SIAM
AMS
proceedings

volume 13

Mathematical Psychology and Psychophysiology

AMERICAN MATHEMATICAL SOCIETY

PROVIDENCE · RHODE ISLAND

Codistributed by LAWRENCE ERLBAUM ASSOCIATES, INC., PUBLISHERS
HILLSDALE, NEW JERSEY and LONDON, ENGLAND

PROCEEDINGS OF THE SYMPOSIUM IN APPLIED MATHEMATICS
OF THE AMERICAN MATHEMATICAL SOCIETY
AND THE SOCIETY FOR INDUSTRIAL AND APPLIED MATHEMATICS

HELD IN PHILADELPHIA
APRIL 15–16, 1980

EDITED BY

STEPHEN GROSSBERG

Prepared by the American Mathematical Society with support from
NSF grant MCS 80-00912

Library of Congress Cataloging in Publication Data

Symposium in Applied Mathematics (1980: New York, N. Y.)
Mathematical psychology and psychophysiology.

(SIAM-AMS proceedings; v. 13)
"Proceedings of the Symposium in Applied Mathematics of the American Mathematical Society and the Society for Industrial and Applied Mathematics held in Philadelphia, April 15–16, 1980"–T. p. verso.
Includes bibliographical references.
Contents: The visual system does a crude Fourier analysis of patterns/Norma Graham–Invariant properties of masking phenomena in psychoacoustics and theoretical consequences/Geoffrey J. Iverson and Michael Pavel–A neural mechanism for generalization over equivalent stimuli in the olfactory system/Walter J. Freeman–[etc.]
1. Psychology, Physiological–Mathematics–Congresses. 2. Psychology–Mathematics–Congresses. I. Grossberg, Stephen, 1939– . II. American Mathematical Society. III. Society for Industrial and Applied Mathematics. IV. Title. V. Series.
QP360.S96 1980 152 81-3500 ISBN 0-8218-1333-1 AACR2

1980 *Mathematics Subject Classification.* Primary 06S25, 08A35, 34A34, 34C15, 34D15, 35B32, 35B40, 39B50, 60H10, 60J60, 62F03, 68G05, 68G10, 92A09, 92A15, 92A17, 92A25, 92A27, 93A13; Secondary 06A10, 06E99, 06F15, 16A72, 16A86, 34C30, 34C35, 34D30, 34K15, 34K25, 39B10, 60B12, 60J70, 60J80, 60K99, 62C99, 62F07, 62F15, 62F35, 62H15, 62H20, 62M02, 62M20, 62P15, 62P20, 68D25, 68D27, 68F15, 90A05, 90A06, 90A12, 90A16, 93A15, 93B20, 93C15, 93C40, 94B60.

Mathematical Psychology
and Psychophysiology

Table of Contents

Preface

Understanding the mind and its neural substrates has long been one of the most challenging and important scientific problems confronting humanity. Experimental and theoretical progress in this area has recently accelerated to the point that our knowledge of brain processes is undergoing a revolutionary transformation. This volume contains articles by the invited speakers at a joint AMS-SIAM Symposium on Mathematical Psychology and Psychophysiology in Philadelphia on April 15–16, 1980 at which several of the theoretical approaches to this area were reviewed.

The articles include contributions to a variety of topics and employ a variety of mathematical tools to explicate these topics. The topics include studies of development, perception, learning, cognition, information processing, psychophysiology, and measurement. Their mathematical substrates include algebraic, stochastic, and dynamical system models and theorems. Despite this diversity, the reader can discover an underlying coherence among the papers. Various concepts and formal laws reoccur in several different subjects. Distinct mathematical tools often probe different levels of the same underlying physical mechanisms.

N. Graham describes the Fourier approach to spatial vision. G. Iverson and M. Pavel discuss an invariance law in psychoacoustics. W. Freeman considers psychophysiological substrates of olfactory coding. D. Willshaw and C. von der Malsburg review a model of retinotectal development.

G. Carpenter analyzes the signal patterns of normal and abnormal nerve cells. S. Geman describes theorems concerning the approximation of stochastic differential equations by deterministic differential equations, and uses these theorems to analyze models of pattern learning and discrimination. S. Grossberg's first article discusses adaptive resonances and competitive dynamics in developmental, perceptual, and cognitive examples, and introduces a new explanation of how depth, brightness, spatial frequency, and filling-in visual computations are related. S. Grossberg's second article analyzes schedule interactions and behavioral contrast as examples of psychophysiological principles.

M. F. Norman's first article discusses some relationships between fitness and genetic mechanisms in a sociobiological context. M. F. Norman's second article describes a stochastic limit theorem whose proof is psychologically motivated.

R. D. Luce and L. Narens review a number of recent results in measurement theory. D. Noreen discusses the psychology and economics of choice. G. Sperling describes models of binocular vision. D. Vorberg analyzes reaction time data using serial self-terminating memory search models.

I would like to thank the other members of the Organizing Committee, W. K. Estes, R. D. Luce, M. F. Norman, H. Simon, and G. Sperling for their thoughtful advice in developing a stimulating Symposium program.

Stephen Grossberg
Boston University
November, 1980

List of Contributors

Professor Gail Carpenter
Department of Mathematics
Northeastern University
Boston, Massachusetts 02115

Professor Walter Freeman
Department of Physiology-Anatomy
University of California
Berkeley, California 94720

Professor Stuart Geman
Division of Applied Mathematics
Brown University
Providence, Rhode Island 02912

Professor Norma Graham
Department of Psychology
Columbia University
New York, New York 10027

Professor Stephen Grossberg
Department of Mathematics
Boston University
Boston, Massachusetts 02215

Professor Geoffrey Iverson
Department of Psychology
Northwestern University
Evanston, Illinois 60201

Professor R. Duncan Luce
Department of Psychology
Harvard University
Cambridge, Massachusetts 02138

Dr. Christopher von der Malsburg
Max Planck Institut für Biophysikalische Chemie
Göttingen, Federal Republic of Germany

Professor Louis Narens
Department of Psychology
University of California
Irvine, California 92717

Dr. David Noreen
Bell Telephone Laboratories
Murray Hill, New Jersey 07974

Professor M. Frank Norman
Department of Psychology
University of Pennsylvania
Philadelphia, Pennsylvania 19104

Dr. Michael Pavel
Department of Psychology
New York University
New York, New York 10012

Professor George Sperling
Department of Psychology
New York University
New York, New York 10012

Professor Dirk Vorberg
Department of Psychology
University of Konstanz
Konstanz, Federal Republic of Germany

Dr. David J. Willshaw
The National Institute for Medical Research
The Ridgeway, Mill Hill
London NW7 1AA, England

SIAM-AMS Proceedings
Volume **13**
1981

The Visual System Does a Crude Fourier Analysis of Patterns

NORMA GRAHAM

Introduction. About a dozen years ago in the Journal of Physiology, John Robson and Fergus Campbell introduced the notion that the human visual system contains multiple spatial-frequency channels–that is, multiple subsystems working in parallel, each of which is sensitive to a different range of spatial frequencies in visual patterns.

At about the same time, in Psychological Review, Jim Thomas made the closely related point that the existence of visual neurons with different sizes of receptive fields has important implications for pattern vision, and, in Science, Allan Pantle and Bob Sekuler suggested the existence of multiple size-selective channels. Since that time, a tremendous amount of psychophysical and physiological work has been inspired by this theoretical notion that there are multiple channels working in parallel to process visual patterns and that each of these channels is sensitive to a different, narrow band of spatial frequencies. Some people have gone so far as to say that the human visual system does a Fourier analysis of the visual scene.

What I do here is review the history of this multiple-channels model of pattern vision and comment on its current status. Some references will be given here, and the reader can find a more extensive bibliography in Graham [**1981**]. Some of the material here is explained more fully at an intuitive level in Graham [**1980**].

First let me point out that in discussing pattern vision, we ignore many dimensions important to vision–color, time, depth–discussing only monochromatic, unmoving, unchanging, flat patterns. We discuss only the initial visual processing, ignoring the higher-order perceptual or cognitive processes that occur, for example, in reading a pattern of letters on a page. In spite of this

---

1980 *Mathematics Subject Classification.* Primary 92A25.

extreme limitation, there would still be too much to cover, so I will further limit it by concentrating on places where mathematics has entered into the development of the multiple-channels model of pattern vision and on places where more mathematics might be useful. (The formal mathematics, which is not presented here in general, can be found in the references.)

Early history. The multiple spatial-frequency channels model developed rather naturally from an earlier model of pattern vision, a single-channel model. In the single-channel model, the important stage of the visual system is a linear system with a two-dimensional input representing the visual stimulus and a two-dimensional output representing the response of the visual system. This model was attractive both because it seemed consistent with known neurophysiology and because it was a very simple model. (People seem implicitly to assume linearity until they have evidence to the contrary.)

Physiological receptive fields. As was discovered a few decades ago, vertebrate retinal ganglion cells, which are the neurons in the retina that send their axons up to the brain, do not respond to single points of light. They respond instead to light in a rather broad area of the visual field. This area is called the "receptive field" of the neuron. Further, a neuron responds differently depending on where in its receptive field the light falls. For example, if light falls in the center of the receptive field, the neuron might respond with increased firing–that is, the neuron is excited. If light falls in an annulus surrounding the center of the field, the neuron responds with decreased firing, that is, the neuron is inhibited. (Some neurons have the opposite arrangement, receptive fields with inhibitory centers and excitatory surrounds.) Importantly, the response of these neurons to the light pattern is approximately linear. If two spots of light are shown, the response to the two is approximately equal to the sum of the responses to each alone. If both spots fall in the center of the receptive field, the neuron responds more to the combination than to either alone. If one spot falls in the center and one in the surround, the responses cancel.

Neurons in the visual cortex of vertebrates were later discovered to have even more complicated receptive fields. Some of these cortical cells are still linear systems with a central excitatory area and adjacent inhibitory areas, but now these areas are rectangular with one or two rectangular inhibitory areas adjacent to the long edge of the rectangular excitatory area. For a review of this physiology, see Robson [**1980**].

Single-channel model. Thus it seemed reasonable to model the important stage of the visual system as an array of many of these neurons. Their receptive fields were assumed to be heavily overlapping and densely distributed across the visual field. (See, for example, Ratliff's delightful book, *Mach Bands* [**1965**].) This conception is called a *Single-Channel Model*. If (1) each neuron responds linearly, (2) the output of this array is taken to be the two-dimensional function giving the response of each neuron as a function of the central position of the neuron's receptive field, and (3) the input to this array is taken to be the two-dimensional function giving light intensity at each point in the visual field, then this array is a linear system.

Behavioral as well as physiological evidence supported this model. For example, the perceived appearance of edges, that is, the existence of Mach bands, is consistent with such a model (Ratliff [**1965**]). I will not go further into this evidence, however, as we are about to discard this model, after we have given it credit for inspiring the use of a new stimulus with which to study vision.

Sinusoidal gratings. This new stimulus was a *sinusoidal grating*. Since the rather prevalent, rather reasonable model of the visual system was a linear system, people who knew about Fourier or linear systems analysis naturally thought of using sinusoidal inputs to the system (Bryngdahl [**1962**], DePalma and Lowry [**1962**], and Patel [**1966**]). What would the appropriate sinusoid be? One candidate, the candidate that was chosen, is a one-dimensional sinusoidal grating like that shown in Figure 1. A sinusoidal grating is a pattern in which the luminance in one direction varies sinusoidally, while the luminance in the perpendicular direction is constant. The *spatial frequency* of a sinusoidal grating is the number of cycles of the sinusoid per unit distance. The *mean luminance* of a grating is the average luminance across the whole grating and is generally kept constant. (One of the attractions of sinusoidal gratings is that the mean luminance of a grating can easily be held constant, keeping the observer in a relatively constant state of light adaptation and thus avoiding manifest early nonlinearities in the visual system, while the contrast and spatial frequency are varied.) The *contrast* of a grating is a measure of the amplitude of the sinusoid; it is usually taken to be one-half the difference between the maximum and minimum luminance divided by the mean luminance.

Fourier analysis. Although I suspect the following will be familiar to most readers, let me review briefly the relevant facts of Fourier analysis. Any two-dimensional function like that describing the luminance at each point in a visual pattern can be Fourier analyzed, that is, the (two-dimensional) Fourier transform of the function can be computed. Further, the Fourier transform can be inverted to give back the original function. Or, to put it in terms of visual patterns, any visual pattern (remember we are only talking about flat, unmoving, uncolored patterns) can be synthesized by adding together sinusoidal gratings of different frequencies and orientations in appropriate phases and contrasts. And there is only one set of sinusoidal gratings which will synthesize any given pattern.

Further, the response of a linear, translation-invariant system, like the single-channel model described above, to a sinusoidal grating is particularly simple. It is a sinusoid of the same frequency and orientation as the grating. Thus only its amplitude and phase need to be specified. The function specifying amplitude and phase for each frequency-orientation combination is known as the transfer function of the system.

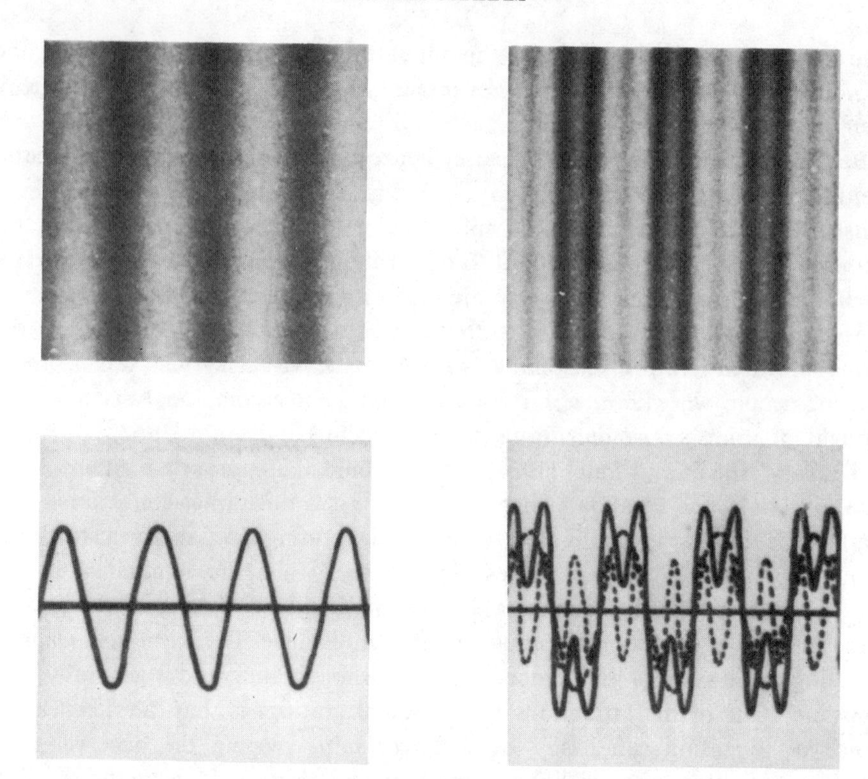

FIGURE 1. A simple sine-wave grating containing one spatial frequency is on the top left, and a compound grating containing two frequencies (one three times the other) is on the top right. The luminance profiles of the patterns are shown underneath each pattern as solid lines. (The dotted lines on the lower right show the luminance profiles of the individual component sine-waves.) From Graham and Nachmias [1971].

By definition, the response of a linear system to any stimulus which is the sum of components is equal to the sum of the responses to the components by themselves. Therefore, according to the linear, translation-invariant single-channel model described above, the response of the visual system to any pattern at all can easily be computed from its response to sinusoids as long as one knows its transfer function.

Application to single-channel model. Thus, if the simple single-channel model introduced above were correct, it would be quite easy to characterize all the parameters of pattern vision. One would just have to know the responses to all sinusoidal gratings (which are a small subset of all possible patterns) and one would know the response to all patterns. Since the response to any sine is a sine of same frequency and orientation, to know the responses to all sinusoidal gratings one would just have to know the amplitudes and phases of the responses.

How does an array of neurons with antagonistic excitatory and inhibitory areas in their receptive fields respond to gratings of different frequencies and orientations? It responds to a limited range of spatial frequencies, responding best to a medium frequency. It responds less well to high frequencies because the nonzero width of the excitatory center smears high frequencies. It responds less well to low frequencies because the inhibitory surround, which is wider than the excitatory center, depresses responses to low frequencies. If the array is made up of neurons with rectangular receptive fields like those in the cortex, it will also respond only to limited ranges of orientation. The orientation it will respond to best is the one where the bars of the gratings are parallel to the long dimension of the rectangular segments in the receptive fields. The phase of the response will depend on the symmetry of the receptive field. Even-symmetric receptive fields like those having equal-sized inhibitory flanks on either side of the excitatory center do not introduce any phase shift.

The simplicity introduced by this simple single-channel model is certainly appealing. But this simple single-channel model can easily be shown to be wrong.

Sine-plus-sine experiments. I am going to describe an experiment that is similar to the original experiment of Campbell and Robson, but allows more telling comparisons between experimental results and theoretical predictions (Graham and Nachmias [**1971**]). Four kinds of patterns were used, all of which were patterns varying in one direction only. See Figure 2, left column, for these patterns' luminance profiles. (A luminance profile is a one-dimensional cut in the interesting direction through the two-dimensional function giving light intensity at each spatial position.) The four patterns were: two simple sinusoidal gratings, one of frequency three times the other, and two compound gratings, each containing both frequencies but in different phases. (Photographs of one of these simple and one of these compound gratings are shown in Figure 1.) For each pattern, the detection threshold (that is, the contrast at which an observer can just tell that a pattern is present rather than a blank field of the same mean luminance) was measured.

Single-channel model predictions. What would the single-channel model predict for this experiment? The middle column of Figure 2 shows the one-dimensional cuts through the two-dimensional output of the channels. These are called response profiles. The response of the channel to the compound grating will just equal the response to the components (in the appropriate phase). To derive the predictions for an observer's threshold, some assumption must be made linking the responses of the model to the behavior of the observer. Here for ease of explanation I will make the simplest assumption. (Many others have been considered over the years, but none has rescued the single-channel model. If one made a sufficiently complicated linking hypothesis, one could undoubtedly rescue the model, but then the interesting part of the model would be in the linking hypothesis, not in the single channel.) We will here assume that an observer detects a pattern whenever the peak-trough difference in the response of the channel (the difference between the largest and the smallest values) reaches some criterion.

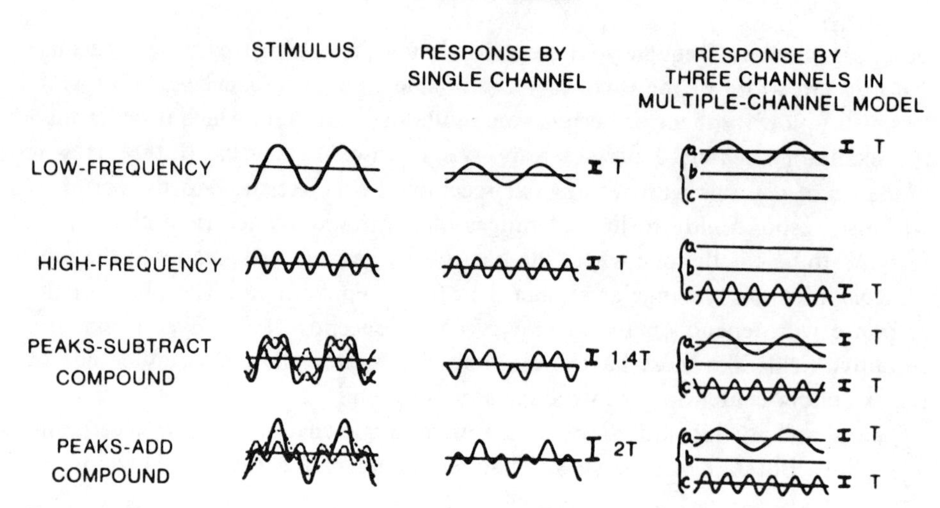

FIGURE 2. Four grating patterns are indicated by their luminance profiles in the left column. (The broken lines show the luminance profiles of the individual components of the compound gratings.) The responses predicted by a single-channel model are shown in the middle column. The responses predicted by three channels of a multiple-channels model are shown in the right column. From Graham and Nachmias [**1971**].

If each component sine-wave's contrast has been set so that it is just at threshold (as illustrated in the first two rows of Figure 2), the single-channel model predicts that the peak-trough difference in the response to each of the compounds will be far above threshold. For the peaks-subtract phase, the peak-trough difference in the response is 1.4 times threshold and for the peaks-add phase 2.0 times threshold. Thus, if the single-channel model were correct, these two compound gratings should be much more visible than their sinusoidal components and the peaks-add much more visible than the peaks-subtract.

For human observers, however, all four patterns are, to a first approximation, equally detectable. This fact is inconsistent with the single-channel model.

Nonuniform single-channel model. A possible variant of the single-channel model that immediately jumps into the mind of many a person is a model in which the size of receptive field changes as one moves from the foveal center to the periphery (although there is still only one size of receptive field at each location). Such a change fits in well with all the phenomena showing that our acuity is better in the middle of the visual field than in the periphery. This modified single-channel model is also wrong, however. A more recent version (Graham, Robson and Nachmias [**1978**]) of the sine-plus-sine experiment described above was done with small patches of grating. (These patches had slow transitions at their edges to avoid introducing too many other spatial frequencies and other edge effects.) These patches were small enough that, according to a nonuniform single-channel model in which receptive field size changes with position at a rate consistent with other visual data, all the patches would have been detected by the same size of receptive field; therefore, the compound would have been more detectable than the components, and the peaks-add compound would have been more detectable than the peaks-subtract. The experimental results, however, were the same as for the full grating, thus ruling out the nonuniform single-channel model.

Multiple-channels model predictions. The experimental results are, however, consistent with a multiple spatial-frequency channels model. Each of the multiple channels might itself be an array of receptive fields all of the same size. Different channels have different sizes of receptive field, a low spatial-frequency channel having larger receptive fields than a high spatial-frequency channel. As far as any individual channel is concerned, the compound grating behaves just like one of its components. For example, in the right column of Figure 2, the channel which is sensitive only to the low-frequency component (channel a) responds to the compound just as if it contained only the low-frequency component. If we assume that a pattern is detectable whenever the response of at least one channel is great enough, and if we continue to ignore, as we have been, the possibility of variability from trial to trial, then this multiple-channels model predicts that all four of the gratings in Figure 2 are just at threshold. In general, a compound grating is exactly as detectable as its most detectable component and phase does not matter. To a first approximation, this is what is found.

Probability summation. But it is only a first approximation. The compound gratings are both, in fact, just a little bit more detectable than their components–not nearly as much more as a single-channel model predicts, but a little. This added detectability came as no surprise to a psychologist, because a favorite notion of psychologists is "probability summation". In all psychophysical experiments, there is a great deal of variability. An observer's response to a particular stimulus is not always the same. A pattern near threshold just does not look the same from trial to trial. Sometimes an observer will see the pattern very clearly; sometimes he will not see it at all. Suppose this variability is due to variability in the responses of the channels, and the variability in different channels is independent. Then, the two channels that respond to a compound grating each have independent chances of detecting the compound grating, but only one of the two channels has a chance to detect a component by itself. Therefore, the compound gratings should be slightly more detectable than either component. A multiple-channels model with probability summation among the channels quantitatively accounts for the exact thresholds of the compound gratings (Sachs, Nachmias, and Robson [**1971**]; Graham, Robson, and Nachmias [**1978**]).

The existence of this probability summation, which is a form of nonlinear summation between channels, greatly complicates calculations from a multiple-channels model.

To further complicate the situation, recent evidence carefully varying the number of bars in a grating (Robson and Graham [**1981**]) strongly suggests there is also probability summation across the spatial extent of each channel. Or, to put it in terms of possible physiology, if the channels are conceived of as arrays of receptive fields, then each receptive field has its own independent variability. Thus, to make predictions from the multiple-channels model, one also needs to take into account this nonlinear probability summation across spatial extent.

Quick pooling model. In order to calculate the predictions of a multiple-channels model with probability summation, the variability in individual receptive fields or channels was originally assumed to be described by a Gaussian distribution. With this assumption, it was extremely tedious to do the calculations of a pattern's threshold from the responses of the multiple receptive fields or multiple channels. Fortunately, a few years ago, Frank Quick **[1974]** pointed out the existence of a function, known in some contexts as the Weibull function, which is a good approximation to the cumulative Gaussian but is much easier to work with as long as it is embedded in a certain psychophysical model.

Let P_{ij} be the probability that the ith receptive field in the jth channel detects the stimulus (that is, the probability that the response of that receptive field is greater than some criterion). Let S_{ij} be the sensitivity of that receptive field to the stimulus (that is, the reciprocal of the contrast necessary to produce detection by the receptive field on half of the trials). Let K be a parameter determining the steepness of the function. Let c be the contrast in the stimulus pattern. Then the function suggested by Quick is

$$P_{ij} = 1 - 2^{-(c \cdot S_{ij})^k}.$$

If one assumes that the observer detects a pattern if and only if at least one of the receptive fields in one of the channels does, then, letting P be the probability of the observer's detecting the pattern, letting M be the number of receptive fields in each channel, and letting N be the number of channels,

$$P = 1 - \prod_{i,j}^{M,N} (1 - P_{ij}).$$

We still need to specify how the observer's probability of detecting a pattern determines his response in an experiment. We will make the simple-minded assumption that, when he does detect the pattern, he always responds correctly and when he does not detect the pattern, he simply guesses. This assumption, sometimes called the high-threshold model of response bias, is known to be wrong (see Green and Swets **[1974]** for example), but perhaps it is a good enough approximation to the truth to be used. (In any case, when it is embedded within the whole model, the resulting predictions seem to account for many results, as will be described below.)

By substituting Quick's psychometric function into the expression for P, the observer's probability of detection, one finds by easy algebraic manipulations that $P = 1 - 2^{-(c \cdot S)^k}$ where S is the sensitivity of the observer to the stimulus (the reciprocal of the contrast necessary for the observer to detect the pattern on half of the trials), and can easily be expressed in terms of the sensitivities of the underlying units

$$S = \left[\sum_{i,j}^{N,M} S_{ij}^{k} \right]^{1/k}.$$

Notice that we have made the implicit assumption that k has the same value for each receptive field, that is the variability in response magnitude is the same for every receptive field. Under this assumption, the function specifying the observer's probability of detection (the so-called psychometric function) has the same form for every stimulus and also the same form as the function specifying a single receptive field's probability of detection. Green and Luce [**1975**] proved that functions of the above form (the base need not be 2) are the only ones having this invariance property.

Empirically, a value for k of about 3.5 seems to describe the psychometric function for a wide variety of stimuli (e.g. Graham, Robson and Nachmias [**1978**], Legge [**1978**], Robson and Graham [**1981**], Watson [**1979**]).

To remind you, the sensitivity of each receptive field S_{ij} is just that of a linear system with the appropriate weighting function (that is, having the size and shape that go with a particular channel's frequency and orientation frequency domain). (Since they are linear systems, the sensitivity of each one to a particular stimulus multiplied by the contrast in the stimulus gives you the response magnitude. Therefore, sensitivity is the same as the response magnitude to a criterion contrast.) In experiments to measure the thresholds of compound gratings containing far-apart frequencies (as in Figure 2) each channel only responds to one frequency so there is no need to actually know the weighting function.

The above expression is easily recognized as a metric in a space where each dimension represents the sensitivity of one receptive field. The sensitivity to a stimulus is just the distance between the origin (which represents the case of no pattern, that is, a blank, homogeneous field) and the point representing that stimulus. (One might extend this approach and let the distance between any two points in the space represent the discriminability of any two stimuli, but almost no work has been done on such an extension so I will not mention it further.) When k is infinity, the observer's sensitivity equals the sensitivity of the most sensitive unit (which makes this version of the above expression equivalent to the model without any probability summation). When k is 1, the sensitivities or responses of different units are just being linearly summed. When k is 2, there is power-summation (which is a model that occurs in many contexts but is false in this context). And when k is about 3.5, this expression quantitatively predicts the thresholds for a wide variety of patterns.

Since this last is an important point, let me elaborate on it a minute. The above expression, with a value for k of about 3.5, produces predictions that agree very closely with the thresholds for a wide variety of patterns including combinations of sines with sines (e.g. Graham, Robson and Nachmias [**1978**], Quick, Mullins and Reichert [**1978**]), patches of sines with various numbers of cycles (Legge [**1978**], Robson and Graham [**1981**]), and some aperiodic stimuli as well as combinations of aperiodic stimuli with sines (e.g. Graham [**1977**]; Bergen, Wilson and Cowan [**1979**]). For the latter predictions, one does need to know the weighting functions. More of that later. Further, one can expand one's perspective and include time as a dimension (varying the length of time a pattern is on, or its temporal frequency content) and this same approach will work (Watson [**1979**], Watson and Nachmias [**1977**]).

In short, a model in which there is probability summation across space (across the receptive fields within each channel) and among channels and in which the variability is described by the function suggested by Quick with a steepness parameter of 3.5 accounts both for the psychometric functions that have been collected for a number of stimuli and for the actual threshold values that have been measured for an even wider collection of stimuli. This is a very impressive feat. However, caution is still in order. The nonlinear pooling across receptive fields and channels exhibited in the equation for S that does such a good job at accounting for thresholds may not actually be due to probability summation (independent noise) but to some other cause, and the agreement between psychometric function steepness parameter and the exponent needed to account for thresholds may be just fortuitious. We know that, in detail, the psychophysical high-threshold theory must be wrong. For discussion of these issues, see Robson and Graham [**1981**].

In any case, the metric in the equation for S has proved its usefulness and has given us insight into the kind of pooling across receptive fields and channels that must be assumed in order to predict thresholds.

Let us now go back to the issue of what the weighting functions for individual receptive fields look like or, to put in another way, what the spatial-frequency and orientation tuning of individual channels is like. Or, to put it still another way, let us go back to the issue of whether the visual system actually does a Fourier analysis of the visual scene.

Strict Fourier analysis? What would it mean if we said that the brain performed a Fourier analysis of the visual scene? We might mean, if we were speaking strictly, that there was a set of neurons that computed the Fourier transform of the visual pattern. The magnitude of the response of a particular neuron in the set would be *completely determined* by the amount (or by the phase) of a particular spatial-frequency/orientation component present in the pattern. In other words, each neuron in the set would respond only to an *extremely narrow* range of spatial frequencies and orientations. Different neurons would respond to different spatial-frequency/orientation combinations so the set as a whole would compute an excellent approximation to a Fourier transform. (It is an approximation because there are only a finite number of neurons and the Fourier transform is a continuous function, but this kind of approximation has no practical consequences.)

There is no such set of neurons anywhere in the brain; at least, there is absolutely no evidence, either physiological or psychophysical, that such a set of neurons exists. The brain, therefore, does not perform a *strict* Fourier analysis of the visual scene. Further, I should say as a historical note, that none of the people (Campbell, Robson, Thomas, Pantle, or Sekuler) mentioned above meant to be implying that the visual system did a strict Fourier analysis, although they, particularly Campbell and Robson, have sometimes been blamed (or credited) with so doing.

Although there is no evidence that the brain performs a strict Fourier analysis, there is an accumulating mass of evidence that the brain performs operations with many of the characteristics of Fourier analysis, operations that could be called crude Fourier analyses.

Evidence for a crude Fourier analysis. When the thresholds for compound gratings containing component frequencies that are close together are properly interpreted (that is, probability summation across space is allowed for, see references given above), they suggest a bandwidth for each channel on the order of $1\frac{1}{2}$ or 2 octaves. Channels of this medium bandwidth could not be said to do a strict Fourier analysis, but nonetheless, the response magnitude of each channel does give you a good idea of how much of the channel's preferred spatial frequency is present in the pattern. So one might want to say that a set of channels with this sort of medium bandwidth does a crude Fourier analysis. Sine-plus-sine experiments are only one kind of evidence, however. Let us look now, very briefly, at the other kinds of evidence that have supported the notion of multiple spatial-frequency channels and at the bandwidths suggested by each kind of evidence.

In *Adaptation and Masking Experiments*, the visibility of test patterns is measured either after the observer has inspected a suprathreshold adapting pattern, or while he is inspecting a masking pattern. The visibility of the test pattern should be affected by the adapting (or masking) pattern if the patterns are processed by the same channel, but not if they are processed by different channels. In fact, when the test pattern and adapting (or masking) pattern contain similar spatial frequencies, the test pattern is affected. But when the test pattern and adapting (or masking) pattern contain very different spatial frequencies or orientations, the test pattern is not affected. These effects occur both at threshold and on some aspects of suprathreshold perceived appearance. That is, the threshold for a test grating, the perceived contrast in a suprathreshold test grating, and the perceived frequency and orientation of a suprathreshold test grating are all altered by previous adaptation to or simultaneous masking by a grating of similar spatial-frequency and orientation. How similar? an octave or so.

In addition to the wide variety of adaptation and masking effects that have been studied there are some less well-studied, higher-order perceptual effects that seem to reveal the action of the spatial-frequency channels.

In *Recognition Experiments*, the observer is asked to say not only whether some pattern is present but what that pattern is (out of some small set of possible patterns). Far-apart frequencies are recognized whenever they are detected (Nachmias and Weber [**1975**]) as if there were spatial-frequency channels and the observer could always tell which of the channels had detected the stimulus. Closer-together frequencies, however, begin to be confused. How close together they need to be to produce confusion may tell something about the bandwidth of channels. The available results suggest medium bandwidths (Hirsch [**1977**], Thomas and Barker [**1977**]).

As an aside off the topic of bandwidth, let me tell you two other interesting recognition results from experiments where compound gratings were used. First, compound gratings containing two frequencies in a ratio of three to one can be far enough above threshold that an observer can always detect both components (can always tell you that both are present rather than just one) and yet be unable to tell you which of two phases the components are in (Nachmias and Weber [**1975**]). This lack of phase discrimination is easy to explain if spatial-frequency channels exist so that the two components are exciting separate channels, and, near threshold, these channels signal nothing about phase so the relative phase of the two components cannot be computed. The second interesting result is that at threshold contrasts, the compound grating is sometimes seen as a compound, sometimes seen as one of its components, sometimes seen as the other component, and sometimes seen as a blank (Hirsch [**1977**]). To a first approximation anyway, the proportions of times it is seen as each one is what you would expect if there were independent variability in each channel (probability summation).

In *Texture Segregation Experiments* (also called grouping or effortless texture discrimination experiments), the observer is asked whether or not he can immediately see that there are two different areas in the pattern that contain two different textures. Julesz originally conjectured that an observer could only see immediately (without scrutiny) that the textures in the two different areas were different if the second-order statistics of the two textures differed. Different second-order statistics implies different autocorrelation functions and hence different amplitude spectra in the Fourier transforms. If there were extremely narrow-bandwidth channels, i.e., if the visual system did a strict Fourier analysis, throwing out the phase information, Julesz's conjecture would amount to saying that textures could only be discriminated when the outputs of these channels were different. Julesz, however, keeps discovering texture pairs for which his conjecture is wrong. Although they have identical second-order statistics and hence identical Fourier amplitudes, they are discriminable. He explains these by postulating a second class of visual mechanism in addition to the one that computes second-order statistics (the one equivalent to extremely narrowband channels). See Julesz [**1981**] for a review of this work. I think it might be better if he did not postulate a second class of visual mechanism but just said that the channels were not extremely narrowband but somewhat broader, as all the other evidence indicates.

Physiology. Receptive fields come in different sizes even at the retinal ganglion cell level where they are concentric. Thus they respond to different ranges of spatial frequencies, although each responds to quite a broad range. In the cortex, recent work (Movshon, Thompson and Tolhurst [**1978**], and DeValois, Albrecht and Thorell [**1977**]) show neurons that respond to very different ranges of spatial frequencies. The bandwidth of these neurons is limited but certainly not extremely narrow. The available physiological evidence, therefore, if you are willing to identify the psychophysical channels with the neurons in the geniculo-cortical pathway, also suggests medium bandwidth channels.

The role of mathematics.

Has mathematics been useful in this work? Historically, knowledge of the theorems about Fourier transforms and linear systems suggested the use of sinusoidal gratings, and the use of sinusoidal gratings has certainly added to our knowledge of the visual system. In particular, the results of experiments using gratings were a major factor in suggesting the multiple spatial-frequency channels model which is now so popular.

If these channels were extremely narrowband, the mathematics of Fourier analysis would have been a very concise, natural description of the system, expressing something fundamental about the way the visual system processed information. The decomposition of a stimulus into a basis set of sinusoidal components by Fourier analysis would have corresponded very closely to a decomposition done by the visual system. Each sinusoidal component would have excited one and only one channel. But the channels are not extremely narrowband.

Even with medium-bandwidth channels, one can often have better intuitions about how the channels will respond by considering the Fourier transforms of the stimuli rather than the stimuli themselves and by considering the transfer functions in the frequency domain of the theoretical channels rather than the weighting functions in the space domain. Formally, one can often calculate the predictions more easily using Fourier transforms (e.g. the calculations for aperiodic stimuli and combinations of sines and aperiodic stimuli in Graham [**1977**]). In this case, however, the mathematics of Fourier analysis is primarily serving as a calculating tool rather than as a natural expression of something fundamental about the visual system.

There is another, more subtle, way in which Fourier analysis may be contributing to our understanding of the visual system. Fourier transforms of stimuli may be enlightening simply as a new description of visual stimuli. Independently of the way the visual system actually works, a new description of visual stimuli may stimulate the human investigators of the visual system and provoke them into creative thought, particularly when the new description is as different from the old description as Fourier transforms are from the original function. In a sense, spatial-frequency descriptions (Fourier transforms) and point-wise descriptions (intensity as a function of spatial position) are opposites. A stimulus consisting of a single spatial frequency is completely localized on the frequency dimension, but infinitely extended in space, whereas a stimulus consisting of a single point is completely localized in space but infinitely extended on the frequency dimension. These two descriptions emphasize very different aspects of the stimulus.

In addition to Fourier analysis, another sort of mathematics seems to have been useful in the work of understanding pattern vision, particularly the thresholds for the detection of patterns. That is the metric expression given above for nonlinear pooling (perhaps probability summation) across receptive fields and channels and the Weibull function as a description of psychometric functions. Again, this mathematics seems primarily to have served as a calculating tool rather than an expression of something fundamental. It does seem, however, to have allowed much greater insight into the possible effects of nonlinearities like those involved in pooling across units.

What would be nice in the way of mathematics?

A substitute for sinusoidal gratings and Fourier analysis. It would be useful to have some mathematics that was as natural a representation of a system involving medium-bandwidth channels (channels with receptive fields that are quite well localized in the visual field and only contain a few excitatory or inhibitory sections) as Fourier analysis is of extremely narrowband channels (channels with a large number of excitatory and inhibitory sections in receptive fields extended across the visual field). There is probably no representation quite as nice as to provide a basis set of stimuli into which any stimulus can be uniquely decomposed and which correspond precisely to the sensitivities of the channels (so that each stimulus in the basis set stimulates one and only one channel). But perhaps some representation exists which is better suited to and thus provides more insight into the case of medium-bandwidth channels. Such a representation might provide a new kind of stimulus (perhaps little patches of something like a sinusoidal grating) which was the optimal kind of stimulus for studying a system of medium-bandwidth channels and might also provide a way of dealing with the responses to those optimal stimuli.

More ways to investigate and describe nonlinearities. Even though the channels are not extremely narrowband, they would be moderately easy to deal with as long as the channels were themselves linear systems and one could examine the properties of one channel at a time. Such, however, is not the case, particularly not for behavioral (psychophysical) phenomena but also not always for physiological data. Thus, investigators of pattern vision, like so many other people, need more flexible and insightful and convenient ways of handling nonlinear systems.

Psychophysics. We have already discussed above the fact that probability summation is necessary to accurately account for the thresholds of patterns. That is, the threshold for a pattern is not determined by the one channel that is most sensitive to the pattern (the one channel that gives, on the average, the largest response to the pattern), but the threshold is always determined by nonlinear pooling of the sensitivities of all the channels that respond at all to the pattern. (Of course, channels that are very insensitive contribute negligibly, but all channels that are almost as sensitive as the most sensitive channel contribute substantially.) Thus, one is always looking at the action of a group of channels rather than a single channel.

Further, even if each receptive field of the channels operative in the pschophysical phenomena does turn out to be a linear system, as is currently assumed, the output of a channel (the output of a whole array of receptive fields of the same size) seems to be the resultant of probability summation across space or some similar kind of nonlinear pooling. So the response of a channel cannot be taken to be the response of a completely linear system.

Although the Quick pooling model presented above does a very good job at accounting for threshold data, we know it cannot be completely correct.

Physiology. If we limit our interest to single neurons, then it is easy to look at only one at a time. Then the only question is whether they are linear or not. Well, many are, or at least linear enough that Fourier analysis works, even in the cortex. But many are not. At all levels in the visual system, a major distinction is being made now between x and y, or sustained and transient, or linear and nonlinear cells. The nonlinear cells are, of course, much more difficult to figure out. Some progress is being made using newly-developed versions of Wiener analysis (Shapely and Victor [**1979**], Victor and Knight [**1979**]).

REFERENCES

Bergen, J. R, H. R. Wilson and J. D. Cowan, 1979. *Further evidence for four mechanisms mediating vision at threshold: Sensitivities to complex gratings and aperiodic stimuli*, J. Opt. Soc. Amer. **69**, 1580–1587.

Bryngdahl, O., 1964. *Characteristics of the visual system: Psychophysical measurements of the response to spatial sine-wave stimuli in the mesopic region*, J. Opt. Soc. Amer. **54**, 1152–1160.

Campbell, F. W. and J. G. Robson, 1968. *Application of Fourier analysis to the visibility of gratings*, J. Physiol. **197**, 551–566.

DePalma, J. J. and E. M. Lowry, 1962. *Sine-wave response of the visual system*. II. *Sine-wave and square-wave contrast sensitivity*, J. Opt. Soc. Amer. **52**, 328–335.

DeValois, R., D. G. Albrecht and L. Thorell, 1977. *Spatial contrast* (H. Spekreijse and L. H. van der Tweel, eds.), North-Holland, Amsterdam, pp. 60–63.

Graham, N., 1977. *Visual detection of aperiodic spatial stimuli by probability summation among narrowband channels*, Vision Res. **17**, 637–652.

______, 1980. *Spatial frequency channels in human vision: Detecting edges without edge detectors*, Visual Coding and Adaptability (C. Harris, ed.), Erlbaum, Hillsdale, N. J.

______, 1981. *Psychophysics of spatial-frequency channels*, Perceptual Organization (M. Kubovy and J. Pomerantz, eds.), Erlbaum, Hillsdale, N. J.

Graham, N. and J. Nachmias, 1971. *Detection of grating patterns containing two spatial frequencies: a test of single-channel and multiple-channels models*, Vision Res. **11**, 251–259.

Graham, N., J. G. Robson and J. Nachmias, 1978. *Grating summation in fovea and periphery*, Vision Res. **18**, 816–825.

Green, D. M. and J. A. Swets, 1974. *Signal detection theory and psychophysics*, Kreiger, Huntington, N. Y.

Green, D. M. and R. D. Luce, 1975. *Parallel psychometric functions from a set of independent detectors*, Psychological Rev. **82**, 483–486.

Hirsch, J., 1977. *Properties of human visual spatial-frequency-selective systems: a two-frequency, two-response recognition paradigm*. Doctoral Dissertation, Columbia University.

Julesz, B., 1981. *Figure and ground perception in briefly presented iso-dipole textures*, Perceptual Organization (M. Kubovy and J. Pomerantz, eds.), Erlbaum, Hillsdale, N. J.

Legge, G., 1978. *Space domain properties of a spatial-frequency channel in human vision*, Vision Res. **18**, 959–969.

Movshon, J. A., I. D. Thompson and D. J. Tolhurst, 1978. *Spatial and temporal tuning properties of cells in areas* 17 *and* 18 *of the cat's visual cortex*, J. Physiology **283**, 100–120.

Nachmias, J. and A. Weber, 1975. *Discrimination of simple and complex gratings*, Vision Res. **15**, 217–224.

Quick, R. F., 1974. *A vector-magnitude model of contrast detection*, Kybernetik **16**, 65–67.

Quick, R. F., W. W. Mullins and T. A. Reichert, 1978. *Spatial summation effects on two-component grating thresholds*, J. Opt. Soc. Amer. **68**, 116–121.

Pantle, A. and R. Sekuler, 1968. *Size-detecting mechanisms in human vision*, Science **162**, 1146–1148.

Patel, A. S., 1966. *Spatial resolution by the human visual system: The effect of mean retinal illuminance*, J. Opt. Soc. Amer. **56**, 689–694.

Ratliff, F., 1965. *Mach bands: quantitative studies on neural networks in the retina*, Holden-Day, San Francisco, Calif.

Robson, J. G., 1980. *Neural images: The physiological basis of spatial vision*, Visual Coding and Adaptability (C. S. Harris, ed.), Erlbaum, Hillsdale, N. J.

Robson, J. G. and N. Graham, 1981. *Probability summation and regional variation in contrast sensitivity across the visual field*, Vision Res. (to appear).

Sachs, M. B., J. Nachmias and J. G. Robson, 1971. *Spatial-frequency channels in human vision*, J. Opt. Soc. Amer. **61**, 1176–1186.

Shapley, R. M. and J. D. Victor, (1979). *Nonlinear spatial summation and the contrast gain control of cat retinal ganglion cells*, J. Physiology **290**, 141–161.

Thomas, J. P., 1970. *Model of the function of receptive fields in human vision*, Psychological Rev. **77**, 121–134.

Thomas, J. P. and R. A. Barker, 1977. *Bandwidths of visual channels estimated from detection and discrimination data*, J. Opt. Soc. Amer. **67**, 140, Abstract.

Victor, J. D. and B. W. Knight, 1979. *Nonlinear analysis with an arbitrary stimulus ensemble*, Quart. Appl. Math. **37**, 113–136.

Watson, A. B., 1979. *Probability summation over time*, Vision Res. **19**, 515–522.

Watson, A. B. and J. Nachmias, 1977. *Patterns of temporal interaction in the detection of gratings*, Vision Res. **17**, 893–902.

DEPARTMENT OF MATHEMATICS, COLUMBIA UNIVERSITY, NEW YORK, NEW YORK 10027

SIAM-AMS Proceedings
Volume 13
1981

Invariant Properties of Masking Phenomena in Psychoacoustics and their Theoretical Consequences[1]

GEOFFREY J. IVERSON AND MICHAEL PAVEL

Introduction. It is one of the ambitions of psychological acoustics to construct a theory of perceived intensity or loudness. There are two directions one may take in attempting to realize this ambition. On the one hand, one can adopt a theoretical language suited to the description and organization of neurophysiological data. On the other hand, another natural theoretical language is that of psychoacoustics, the study of loudness judgments. In this work we discuss psychoacoustic data only.

With the rare exception, one cannot expect to gain much insight into the problem of loudness from data collected within any one experimental context. It is when the data from one empirical paradigm share common, nontrivial features with those of another that advances are made. Examining data of our own, as well as those from a number of other studies, we have noted that such common features do occur, and we have abstracted them in terms of so-called "homogeneity" laws. These laws express the way in which a large class of data involving loudness judgments transforms under change of unit of the physical intensity scales used in measuring the data.

We briefly review our work on homogeneity laws in this paper, and in doing so we relate data from loudness discrimination, detection, partial masking and forward masking paradigms. While homogeneity laws are functional equations of great simplicity, they provide powerful constraints on any theory of loudness. If nothing else, we hope to make this point clear throughout.

Partial masking. Many real life situations illustrate the phenomenon of partial masking. For instance, it is often difficult, if not impossible, to carry on a

1980 *Mathematics Subject Classification.* Primary 92A27.

[1]This work was supported in part by National Science Foundation Grant No. 77-16984, awarded to New York University.

conversation in a moving subway car without raising one's voice considerably. Provided one is willing to do so however, speech intelligibility is not dramatically impaired.

Partial masking can be studied in the laboratory using simple stimuli that are easily measured and controlled–in particular, pure tones and broadband white Gaussian noise. If one listens to a pure tone in a "quiet" environment such as an acoustically insulated chamber, perceived loudness grows rather more slowly than linearly with physical sound pressure, except possibly at very low levels. If noise is now added as a constant background, the tone is partially masked: at low levels the tone will not be heard until it is raised to a just audible intensity considerably in excess of the corresponding value in quiet, while at high tone levels, the presence of the noise appears to have virtually no effect. Between these two extremes occurs a region of "abnormally rapid" loudness growth, the signature of partial masking.

This description of pure tone partial masking is so subjective as to be of intuitive value only. More precise and objective empirical techniques for measuring the partial masking effect are not difficult to come by, however. A common technique involves a loudness matching task in which a masked and an unmasked tone are adjusted in intensity until they are perceived equally loud. For a fixed noise level, a plot of the unmasked tone intensity as a function of the masked tone intensity is called a (partially masked) loudness matching function. A typical example of such a function is shown in Figure 1. Departure of the data from the leading diagonal is due to the effect of the noise, and this difference may legitimately be regarded as an alternative way to characterize the masking effect.

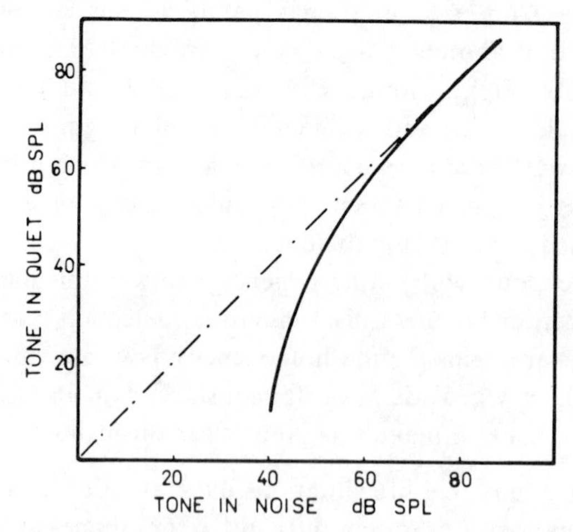

FIGURE 1. A hypothetical loudness matching curve. Note that the physical intensities are measured in decibels, i.e. both coordinates are logarithmic functions of sound pressure.

Consider now the change induced in a matching function as the noise level is increased. Clearly the matching function moves to the right, in some fashion. We are interested in describing such transformations in detail. This task turns out to be remarkably simple. If one compares two distinct matching functions, it turns out that they are merely displaced versions of one another, the displacement being made in a direction parallel to the leading diagonal (see Figure 2). We call this empirical property *shift invariance*.

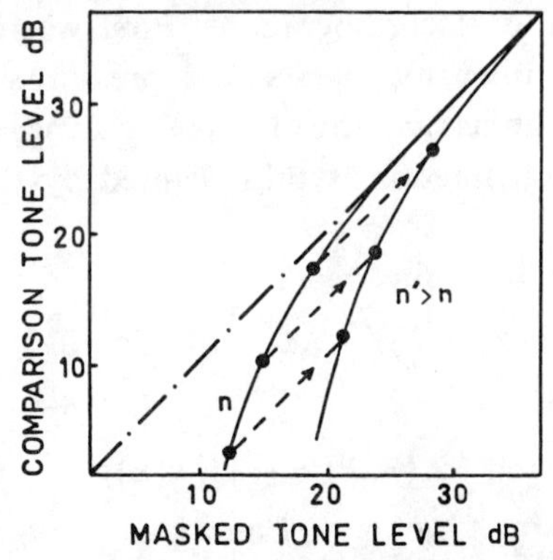

FIGURE 2. Hypothetical loudness matching functions illustrating shift invariance. The right-most function is identical to the left-most except for a rigid translation in a diagonal direction. Arrows connect corresponding points following an increase in masker level from n to n'.

To verify shift invariance empirically requires data appropriate to the task. We have investigated shift invariance in a careful empirical study of our own (Pavel and Iverson [**4**]), and in addition we have reanalyzed data from a number of other laboratories. In all cases, the invariance hypothesis is extremely well supported. Perhaps the most extensive set of partial masking data is that of Stevens and Guirao [**6**]. Their data are replotted in the following figure, along with a family of curves satisfying shift invariance.

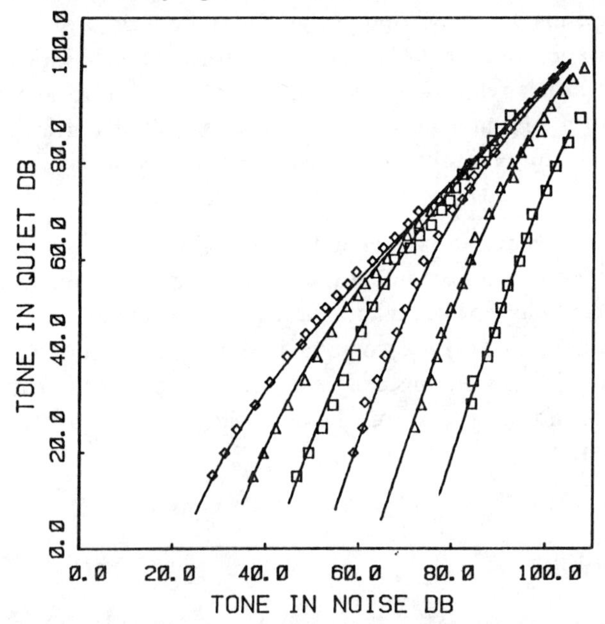

FIGURE 3. Loudness matching data replotted from Stevens and Guirao [**6**]. The family of solid curves satisfies shift invariance. Adjacent sets of data correspond to equal 10 decibel increments in noise level.

It is clear that these curves describe the data quite well. The parametric form of this family of curves is obtained by a theoretical argument which will be sketched below.

Let us turn to theoretical considerations. First we need some notation: we denote by $\phi(x, n)$ the intensity (*rms* sound pressure)^2 of an unmasked tone required to match, in loudness, a tone of intensity x masked by noise of intensity n. A loudness matching function is thus defined by the correspondence $x \mapsto \phi(x, n)$.

In this notation, shift invariance reads

$$\phi(\lambda x, K(n, \lambda)) = \lambda\phi(x, n) \quad \text{for all } \lambda > 0. \tag{1}$$

It is easily shown that

$$K(n, \lambda) = \kappa^{-1}(\lambda\kappa(n)) \tag{2}$$

for some strictly increasing function κ. Empirically, a somewhat stronger version of (1) holds: writing $e(x, n) = \phi(x, n)/x$, we have in fact

$$e(x, n) = e(x', n') \quad \text{iff} \quad e(\lambda x, \mu n) = e(\lambda x', \mu n') \quad \text{any } \lambda > 0, \mu > 0 \tag{3}$$

and (3) is, under natural side conditions, sufficient to determine the function κ in (2) as a power function. In other words, shift invariance amounts to the functional equation

$$\phi(\lambda x, \lambda^\theta n) = \lambda\phi(x, n), \qquad \lambda > 0, \tag{4}$$

where θ is a constant measured empirically to be less than 1. We call properties such as (3) or (4) psychophysical *homogeneity laws*.

From (4) we derive $\phi(x, n) = xf(x^\theta/n)$ so that the "masking effect" $e(x, n) = \phi(x, n)/x$ depends only on a kind of signal to noise ratio, x^θ/n.

Shift invariance does not determine the form of any individual matching function, but rather relates any two such functions, one to another. To obtain the analytic form of a matching function requires a model of the partial masking effect. We conjecture that the following representation holds for $\phi(x, n)$.

$$\phi(x, n) = F(g(x)/(h(x) + k(n))) \tag{5}$$

in which F, g, h, k are strictly monotone functions. The intuitive idea behind this representation is, at least under the present empirical circumstances, that the auditory system behaves like an automatic gain control device.

When equations (4), (5) are combined with natural smoothness hypotheses, parametric families of solutions can be determined (see Iverson and Pavel [2] for details). There are just two of these families which satisfy the empirical "boundary condition"

$$\lim_{n\to 0} \phi(x, n) = Ax, \qquad A \text{ a constant} \simeq 1.$$

These are given by

$$\phi(x, n) = A\left(\frac{x^\alpha}{x^\beta + Bn^{\beta/\theta}}\right)^{1/(\alpha-\beta)}, \tag{6a}$$

and

$$\phi(x, n) = A\left(x^{\alpha}\left(x^{\beta} - Bn^{\beta/\theta}\right)\right)^{1/(\alpha+\beta)}, \tag{6b}$$

where in each case $\alpha > \beta > 0$, $A > 0$, $B > 0$. The first of these families was used to generate the curves displayed in Figure 3.

It is natural to identify the ratio $g(x)/(h(x) + k(n))$ as a masked loudness scale $L(x, n)$. In these terms, (5) reads $L(\phi(x, n), 0) = L(x, n)$, in which, employing equation (6a),

$$L(x, n) = Ax^{\alpha}/\left(x^{\beta} + Bn^{\beta/\theta}\right).$$

In particular, $L(\cdot, 0)$ is a power function, and for small values of x the condition $L(x, n) =$ constant amounts to $x^{\alpha}/n^{\beta/\theta} =$ constant. These two special circumstances give a good account of discrimination and detection data, as we now discuss.

Detection. In detection studies, "threshold" (i.e. just audible) levels of a pure tone are determined for each of a number of noise intensities. Such data (Hawkins and Stevens [**1**]) are usually characterized by a constant signal to noise ratio; in the notation established above this means $x/n =$ constant. (Actually there is some evidence in detection data of our own as well as in other published studies that the tone signal needs to be increased about 11 dB for each 10 dB increment of noise.) Such a signal to noise invariance is consistent with the ideas promoted at the end of the prior section, provided only that $\alpha \simeq \beta/\theta$, or equivalently, $\alpha\theta/\beta \simeq 1$. We have determined values of β/α and θ from seven distinct sets of data. These parameters vary a little from study to study, subject to subject; values of β/α range uniformly from .70 to nearly 1, the average, over all determinations being .85. Likewise, values of θ range from .61 to .90 with a mean of .77. The ratio $\beta/\alpha\theta$ ranges from 1.04 to 1.20 with a mean value of 1.10.

Discrimination. Discrimination of pairs of both unmasked and masked pure tones has been the subject of considerable recent research. We shall confine our remarks to pure tone discrimination in quiet (or at least in very low levels of background noise).

A major reason for the recent activity in discrimination has to do with the somewhat surprising fact that loudness discriminability improves in an apparently strictly monotone fashion from medium intensities and beyond. That is, practically speaking, no asymptotic performance level is ever reached. This phenomenon, which is a "small" but consistent effect, has been termed the "near miss" to Weber's law.

Discrimination data can be collected in two ways. Estimates may be obtained of P_{xy} the probability of choosing a tone of intensity x as louder than a similar tone of intensity y. A far more popular technique, however, consists of measuring *Weber* functions. A π-Weber function, Δ_{π}, is obtained by determining the value of $x = y + \Delta_{\pi}(y)$ required to be judged louder than y with a fixed probability π. That is

$$P_{x,y} = \pi \quad \text{iff} \quad x = y + \Delta_\pi(y).$$

In a study of Jesteadt et al. [3], Weber functions were carefully estimated. The data of these authors are extremely orderly, and indicate that for all but very low levels of intensity,

$$\Delta_\pi(x)/x \simeq \text{const } x^{-\delta}, \tag{7}$$

with a value of δ of about .075. (Other studies measure the exponent as $\simeq$ .1.) Here π is fixed at a value of about 0.7.

In the context of the simple one-dimensional account of loudness introduced above, it seems natural to investigate the discrimination theory

$$P_{x,y} = F[L(x,0) - L(y,0)] \tag{8a}$$

$$= F(x^{\alpha-\beta} - y^{\alpha-\beta}), \tag{8b}$$

F some strictly increasing function. The prediction of (8) for the form of Weber fractions is straightforward. We derive (with $F^{-1} = \gamma$)

$$\frac{\Delta_\pi(x)}{x} = \left(1 + \frac{\gamma(\pi)}{x^{\alpha-\beta}}\right)^{1/(\alpha-\beta)} - 1 \simeq [\gamma(\pi)]^{1/(\alpha-\beta)}/x \quad \text{for small } x$$

$$\simeq \frac{\gamma(\pi)}{\alpha-\beta} x^{-(\alpha-\beta)} \quad \text{for large } x.$$

This latter prediction is in accord with the empirical finding summarized in (7).

From our analysis of partial masking data we have a number of estimates of $\alpha - \beta$. These range from near zero to .2, with a mean of .11, in fair agreement with the estimates of Jesteadt et al. [3] and others.

We remark that the representation (8b) is equivalent, in the context of (8a) (for any increasing function $L(x, 0)$), to the homogeneity law

$$P_{x,y} = P_{x',y'} \quad \text{iff} \quad P_{\lambda x,\lambda y} = P_{\lambda x',\lambda y'} \tag{9}$$

provided $\lambda \mapsto P_{\lambda x,\lambda y}$ is strictly increasing for all $x > y$. While this latter condition is likely to hold empirically, no direct test of (9) has ever been carried out. This is not surprising in view of its consequence (8b), which presents rather less of a problem to test.

In this brief review of partial masking, detection and discrimination, we have shown that empirically verified homogeneity laws lead to a simple and coherent account of the data of all three paradigms in terms of a one-dimensional theory of loudness. While further research may demonstrate the shortcomings of this simple account, homogeneity laws are likely to retain a prominent role in structuring any theoretical improvements.

Our discussion thus far has been concerned with stimuli which, as far as the auditory periphery is concerned, can be regarded as "steady state". We now examine temporal, i.e. transient, behavior of the auditory system.

Forward masking. The detectability of a brief auditory signal (such as a "click") is decreased by the presence of a simultaneously presented noise. In fact, it is well known that detection performance is characterized by a signal to noise ratio, just as for steady pure tones. This result is not particularly constraining. It is a prediction of many dynamical systems. But a question arises in this connection which offers the possibility of a far more powerful constraint. Suppose the click is delayed in time relative to the offset of a burst of noise. Contrary to the behavior of a system with complete knowledge, masking still occurs. This is called forward masking. We ask: Is detection performance under forward masking conditions dictated by a signal to noise ratio for each delay?

Curiously, forward masking has not been investigated in the past with this question in mind. Recently we have carried out a study (Pavel and Iverson [5]) to fill this gap. Some preliminary results are presented here.

We do not go into details of experimental method. Suffice it to say that one task involved the determination of detection thresholds of a 10 msec burst of a 1000 Hz tone placed either simultaneously with or at one of a number of intervals following a 400 msec burst of white noise. The data of each subject were so similar that averages over subjects were taken. These averaged data are shown in the following figure, plotted separately for each delay–a departure from the custom of plotting signal level against delay.

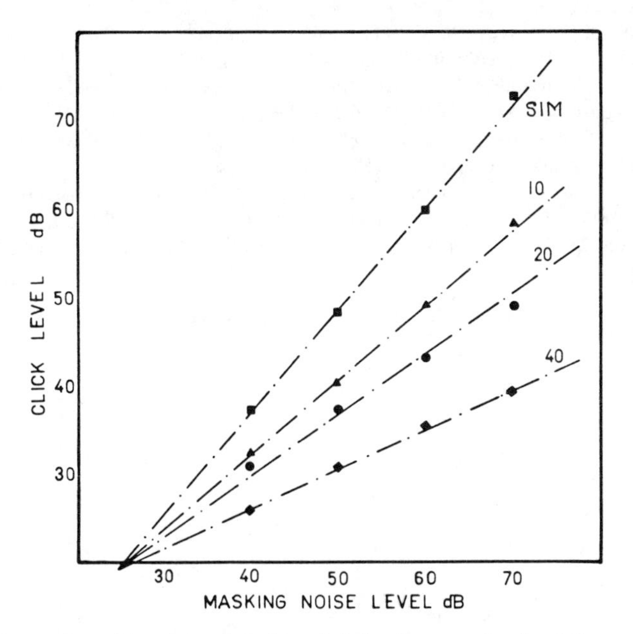

FIGURE 4. Forward masking data plotted as click level vs. masker level for each of four delays (0, 10, 20, 40 msec).

As is evident from a glance at Figure 4, click level is, for each delay, a linear function of masker level in these logarithmic coordinates. This fact is captured by the following homogeneity law written in terms of $P(x, n, \tau)$, the probability of detecting a signal of intensity x delayed τ msec relative to the offset of a noise mask of intensity n.

$$P(x, n, \tau) = P(x', n', \tau) \quad \text{iff} \quad P(\lambda x, \mu n, \tau) = P(\lambda x', \mu n', \tau). \tag{10}$$

Actually, the data suggest an even stronger hypothesis. To an excellent approximation, the data appear to fall on a pencil of lines (see Figure 4). This leads to the very strong conclusion

$$P(x, \tau, n) = F(x/n^{e(\tau)}) \tag{11}$$

for some strictly increasing function F, independent of delay. The question that we set forth above has thus been answered affirmatively.

It is of interest to seek dynamical systems which predict detection performance in accord with equation (11). Interestingly, our own efforts in this regard have proved unsatisfactory. Standard systems involving e.g. feedforward stages (as in gain control devices) appear to be incapable of performing in the desired manner. The apparent difficulty of this problem serves once again to emphasize the power of homogeneity laws abstracted here in (10), (11). We leave this issue unresolved as a challenge to the reader.

References

1. J. E. Hawkins and S. S. Stevens, *The masking of pure tones and of speech by white noise*, J. Acoust. Soc. Amer. **22** (1950), 6–13.

2. G. J. Iverson and M. Pavel, *On the functional form of partial masking functions in psychoacoustics*, J. Math. Psych. (1980) (submitted).

3. W. Jesteadt, C. C. Weir and D. M. Green, *Intensity discrimination as a function of frequency and sensation level*, J. Acoust. Soc. Amer. **61** (1977), 169–177.

4. M. Pavel and G. J. Iverson, *Invariant characteristics of partial masking: Implications for mathematical models*, J. Acoust. Soc. Amer. **69**(3) (1981).

5. ———, *Temporal properties of complete and partial forward masking*, Mathematical Studies in Perception and Cognition, No. 81–2, New York University, Department of Psychology, 1981.

6. S. S. Stevens and M. Guirao, *Loudness functions under inhibition*, Perception and Psychophysics **2** (1967), 459–465.

Department of Psychology, New York University, New York, New York, 10003

SIAM-AMS Proceedings
Volume **13**
1981

A Neural Mechanism for Generalization over Equivalent Stimuli in the Olfactory System[1]

WALTER J. FREEMAN

Introduction. The olfactory system in mammals poses an interesting problem in stimulus equivalence. On the one hand the input to the system is provided by an immense number of receptors (Le Gros Clark [**1957**]), roughly 10^8, embedded in an intricately folded membrane exposed to the air in the nasal passages. The number of types of receptor specificity for odor quality has been estimated (Amoore [**1971**]) to lie between 10 and 10^2, suggesting that there may be 10^6 or more receptors that are sensitive to any one odor. On the other hand the remarkable sensitivity of this system indicates that excitation of a small number of receptors, say 10^1 to 10^2, suffices for detection by a trained animal of the presence of an odor. The detection is consistent over multiple presentations, as shown by the performance of a tracking dog. Ccnsidering the turbulence of air flow through the nose, it is unlikely that an odorous substance falls on the same small subset of receptors on any two or more sniffs. In order to explain the perceptual invariance that is implied by the animal behavior, a neural mechanism must be postulated that gives a fixed output for all samples of an odor given to varying subsets of receptors among the set that is responsive to that odor.

A mechanism that can do this is based on the postulate that an animal forms a template of connections among those neurons in the olfactory bulb that are activated by a given odor during a training or familiarization experience. This postulate stems from observations on the spatial patterns of electrical waves generated by the olfactory bulb in cats and rabbits that were trained to identify particular odors.

1980 *Mathematics Subject Classification.* Primary 92A09.
[1]Supported by Grant MH06686 from the National Institute of Mental Health, USPHS.

Anatomical and physiological observations. The neurons of the bulb form layers in a curved sheet, roughly hemispherical, 2.5 mm in average radius, and 1.5 mm in thickness. The receptors send axons onto the surface in a relatively homogeneous distribution over the entire surface. Each axon ends synaptically onto one or a small number (1–10) of bulbar projection neurons totalling about 10^5, which send their axons through the interior of the bulb to other parts of the brain. These projection neurons also form reciprocal synaptic connections with bulbar interneurons (Rall and Shepard [**1968**]) numbering on the order of 10^8.

Neural activity in the bulb is manifested by a dipole field of potential; the zero isopotential surface is also roughly hemispheric and conforms to the mid-region of the layers of bulbar neurons (Freeman [**1975**]). The spatial pattern of potential is observed over one pole by placing arrays of electrodes onto the surface of the bulb where the receptor axons end, and recording the potential from each electrode with respect to a part of the skull well removed from the bulb. The time course of this activity takes the form of repeated brief bursts of oscillation in electrical potential lasting about 10^{-1} sec, with a frequency of oscillation in the range of 40–80 Hz, and a recurrence rate with each inspiration at 3–6 Hz in rabbits and 1–3 Hz in cats (Figure 1).

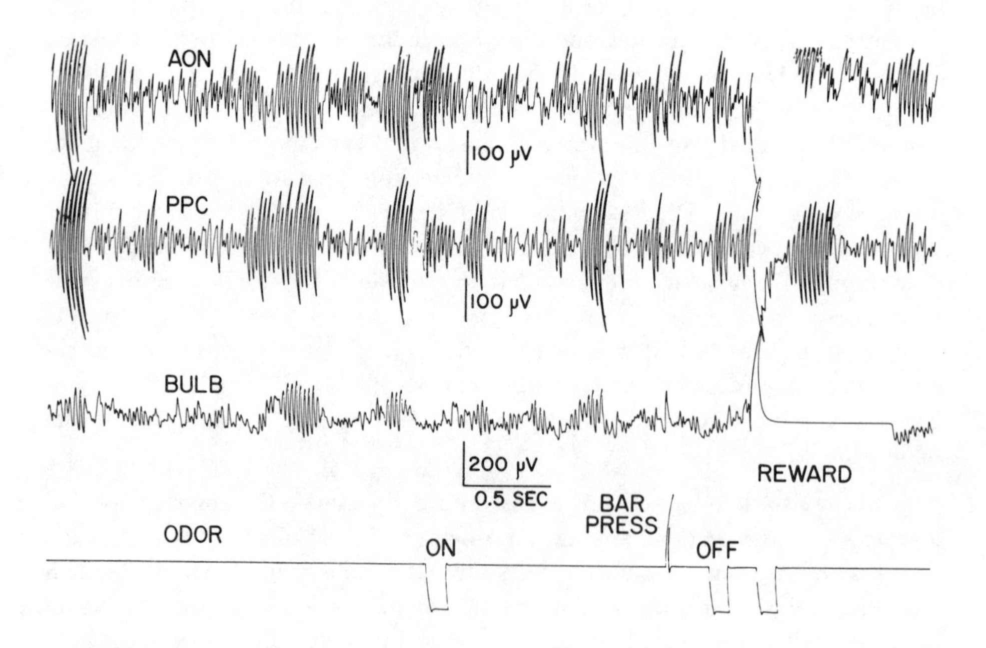

FIGURE 1. The elemental event in the neural activity pattern of the olfactory system is a brief oscillation called a wave packet (Freeman [**1975**]), which is revealed by an oscillation in brain potential called a burst. Examples are shown from three parts of the olfactory system of a cat (AON, anterior olfactory nucleus; PPC, prepyriform cortex; BULB, olfactory bulb) trained to press a bar for milk as a reward whenever a particular odor is delivered to the animal as a signal. Bursts occur with inspiration or sniffing.

The instantaneous frequency and duration of the burst are everywhere the same in the bulb. For this reason the bulbar activity can be time-parsed into a train of discrete events, and each event (burst) can be described by a two-dimensional vector sheet. The value of the vector at each surface recording point is given by the root mean square (rms) amplitude of the burst and the phase with respect to the ensemble average from a set of simultaneous records. The rms amplitude and phase are displayed in contour plots (Freeman **[1978a]**, **[1978b]**).

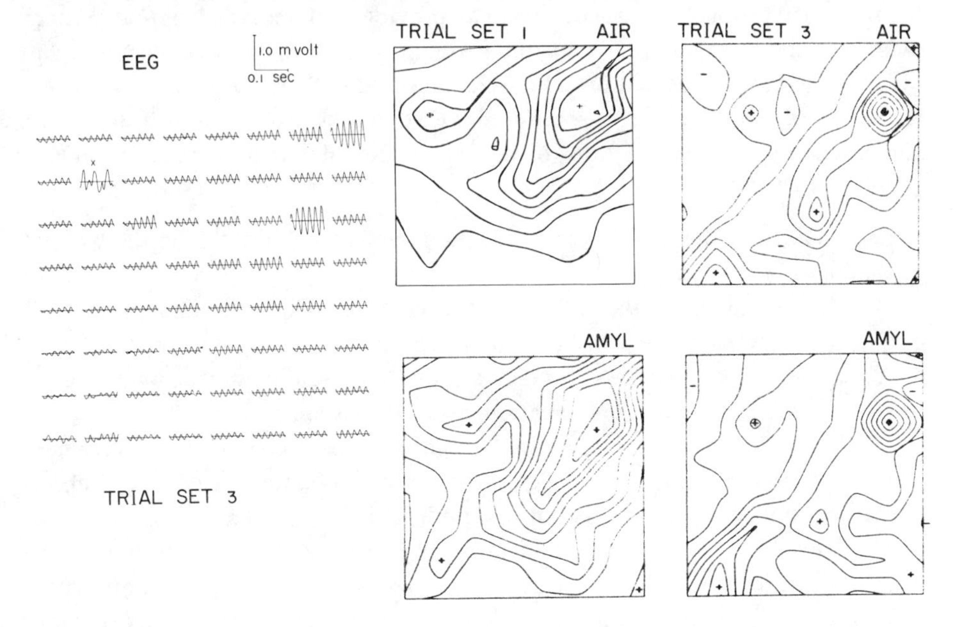

FIGURE 2. Each wave packet occupies a local region of the bulb and is manifested by a collection of burst recordings from an array of electrodes on the bulb. The example at left shows 64 traces from one burst recorded with an 8 × 8 array (3.5 × 3.5 mm). At the right the four frames show contour plots of burst amplitude averaged over ten trials. The upper frames are with background odor (AIR), and the lower frames are with a warning odor (AMYL acetate). The left frames are from the 1st day of training and show a difference in spatial pattern between AIR and AMYL. By the 3rd day of training the difference has vanished (right frames).

An example is shown in Figure 2 (left) of a set of 64 simultaneous burst recordings from an 8 × 8 array of electrodes. Examples of contour plots of rms amplitude are shown in Figure 2 (right). These contour plots do not change when an odor is presented to animals, unless the odor is paired with an aversive or rewarding stimulus (Freeman **[1978a]**). If this is done so as to establish a conditioned emotional response (CER) to one odor A, which is paired with an aversive shock, and not to another odor B, which is not paired with the shock, then during the first day of training the rms amplitude shows a new pattern during odor A presentation (left lower frame, AMYL), which is not present with odor B or with no odor (left upper frame, AIR).

In the next training session in which the CER is consolidated and thereafter, a new pattern is present whether or not the odor A or any other odor is present.

This pattern persists until retraining is undertaken with a new odor, and the above sequence repeats. In brief, the rms pattern does not depend on the odor given, but it appears to reflect the expectation by an animal of what significant odor it is to receive (Freeman [**1978b**]).

It is postulated that the rms amplitude manifests a template of neural connections in the bulb, which is formed during training to identify an odor, and which serves thereafter as a selective filter for searching for the odor. The template serves also as a mechanism to provide a stereotyped output pattern over a set of equivalent inputs, i.e. the varieties of combinations of receptors that are activated over repeated presentations of an odor. This postulate is expressed in a set of nonlinear integrodifferential equations, the solutions to which simulate the space-time patterns of the bulbar neural activity (Freeman [**1979a**], [**1979b**], [**1979c**]).

Construction of the model. The following premises are adopted on the basis of physiological analysis.

1. The dynamics of the bulb can be represented in time and in the two surface dimensions of the array of bulbar neurons.

2. All neural properties are time-invariant and spatially homogeneous, except in relation to input, template formation, and habituation.

3. The bulbar neural mass can be spatially divided into any convenient number of contiguous local neighborhoods or columns called *KII* subsets containing both excitatory (projection) and inhibitory (inter) neurons.

4. The subsets of excitatory and inhibitory neurons in every column can be each divided into two mirror-image subsets, respectively KO_{e1}, KO_{e2}, and KO_{i1}, KO_{i2} (Figure 3). They have reciprocal connections that are mutually excitatory (k_{ee}), mutually inhibitory (k_{ii}), excitatory to inhibitory (k_{ei}), or inhibitory to excitatory (k_{ie}).

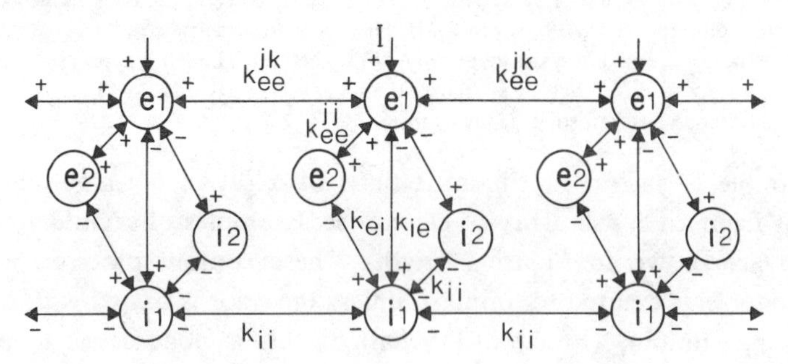

FIGURE 3. Each element is a *KII* subset that consists of two subsets of excitatory neurons (KO_{e1} and KO_{e2}) that by mutual excitation (+) form a KI_e subset (Freeman [**1975**]) and two subsets of inhibitory neurons (KO_{i1} and KO_{i2}) that by mutual inhibition (−) form a KI_i subset. Negative feedback coupling (k_{ei}, k_{ie}) is only within *KII* subsets and is fixed at the same level throughout the *KII* set representing the bulb. Excitatory to excitatory coupling (k_{ee}^{jk}) between and within *KII* subsets is modifiable; inhibitory to inhibitory coupling (k_{ii}) is fixed; the superscripts denote modifiability. The *KII* subsets are arranged in a two-dimensional array with a toroidal boundary condition. From Freeman [**1979c**].

5. Input to each column is given only to one subset, KO_{e1} or KO_{i1}, and not to both; output is from both subsets.

6. Every column is connected to every other column by mutually excitatory and mutually inhibitory connections, but not by negative feedback connections (k_{ei}, k_{ie}). To be consistent with Premise 2 a toroidal boundary condition is invoked.

7. The operations performed by each neural subset can be represented by a linear, time-invariant differential equation and a static nonlinearity.

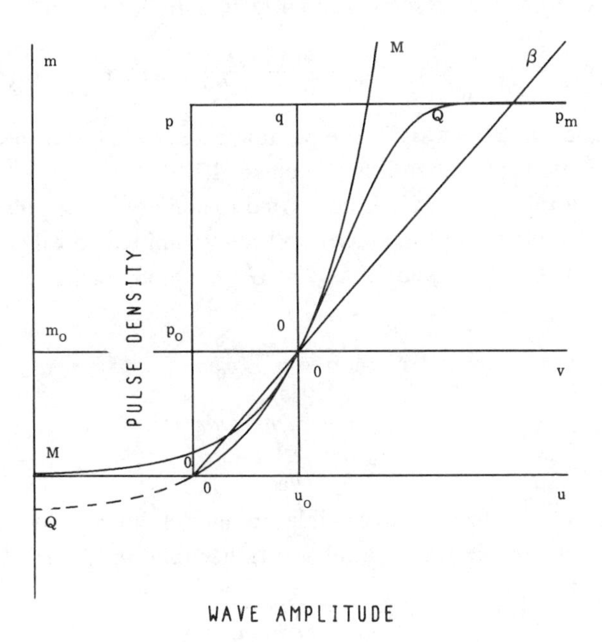

FIGURE 4. Pulse density p in a KO subset is related to wave density u through an intervening state variable m that can represent the mean level of sodium activation in the subset. The coordinate axis for m is orthogonal to those for u and p but is shown as coplanar for convenience. M, Q, and v are normalized variables. From Freeman [**1979a**].

8. The state of each KO subset can be represented by three state variables (Figure 4) that are continuous in time: axonal pulse density p, wave density (mean local dendritic current) u, and an intervening variable m that relates p to u, which can be thought of as the average tendency of neurons in the subset to give pulses, or as the average subthreshold value in the subset of the sodium activation factor defined in the Hodgkin-Huxley equations (Hodgkin and Huxley [**1952**]).

9. The variable m increases exponentially with increasing v.

10. Pulse density p approaches a maximum p_m asymptotically as m increases.

11. In normal operation a steady state is posited at (u_0, m_0, p_0); these steady state values are linearly proportional to each other.

Each KO subset receives input in the pulse mode from other subsets, which is delayed by axonal conduction, and which is converted by synapses to activity in the wave mode in the dendrites. This operation is linear insofar as normal (nonepileptic) activity is concerned and is represented by a coefficient ζ.

The wave activity undergoes delay owing to the cable properties of the dendrites and to the passive membrane resistance and capacitance of the neurons. The axonal conduction, synaptic, and dendritic cable delays can be lumped into a single term, so that the time-dependent properties of the nth subset can be expressed by a second order differential equation.

$$F[v_n] \stackrel{\Delta}{=} \frac{1}{ab}\frac{d^2}{dt^2}[v_n(t)] + \frac{a+b}{ab}\frac{d}{dt}[v_n(t)] + \frac{1}{ab}[v_n(t)] \qquad (1.1)$$

where $\stackrel{\Delta}{=}$ means "is defined by". The parameters $a = 220/\mathrm{sec}$ and $b = 720/\mathrm{sec}$ have been evaluated experimentally (Freeman [**1975**]).

The activity in the wave mode is converted to activity in the pulse mode at the trigger zones of neurons in the subset and is transmitted to other subsets. This nonlinear conversion is designated as $p = G(v)$. Three normalized variables are defined:

$$Q \stackrel{\Delta}{=} (p - p_0)/p_0, \qquad (1.2)$$

$$M \stackrel{\Delta}{=} (m - m_0)/m_0, \qquad (1.3)$$

$$v \stackrel{\Delta}{=} u - u_0. \qquad (1.4)$$

Because u_0 is not directly accessible to measurement, the equations are developed in terms of the wave variable v rather than u. Under Premise 9, and with proper scaling

$$dm/du = m. \qquad (2.1)$$

Then

$$\int_{m_0}^{m} \frac{dm'}{m} = \int_{u_0}^{u} du', \qquad (2.2)$$

or, by definitions (1.3) and (1.4),

$$M = e^{v} - 1. \qquad (2.3)$$

Under Premise 10,

$$\frac{dp}{dm} = \frac{p_m - p}{p_m - p_0}. \qquad (3.1)$$

By Premise 11

$$p_0 = m_0. \qquad (3.2)$$

Then

$$\int_{p_0}^{p} \frac{dp'}{p_m - p'} = \frac{1}{p_m - p_0}\int_{m_0}^{m} dm', \qquad (3.3)$$

and from definition (1.2),

$$Q = Q_m[1 - \exp(-M/Q_m)], \tag{3.4}$$

where Q_m is the maximal pulse rate normalized from p_m by equation (1.2). Empirically it is obvious that $p \geqslant 0$ and therefore $Q \geqslant -1$. From equations (2.3) and (3.4), and from definition (1.4), and by setting $u = 0$ at $p = 0$,

$$Q = Q_m\{1 - \exp[-(e^v - 1)/Q_m]\}, \qquad v > -u_p, \tag{4.1}$$

$$Q = -1, \qquad v \leqslant -u_0, \tag{4.2}$$

$$u_0 = -\ln[1 - Q_m \ln(1 + 1/Q_m)]. \tag{4.3}$$

From Premise 11,

$$p_0 = \gamma u_0. \tag{4.4}$$

Experimental analysis shows that with proper scaling of variables, $\gamma = 1$. From definition (1.2)

$$p = u_0(Q + 1). \tag{4.5}$$

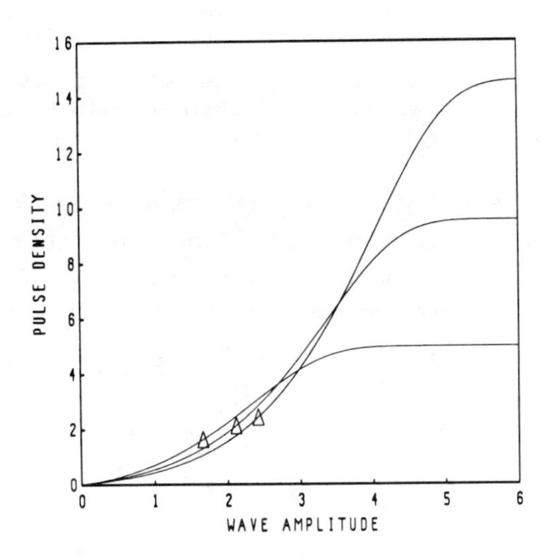

FIGURE 5. Three examples are shown for the static nonlinear wave to pulse function $p = G(v)$ for three values of Q_m. The triangles show the steady state values (p_0, u_0). From Freeman **[1979a]**.

The desired function $p = G(v)$ is shown in Figure 5 for three values of Q_m. The selection of Q_m over u_0 or p_0 as the single controlling parameter is based on its accessibility to experimental evaluation (Freeman **[1975]**, **[1979a]**).

The nonlinear gain is given by the derivative

$$\frac{dp}{dv} = \frac{dp}{dQ}\frac{dQ}{dM}\frac{dM}{dv}. \tag{5.1}$$

From equations (1.4), (2.3), (3.4), (4.5), and (5.1),

$$\frac{dp}{dv} = u_o \exp\left[v - \frac{e^v - 1}{Q_m}\right]. \tag{5.2}$$

The gain approaches zero as v approaches $\pm\infty$ and reaches a maximum at

$$v = \ln(Q_m). \tag{5.3}$$

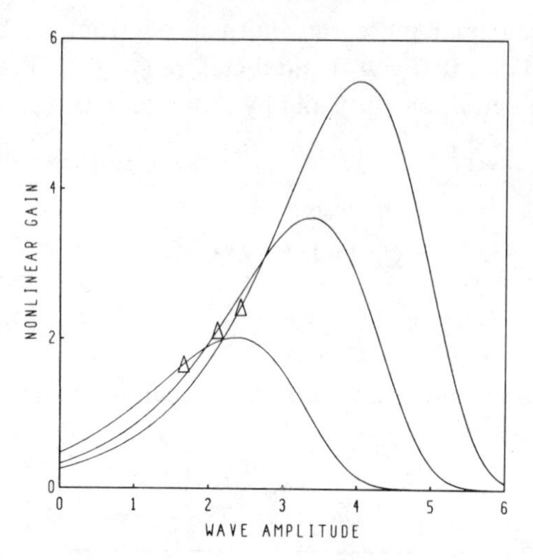

FIGURE 6. Three examples are shown of the gain for nonlinear wave to pulse conversion at three values of Q_m. Excitatory wave input causes increased pulse output and a nonlinear increase in gain. This nonlinearity is the key property that underlies burst formation. From Freeman [1979a].

As shown in Figure 6, a small positive (excitatory) input to a *KO* subset not only gives positive output; it increases the gain of the subset. The amount of gain increase is determined by the steady state dendritic depolarization u_0 that relates to p_0 by equation (4.4) and the maximal pulse density Q_m through equation (4.3). From the connections given in Figure 3 and from equations (1.1) and (4.5),

$$F(v_{e1,j}) = \zeta_e^j k_{ee}^{jj} p_{e2,j} - \zeta_e^j k_{ie}(p_{e1,j} + p_{i2,j}) + \sum_{k \neq j}^{N} \zeta_e^j k_{ee}^{jk} p_{e1,k} + I_j,$$

$$F(v_{e2,j}) = \zeta_e^j + k_{ee}^{jj} p_{e1,j} - \zeta_i k_{ie} p_{i1,j},$$

$$F(v_{i2,j}) = \zeta_e^j k_{ei} p_{e1,j} - \zeta_i k_{ii} p_{i1,j},$$

$$F(v_{i1,j}) = \zeta_e^j k_{ei}(p_{e1,j} + p_{e2,j}) - \zeta_i k_{ii} p_{i2,j} - \zeta_i k_{ii} \sum_{k \neq j}^{N} p_{i1,k}, \tag{6.1}$$

where the superscripts denote pulse-wave conversion coefficients ζ_e^j and modifiable synapses k_{ee}^{jk} from the kth to the jth subset. The equations are solved by numerical integration with zero initial conditions and a toroidal boundary condition. The form of the input for each element I_j is structured to simulate the time-dependency of input in the olfactory system; the input amplitude p_j is either zero or unity for each element. In the procedure for evaluation of the model all values for ζ and k are set provisionally at unity, until the proper range of values is found for Q_m, such that the EEG burst is simulated. The performance is then optimized by setting $k_{ei} = 5.0$ and $k_{ie} = 0.2$, reflecting the smaller number and greater size of projection neurons (Freeman [1979b]). From stability considerations (Freeman [1979c]),

$$k_{ii} = k_{ee}^{jj} = 0.25, \tag{6.2}$$

$$\zeta_i = 1.64 \exp(-0.53\, Q_m), \tag{6.3}$$

$$\zeta_e^j = 1.64\, \beta_j \exp(-0.53\, Q_m), \tag{6.4}$$

where β_j is a parameter representing the effects of habituation that ranges downwardly from 1.0 to 0.6 with increasing habituation. These relations have the effect of reducing the nonlinear gain to an "on-off" parameter, such that for $Q_m \geqslant 2$ a stereotypic burst is generated for appropriate input, whereas for $Q_m < 2$ it is not. An example is shown in Figure 7 for simulated bursts from a (10×1) array of elements. The output for each element is taken as the rms amplitude V_j of $(v_{e1,j} + v_{e2,j})$ and the phase Φ_j with respect to the ensemble average over all elements.

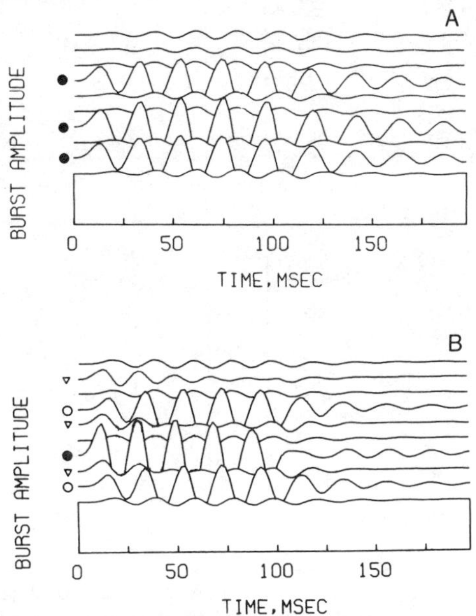

FIGURE 7. Two examples are shown of simulated bursts from a 1×10 array. A. A template exists for the 3 elements shown by dots, and input is given to all 3. B. Input is given only to one element in the template (a "hit") and not to the other two elements (circles). Noise input is given to the 3 elements marked by triangles. From Freeman [1979c].

Formation and testing of templates. A template is formed over a sequence of simulated bursts, or most quickly over one burst. All elements (selected at random) that are to be included in the template are given unit input, $p_j = 1$, $j = 1, \ldots, A$, where A is the number of template elements $< N$, the number of elements. All other inputs are zero. The simulated burst of each element (Figure 7, upper frame) is cross-correlated with every other burst to form an $(N \times N)$ product-moment correlation matrix $r(j, k)$. The matrix is weighted by an empirical coefficient $\beta_a = 1.15$. If, for any pair of elements j and k, $r(j, k)\beta_a > 0.25$, and if the geometric mean of V_j and V_k exceeds E_{rms}, the rms amplitude of the ensemble average, then k_{ee}^{jk} is replaced by $r(j, k)\beta_a$, k_{ee}^{jj} by $r(j, j)\beta_a$ and k_{ee}^{kk} by $r(k, k)\beta_a$; otherwise it is unchanged. The set of modified k_{ee}^{jk} coupling coefficients among the A elements constitutes the template.

The modified array of elements is tested by giving unit input to elements within a template (each such input is referred to as a "hit"), and to other elements ("noise"). The output is in two forms: a contour plot of the rms amplitudes for all of the elements V_j, and the rms amplitude of the ensemble average E_{rms}.

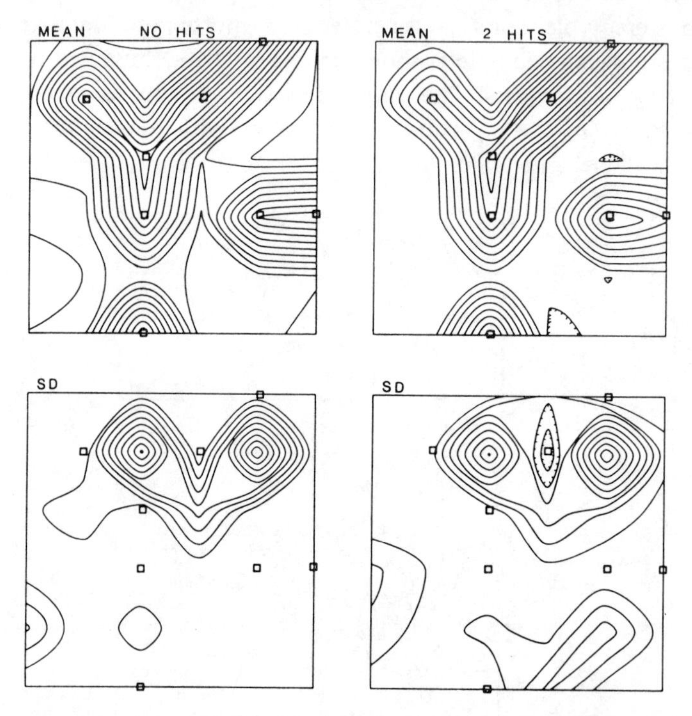

FIGURE 8. The upper frames show the rms amplitudes in contour plots for a 6 × 6 array of interconnected *KII* subsets. The 8 small squares show the locations of template elements. Each upper frame shows the mean of 5 simulated bursts. The lower frames show the standard deviations (S.D.). For the left frames the inputs consist of noise given at random to 8–16 nontemplate elements. For the right frames the inputs consist of the same noise plus 1–4 "hits" on each run at randomly differing template sites. Output image invariance in respect to input (Grossberg [1980]; Babloyantz and Kaczmarek [1979]) is shown by the displacement of maximal variance from maximal rms amplitudes. From Freeman [1979c].

It is found that the contour plot is dominated by the template pattern, irrespective of the number or location of "hits" and the presence of "noise". An example for a 6 × 6 array is shown in Figure 8 (left) of the mean and standard deviation (S.D.) of 5 contour plots with no "hits" on any of the 8 template elements (small squares), but with "noise" given to 8 randomly selected elements. In Figure 8 (right) the mean and S.D. are given for the same noise patterns but with 2 "hits" on each trial at differing template locations. There are no significant differences between the contour plots for "hits" and no "hits". The stability of the spatial image is shown by the finding that the maximal values for the S.D. are not on the template. This replicates the relations observed in the bulbar field potentials (Figure 9), in which the locations of maximal variance were found to be eccentric to the foci of maximal amplitude.

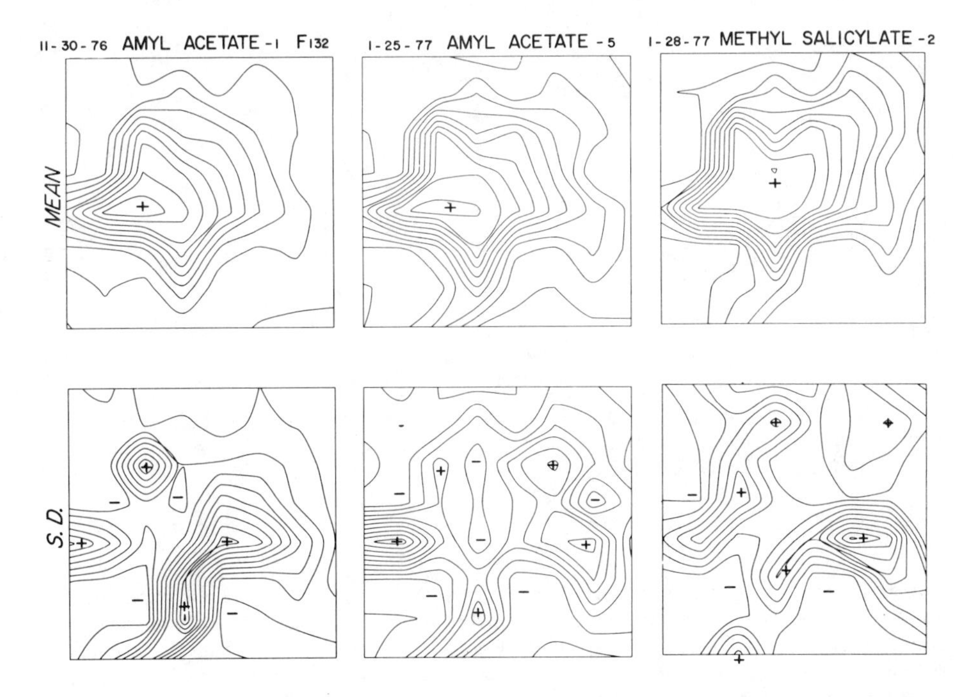

FIGURE 9. The three upper frames show the contours of mean rms amplitudes over 10 bursts from the olfactory bulb of a rabbit on presentation of warning odors used as conditioned stimuli. The lower frames show the standard deviations (S.D.) of the rms amplitudes. EEG image stabilization is reflected in the spatial disparity between maxima for mean and S.D.

The E_{rms} value is found to exceed the background level determined by "noise" for any one or more "hits", provided that the number of elements H receiving "noise" does not exceed A, the number of template elements (Figure 10). This restriction is entirely removed if the elements receiving noise undergo habituation by setting $\beta_j = 0.6$ in equation (6.4) for all elements receiving "noise" p_j, $j = 1, \ldots, H$. This operation also permits partial or complete suppression of templates.

The increase in E_{rms} with one or more "hits" is not usually due to an increase in V_j for the template elements that receive "hits". Globally it is due either to an increase in V_j of nontemplate elements (Figure 10D) or to a decrease in phase dispersion of the Φ_j's (Figure 10B). The output of each element can be represented by a vector. The outputs of similar elements tend to cluster and can be summed as vectors representing the outputs of template (T), noisy (N), and unstimulated (U) elements.

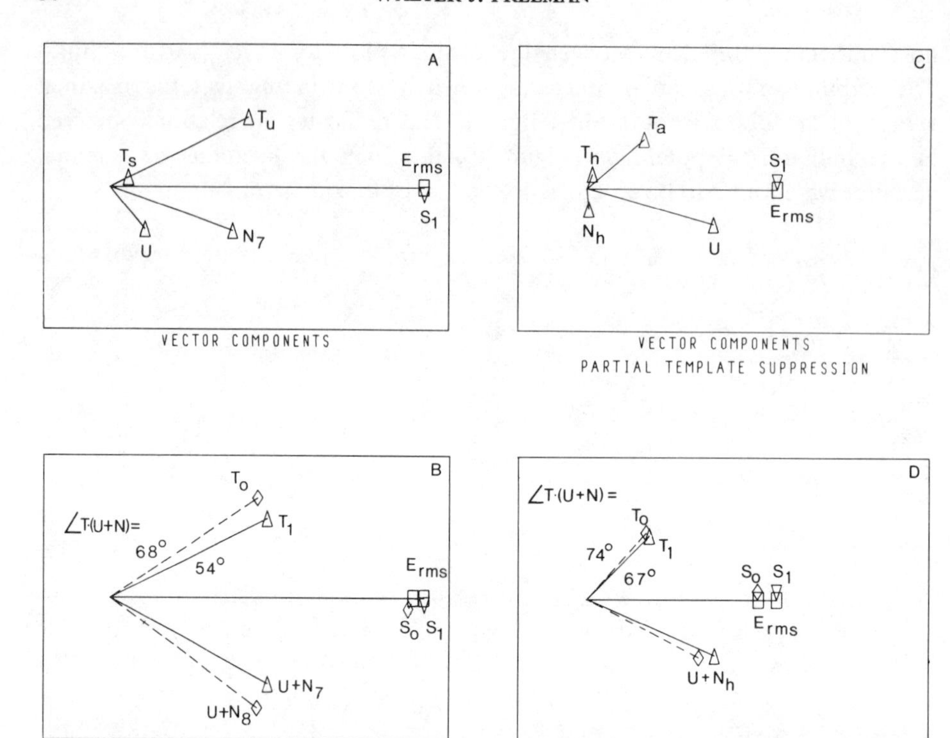

FIGURE 10. The output of each *KII* subset can be expressed as a vector having rms amplitude and phase with respect to the ensemble average E_{rms}. The outputs from subsets that form groups are added vectorially to give group vectors, and these are added to give the total vector S. Examples of summated output vectors are shown for the two-dimensional case with an 8-element template and noise input to 7 or 8 elements (no matched filter). A. There are 7 elements without input (T_u), one "hit" (T_s), 7 elements with noise (N_7), and 21 unstimulated elements (U). B. The summated vectors are shown for template elements with no "hit" (T_0) and with one "hit" (T_1), and for the 28 nontemplate elements with 7 noise inputs ($U + N_7$) and with 8 noise inputs ($U + N_8$). The increase in total output (S_0 to S_1) takes place with no difference in total input, and can be ascribed to the decrease in angle $\angle T \cdot (U + N)$ between summated vectors. The nontemplate element that no longer receives input has an increase in the phase lag of its output from $E(t)$ of about −30°. C. A template is formed with 8 elements, and then 4 of the 8 elements are suppressed with a matched habituation filter ($\zeta_e^j = 0.33$). Input is given to 3 of those 4 (T_h), to 1 of the 4 active elements (T_1), and to 8 nontemplate elements (N_h) with a matched habituation filter ($\zeta_e^j = 0.33$). There are 20 remaining unstimulated elements (U). D. The changes in summated template vectors (T_0 and T_1) and nontemplate vectors ($U + N_h$) are shown, when one "hit" is changed from a suppressed template element (T_0) to an active template element (T_1). The angle $\angle T \cdot (U + N)$ decreases by 6°, and E_{rms} increases by 11%.The mean values for $\angle T \cdot (U + N)$ and its change with one "hit" are 66° ± 8° and 10.5° ± 3.3°, over 6 computer runs with variations in matched habituation filter strength. These values simulate the average EEG phase differences observed in trained rabbits (67° ± 4°) and the average magnitude of change (8.5° ± 2.9°) on presentation of an expected odor (Freeman [**1980**]). From Freeman [**1979c**].

Preliminary physiological evidence (Freeman [**1980**]) shows that a decrease in phase dispersal within a burst occurs in trained animals when an expected odor arrives. The requisite spatial integration is known to take place in the brain structure to which the bulb sends its output. These results imply that the detection of an expected event is a global property of the array, and that it is not manifested in the elements where "hits" occur, nor is that information contained in the output.

Conclusion. The dynamics of this model indicate that the operation of the olfactory system depends heavily on centrifugal controls, i.e. signals carried by neural fiber bundles that run from the brain to the olfactory bulb and cortex in the direction opposite to the main olfactory information flow. These signals can be expected to give nonspecific commands to the bulb, such as "turn on", "form template to whatever is there", "activate template to whatever is there", "suppress all activated templates", "habituate to whatever is there", "suppress all habituation", and so on. There are known anatomically to be ten or more centrifugal paths to the bulb and yet others to the olfactory cortex, although their functions have not yet been physiologically sorted out.

According to the model, when an animal is aroused by a novel motivating stimulus in any sensory modality, it will orient to the on-going odor input and thereby form a compound template to a test odor in association with background odors. Presentation of background odors subsequently may suffice for selective activation of the compound template, and selective habituation "to whatever is there" may suppress the background components, leaving the component of the template for the test odor as a matched filter for that odor. The neural active state manifested in the EEG can be viewed as a neural search image for an expected odor, or as a hypothesis to be tested against sensory input, or as the objective substrate of a mental image or recollection of the odor.

The applicability of this model to the visual somesthetic and auditory systems has not yet been determined; too little is known about their central physiology, although the requisite gating and carrier frequencies are abundantly manifest in their EEG's. Certainly neocortical dynamics are more complex than the mechanisms of the olfactory forebrain, and the olfactory perceptual apparatus together with the archicortex phylogenetically preceded the emergence of neocortex in mammals. The manner of growth and the need for commonality of coding among the several modalities give reason to believe that neocortical mechanisms may be adaptations and elaborations of the perceptual processes that first appeared in the sphere of olfaction, such that the findings reported here may have a certain generality.

REFERENCES

Amoore, J. E., 1971. *Olfactory genetics and anosmia*, Handbook of Sensory Physiology. IV, Chemical Senses, Part 1, Olfaction (L. M. Beidler, ed.), Springer-Verlag, Berlin, Chapter 11.

Babloyantz, A. and L. K. Kaczmarek, 1979. *Self-organization in biological systems with multiple cellular contacts*, Bull. Math. Biol. **41**, 193–201.

Freeman, W. J., 1975. *Mass action in the nervous system*, Academic Press, New York.

______, 1978a. *Spatial properties of an EEG event in the olfactory bulb and cortex*, Electroencephalogr. Clin. Neurophysiol. **44**, 586–605.

______, 1978b. *Models of the dynamics of neural populations*, Electroencephalogr. Clin. Neurophysiol. Suppl. No. **34**, 9–18.

______, 1978c. *Spatial frequency analysis of an EEG event in the olfactory bulb*, Multidisciplinary Perspectives in Event-Related Brain Potential Research (D. A. Otto, ed.), U.S. Govt. Printing Office, EPS-600/9-77-043, pp. 531–546.

______, 1979a. *Nonlinear gain mediating cortical stimulus-response relations*, Biol. Cybernet. **33**, 237–247.

______, 1979b. *Nonlinear dynamics of paleocortex manifested in the olfactory EEG*, Biol. Cybernet. **35**, 21–37.

______, 1979c. *EEG analysis gives model of neuronal template-matching mechanism for sensory search with olfactory bulb*, Biol. Cybernet. **35**, 221–234.

______, 1980. *Use of spatial deconvolution to compensate for distortion of EEG by volume conduction*, IEEE Trans. Biomed. Engin. **BME-27**, 421–429.

Grossberg, S., 1980. *How does a brain build a cognitive code?*, Psychol. Rev. **87**, 1–51.

Herrick, C. J., 1963. *Brains in rats and men*, Hafner, New York.

Hodgkin, A. L. and A. F. Huxley, 1952. *A quantitative description of membrane current and its application to conduction and excitation in nerve*, J. Physiol. **117**, 500–544.

Le Gros Clark, W. E., 1957. *Inquiries into the anatomical basis of olfactory discrimination*, Proc. Roy. Soc. London **146B**, 299–319.

Rall, W. and G. M. Shepherd, 1968. *Theoretical reconstruction of field potentials and dendrodendritic synaptic interactions in olfactory bulb*, J. Neurophysiol. **31**, 884–915.

DEPARTMENT OF PHYSIOLOGY-ANATOMY, UNIVERSITY OF CALIFORNIA, BERKELEY, CALIFORNIA 94720

SIAM-AMS Proceedings
Volume **13**
1981

Differential Equations for the Development of Topological Nerve Fibre Projections

CH. VON DER MALSBURG AND D. J. WILLSHAW

We are concerned with the period in vertebrate ontogenesis when the nerve cell layer of the eye, the *retina*, develops output cells, *ganglion cells*. Each of these puts out an *axon* which grows over the surface of the retina and through the optic stalk into the optic pathway to the brain, where it establishes connexions, *synapses*, with cells of the developing nervous system. In lower vertebrates, such as frog, toad or fish, which are the animals that have been most studied, the main central target structure is the *optic tectum*. The initial connexions made by the optic nerve fibres are, in some cases, quite haphazard. After some time, however, a final configuration is reached in which there is an ordered map of retina onto tectum. Such maps are called retinotopic. The problem that has been studied in this field for several decades is: how do the fibres manage to find their proper termination area? A large number of experiments has been performed to gather constraints for possible theories. Many of these are reviewed in **[1]**, **[2]**.

This intensive work on the problem of retinotopy has been motivated by the hope of having a paradigm problem which is both conceptually very simple and provides insight into the way in which specific patterns of nerve connexions are constructed in the brain during ontogenesis and possibly also during learning.

Now that many different theories have been formulated for the problem of retinotopy, a certain consensus is emerging for what might be the important ingredients. At least two mechanisms seem to be acting in cooperation.

(a) A *local* fibre ordering mechanism ensures that each retinal fibre preferentially establishes synapses at those places on the tectal surface where other fibres coming from the same retinal region have their synapses.

(b) A *global* mechanism provides just enough information to specify the orientation of the final mapping between retina and tectum.

1980 *Mathematics Subject Classification*. Primary 92A09, 34H05.

The local mechanism logically requires the existence of a *marker* system to encode the neighborhood relationships of the retinal ganglion cells. The nature of this marker system, the relative importance of the two mechanisms as well as other details are still under discussion. But an overall mathematical framework can already be given.

The most obvious mathematical formulation for the retinotopy problem is in terms of differential equations for the locations and the strengths of the contacts made by the retinal fibres on the tectal surface. The difficulty with this approach, as often in neurobiology, is that no precise analytical form for the different terms offers itself in any experimentally testable way. In our earlier papers **[2]**–**[4]** we described, in the form of a computer program, a set of differential equations which seemed to us to represent the biological processes most naturally. We characterized the behaviour of the system in terms of rules derived from numerical solutions to these equations; in this paper we discuss their mathematical properties. We must stress that many of the details were formulated on intuitive grounds as detailed experimental evidence is lacking.

The signal code. We have discussed two possibilities for the physical nature of the markers carried by the ganglion cell axons to encode their neighborhood relationships. One of them consists of electrical pulse activity in the fibres **[4]**. The other is that there is a relatively small number of marker molecule types which are carried in different proportions by the various ganglion cells and axons **[3]**. What we know of action potentials on the one and of molecules on the other hand suggests slightly different types of 'boundary conditions' for the two cases, so that it may be worth discussing and comparing them here.

The electrical activity code is imagined to express neighborhood relationships in the retina in the form of correlations between spike sequences in different cells: Isolated ganglion cells have random spontaneous activity. They are coupled within the retina by lateral connexions by which correlations and anticorrelations are introduced between the individual spike trains. Short range connexions are excitatory, leading to correlations between the activity of neighboring cells; long range connexions are inhibitory, so that corresponding anticorrelations are produced between the activity of cells a long way apart. It is convenient to think in terms of the extreme case in which at each moment in time retinal activity is concentrated into one small island of a few cells, the rest of the retina being silent. Over time, activity will occur in all possible islands on the retina. When considering two fibres which terminate on the tectum, one can decide whether their ganglion cell bodies are neighors in the retina by looking for correlations in their spike activity. The activity pattern does not, however, tell us anything about the absolute location of the cells in the retina.

In the other alternative, the molecular signal code, each retinal cell and fibre carries a vector which records the concentrations of a fixed set of a small number of molecule types. A suitable marker system for the retina would be one in which there is one isolated point source for each molecule type. Given that there are also lateral transport and decay of molecules, the steady state retinal distribution of the concentration of each molecule will have a single peak, and a monotonical decrease in all directions from the peak.

The concentration c_ρ^m of marker type m in ganglion cell ρ is the solution of the diffusion equation

$$\frac{\partial c_\rho^m}{\partial t} = -\alpha c_\rho^m + \sum_{\rho'} D_{\rho\rho'} c_{\rho'}^m + Q_\rho^m$$

where α is the decay constant and D is a matrix of diffusion constants specifying the exchange of molecules between neighboring cells. Q_ρ^m describes the production of molecule m. It has a positive value for one cell only and is otherwise zero. Proximity of cells is expressed by the similarity between concentration vectors. If we consider n different types of molecules, similarity is measured by some function s: $V^n \times V^n \to [0, 1]$. s has the value 1 if the two concentration vectors are the same, 0 if the two vectors are normal to each other. Similarity should be a monotonically decreasing function of distance between ganglion cells.

From the mathematical point of view the molecular alternative is much more convenient because in this case signal processing is entirely linear, whereas processing of electrical activity of cells is known to be a highly nonlinear phenomenon. However, both alternatives could be realized in nature.

Signal coupling between retina and tectum. Retina and tectum are coupled by contact structures, *synapses*, between retinal *fibres* and tectal *cells*. Each synapse may be characterized by a number, called *synaptic strength* or *weight*.

At each time, the set of synaptic strengths defines the state of the rectinotectal projection. The transformation of an initially disordered projection into a retinotopic map shall be described by rules for the plastic changes of synaptic strengths, which take place in response to the locally available signals.

The topological structure of the tectum impresses itself on the system by there being the same kind of lateral transport of signals as in the retina: excitatory and inhibitory connexions for electrical signals, or lateral diffusion for molecules.

For each synapse one can evaluate the similarity between the signals in the retinal fibre and the tectal cell, by the function described earlier for evaluating similarity in the retina. The similarity between the marker in the retinal cell with index ρ and the tectal cell with index τ will here be called $s_{\tau\rho}$. It is these quantities which control the growth of synapses.

The type of solution to the retinotopy problem we are examining is one where the signals in the tectum are themselves supplied from the retina through the synapses already made. Thus each synaptic strength determines the degree to which the signal from the fibre influences the signal on the tectal cell. In the case of molecular markers, the synaptic strength may be regarded as a diffusion constant for the passage of retinal markers into the tectal cell. Thus the tectal markers are not preprogrammed as suggested by Sperry [**5**] but are *induced* by the retinal fibres. Such an inductive scheme solves the problem of how to produce two matching sets of markers in two independently developing structures, retina and tectum. Moreover, only such a scheme is compatible with the experimental findings, as discussed in [**2**].

The mathematically most convenient assumption to make about the distribution of synapses is that at all times all possible synapses are present, but many of them have zero strength and so are silent. We found it biologically more plausible to imagine that only a selection of synapses is present at any given time. Then in order that the synaptic distribution can change over time, a facility for the creation of new synapses is required. We supposed that each fibre has the facility to continually put out branchlets (sprouts) from each tectum cell which it already contacts onto the neighboring ones which it does not yet occupy. Conversely, contacts which have reached zero strength are assumed to be no longer in existence.

Plasticity. Let us call $W_{\tau\rho}$ the strength of the synapse between retinal cell ρ and tectal cell τ. The basic idea is to let the growth of $W_{\tau\rho}$ be controlled by the similarity $s_{\tau\rho}$, so that a larger $s_{\tau\rho}$ leads to a faster growth of $W_{\tau\rho}$. For this purpose we introduce the term $F_{\tau\rho}$ which specifies the rate of growth of synapse $\tau\rho$. In addition to its dependence on $s_{\tau\rho}$, $F_{\tau\rho}$ may also depend on the $W_{\tau\rho}$'s (see below). In symbols:

$$\dot{W}_{\tau\rho} = F_{\tau\rho}. \tag{1}$$

(We are assuming that this growth rule applies only to those synapses which are existing, $\dot{W}_{\tau\rho}$ being zero otherwise.) Since this law would lead to unlimited growth, the strengthening of some 'successful' synapses has to be balanced by the weakening of others. This is most conveniently enforced by keeping the sum of synaptic strengths available to each retinal cell constant, i.e.

$$\sum_{\tau'} W_{\tau'\rho} = W, \tag{2}$$

where this sum is (as are all future sums over tectal cells) over only those tectal cells with which fibre ρ has contacts. Since these rules cannot both be obeyed simultaneously they must be modified slightly. The procedure we proposed earlier [**2**], [**3**] was the following: In each small time interval Δt, let $W_{\tau\rho}$ increase in proportion to $F_{\tau\rho}$. Its new value is $W'_{\tau\rho} = W_{\tau\rho} + \Delta W_{\tau\rho}$ where $\Delta W_{\tau\rho} = F_{\tau\rho}\Delta t$. Then renormalize it multiplicatively to reinstall (2):

$$W'_{\tau\rho} \to W''_{\tau\rho} = W'_{\tau\rho} \cdot W / \sum_{\tau'} W'_{\tau'\rho}.$$

The differential equation for $W_{\tau\rho}$ is then defined as

$$\frac{dW_{\tau\rho}}{dt} = \lim_{\Delta t \to 0} (W''_{\tau\rho} - W_{\tau\rho})/\Delta t$$

which by (2) is:

$$= \lim_{\Delta t \to 0} \left[\frac{W_{\tau\rho} + \Delta W_{\tau\rho}}{\left(1 + \sum_{\tau'} \Delta W_{\tau'\rho}/W\right)} - W_{\tau\rho} \right] / \Delta t$$

$$= \lim_{\Delta t \to 0} \left[(W_{\tau\rho} + F_{\tau\rho}\Delta t)\left(1 - \Delta t \sum_{\tau'} F_{\tau'\rho}/W\right) - W_{\tau\rho} \right] / \Delta t$$

or

$$\dot{W}_{\tau\rho} = F_{\tau\rho} - W_{\tau\rho} \sum_{\tau'} F_{\tau'\rho}/W. \tag{3}$$

This is the basic form of the differential equation that we have used for the strengths of the retinotectal synapses. (To be precise, this differential equation would have to be modified slightly by an additive term by which new synapses are brought into existence by sprouting.) For its solution the differential equation has to be supplemented by equations for the $F_{\tau\rho}$, which depend on the signals in retina and tectum, some of which in turn depend on the synaptic weights.

Discussion of the differential equation. We have not yet decided precisely how $F_{\tau\rho}$ should depend on the synaptic weights. Since there is no direct experimental information nor even a generally accepted view on the issue, the following discussion might help. Consider the synapses made by one retinal fibre ρ. Omitting the index ρ, we are left with the equation $\dot{W}_\tau = F_\tau - W_\tau \Sigma_{\tau'} F_{\tau'}/W$.

In the stationary case, when $\dot{W}_\tau = 0$,

$$W_\tau / W = F_\tau / \sum_{\tau'} F_{\tau'}, \tag{4}$$

i.e. the distribution of the synaptic weights made by one retinal fibre mirrors the distribution of the corresponding F_τ's.

Insight into the significance of the basic differential equation (3) may be gained by noticing the strong similarity it bears to the differential equations used to describe evolution. The quantity W_τ can be interpreted as the *population number*, the number of individuals, in the species τ, F_τ as the Darwinian fitness of that species. The subtractive term acts to keep the total population number constant. This equation has been extensively discussed in **[6]**; see equation 37. The process of creation and destruction of synapses by sprouting and retraction finds its analogy in the evolutionary case in mutation, by which a new species is created, and in extinction of a species.

Let us suppose that F_τ depends on W_τ in the following way:

$$F_\tau = W_\tau^\mu s_\tau \tag{5}$$

where μ is a positive exponent. We first consider the case when s_τ is a constant. (For the retinotopy problem this simplifying assumption of constant s_τ is not too unrealistic, since $s_{\tau\rho}$ is a function of all $W_{\tau\rho}$'s and the dependence on any particular fibre is weak.) The differential equation is now:

$$\dot{W}_\tau = W_\tau^\mu s_\tau - W_\tau \sum_{\tau'} W_{\tau'}^\mu s_{\tau'} / W. \tag{6}$$

The behaviour of (6) will depend on the chosen value of μ, as discussed in [**6**] for the evolution case. There, s_τ is determined solely by the structure of the species τ. We will now discuss (6) in terms of the evolution analogy.

Case $\mu = 1$. This case is referred to as Darwinian evolution in the narrow sense. We obtain the special form

$$\dot{W}_\tau = W_\tau(s_\tau - \bar{s}) \tag{7}$$

where $\bar{s} = \Sigma_{\tau'} W_{\tau'} s_{\tau'} / \Sigma_{\tau'} W_{\tau'}$ is the weighted mean of the fitnesses of all species. This differential equation has the nice property that precisely those species grow which have higher than average fitness, irrespective of their population number (as long as $W_\tau \neq 0$). In the course of evolution $\bar{s}$ grows monotonically, pushing more and more species below the threshold of growth. In the final stationary state, $W_\tau = 0$ for most species. For the survivors $s_\tau = \bar{s}$, i.e. the survivors have maximal fitness. In general there will be only one survivor.

Case $\mu = 2$. To compare this case with the previous one let us write (5) in the form $F_\tau = W_\tau \cdot (W_\tau s_\tau)$; the similarity s_τ is replaced by $W_\tau s_\tau$. This means that low s_τ can be compensated by high W_τ. So a species will win if it happens to start out with a high enough population number. For each species there is a possible (stable) stationary state, in which only that species survives and all the others have died out. The case $\mu = 2$ would be a bad choice for the retinotopy problem, as the final state depends heavily on the initial conditions; the behaviour of the system is not dominated by the s_τ, as it should.

Case $\mu = 0$. In the evolution case, no true selection occurs. Equation (6) turns into

$$\dot{W}_\tau = s_\tau^* - W_\tau \sum_{\tau'} s_{\tau'}^* / W. \tag{8}$$

($s_\tau^* = \theta(W_\tau) s_\tau$ with

$$\theta(W_\tau) = \begin{cases} 1 & \text{for } W_\tau > 0, \\ 0 & \text{otherwise,} \end{cases}$$

i.e. nonexistent synapses cannot grow, as was stated above. With $\mu > 0$ this is taken care of automatically.) The stationary state is

$$W_\tau / W = s_\tau^* / \sum_{\tau'} s_{\tau'}^*. \tag{9}$$

In other words, in the stationary state the population number of any species is proportional to its fitness.

Applying this discussion of the evolution equations to the case when the similarities can vary, the case of $\mu = 1$ seems to be favoured. In this case the F_τ-values have just the right degree of dependence on the W_τ's to sharpen the fitness profile from an initial broad distribution to a final narrow one but not to give the large W_τ-values too strong an advantage to make the system insensitive to the s_τ's and too dependent on the initial conditions (as in the case $\mu = 2$). Unfortunately, if the dependence of the similarities on the variable tectal signals is taken into account, the case $\mu = 1$ turns out to have the same drawback as $\mu = 2$. The strong synapses are able, by locally inducing their markers into the tectum, to increase their s_τ, thus creating an effective μ that is bigger than 1. This leads to the establishment of nonoptimal mappings for many initial synaptic configurations. Therefore the case of $\mu = 0$ had to be adopted.

The adoption of $\mu = 0$ leads to another difficulty. In the final state each retinal fibre should project to a small region of tectum. As the profile of W_τ across the tectum is controlled, according to (9), by the profile of s_τ, this would mean that s_τ in the final state has to have a narrow profile. In other words, if the marker of one retinal point is projected into one tectal spot by means of a narrow W_τ distribution, then the presence of this marker can be felt only in a small region of tectum (narrowly distributed s_τ!). This can easily be arranged by giving markers in tectum a small diffusion length and by letting s be a sensitive function of the concentration vectors. However, such choice of diffusion length and s-function means that there is only a small distance over which neighboring retinal fibres can attract each other with their synapses in the tectum. The difficulty, now, is that in such a system with short interaction length between fibres in the tectum the achievement of ordered mappings is possible only for very special initial distributions of synapses.

It is possible to have a long interaction length *and* a narrow effective s_τ finally by subtracting from each value s_τ a threshold similarity ϕ, which can vary over time. When ϕ is chosen to be the arithmetic mean $\bar{s} = n^{-1}\Sigma_{\tau'} s_{\tau'}$, averaged over the fibre's n synapses, (8) becomes $\dot{W}_\tau = (s_\tau - \bar{s})\theta(W_\tau)$. This equation has the property that all those synapses with above average similarity, even the new weak ones, will grow and the below average similarity synapses will eventually disappear. This is the basic form of the equation we have used. The factor $\theta(W_\tau)$ ensures that only existing synapses can grow and that no synaptic weights can fall below zero.

Many different similarity functions are possible, and the one we used is calculated as follows:

$$s_{\tau\rho} = \sum_{m \in M} \text{minimum}(\hat{c}_\rho^m, \hat{c}_\tau^m),$$

where $\hat{c}_\rho^m$, $\hat{c}_\tau^m$ are normalised concentrations calculated over the selection M of molecule types possessed by both cells ρ and τ:

$$\hat{c}_\rho^m = c_\rho^m / \sum_{m' \in M} c_\rho^{m'}, \qquad \hat{c}_\tau^m = c_\tau^m / \sum_{m' \in M} c_\tau^{m'}$$

so that $\Sigma_{m' \in M} \hat{c}_\rho^{m'} = \Sigma_{m' \in M} \hat{c}_\tau^{m'} = 1$. From this it follows that $s_{\tau\rho}$ can take values between 0 and 1. Similarity between two retinal fibres is measured in an identical fashion.

The concentrations in a given tectal cell τ are a mixture of those induced by the retinal fibres and those supplied by lateral transport within the tectum. A desired property of the system is that fibres should concentrate their contacts in compact regions of the tectum, and that tectum cells should receive contacts from compact regions of the retina. The first is ensured by the *competition* that is a consequence of the sum rule (2); the second is the *interference* realized by calculating the similarities as a function of the normalized concentrations. The calculation of similarity in this way has a plausible physiological basis; for more details see [**2**].

A final remark, regarding the form of differential equation (3), may be added here. Grossberg [7] discusses the differential equation:

$$\dot{x}_i = -\alpha x_i + (B - x_i)I_i - x_i \sum_{i' \neq i} I_{i'}, \tag{10}$$

in which x_i is the degree of excitation of cell i in a field of cells, B an upper bound to x_i, α a decay rate constant and I_i an input impinging on cell i. Grossberg refers to the last two terms as an on-center off-surround shunting interaction. The equation has the pleasing property that in the stationary case of $\dot{x}_i = 0$, $x_i = BI_i/(\alpha + \Sigma_{i'} I_{i'})$, i.e. the ratio of two x_i's depends on the ratios of the corresponding I_i's, irrespective of the total input $\Sigma_{i'} I_{i'}$. In the special case of $\alpha = 0$, (10) becomes:

$$x_i = BI_i - x_i \sum_{i'} I_{i'}, \tag{11}$$

which has the same form as (3). This shows that the special case of (10) with $\alpha = 0$ can be derived from the 'rules' $\dot{x}_i = BI_i$ and $\Sigma_{i'} x_{i'} = B$ in the same way as (3) was derived from (1) and (2). The case of $\alpha \neq 0$ cannot be introduced into the differential equation in this way, as modification of the first 'rule' to $\dot{x}_i = -\alpha x_i + BI_i$ still leads to the same equation (11). It may be regarded as a very interesting fact that the three differential equations, (3), (10), and equation (37) of [**6**], although having very different meaning and interpretation, bear so close formal relationships.

ACKNOWLEDGEMENT. We would like to thank Dr. A. Häussler, who proposed important improvements to the manuscript.

REFERENCES

1. R. M. Gaze, *The problem of specificity in the formation of nerve connections*, Specificity of Embryological Interactions (D. Garrod, ed.), Chapman and Hall, London, 1978.

2. D. J. Willshaw and C. von der Malsburg, *A marker induction mechanism for the establishment of ordered neural mappings: its application to the retinotectal problem*, Philos. Trans. Roy. Soc. London Ser. B **287** (1979), 203–243.

3. C. von der Malsburg and D. J. Willshaw, *How to label nerve cells so that they can interconnect in an ordered fashion*, Proc. Nat. Acad. Sci. U.S.A. **74** (1977), 5176–5178.

4. D. J. Willshaw and C. von der Malsburg, *How patterned neural connections can be set up by selforganization*, Proc. Roy. Soc. London Ser. B **194** (1976), 431–445.

5. R. W. Sperry, *Chemoaffinity in the orderly growth of nerve fibre patterns and connections*, Proc. Nat. Acad. Sci. U.S.A. **50** (1963), 707–709.

6. M. Eigen and P. Schuster, *The hypercycle, a principle of natural self-organization*. B, *The abstract hypercycle*, Naturwissenschaften **65** (1978), 7–41.

7. S. Grossberg, *On the development of feature detectors in the visual cortex with applications to learning and reaction–diffusion systems*, Biol. Cybernet. **21** (1976), 145–159.

Max-Planck-Institute for Biophysical Chemistry, Department for Neurobiology, Göttingen, Federal Republic of Germany

The National Institute for Medical Research, Mill Hill, London, England

SIAM-AMS Proceedings
Volume 13
1981

Normal and Abnormal Signal Patterns in Nerve Cells

GAIL A. CARPENTER[1]

1. Introduction. A fundamental problem of psychophysics is an inverse problem: induce underlying cellular or network mechanisms from signal patterns. The patterns may be observed at various levels: from EEG, extracellular, or intracellular recordings or from psychological experiments. In each case the recorded pattern is an ensemble of events taking place at a finer, and usually unobservable, level.

At the center of the work presented in this article is the following simple question:

Which intracellular signal patterns are the result of basic single-cell membrane mechanisms and how are these patterns altered as membrane parameters change?

In other words, we will be considering the inverse problem at the level of the single cell.

At the heart of all models of the nerve impulse is the ionic hypothesis, which assumes that the main factor in transmembrane voltage shifts is the flow of selected ions into and out of the nerve cell. §2 contains an outline of the ionic hypothesis and its classical realization, the Hodgkin-Huxley equations (Hodgkin and Huxley [**1952**]). The latter is a nonlinear diffusion equation coupled with three first-order ordinary differential equations. The model was constructed partly from classical electrical theory and partly from a careful series of experiments which allowed Hodgkin, Huxley, et al., to separate the various components of current and to model them empirically. These experiments were performed on the fortuitously large giant axon of the squid. Hodgkin-Huxley (HH) models of other experimental preparations have been obtained and a large

---

1980 *Mathematics Subject Classification*. Primary 92A09.

[1]Supported in part by the National Science Foundation under Grant MCS-80-4021 and by the Northeastern University Research and Scholarship Development Fund.

number of variations on the model's themes have been proposed (Adelman and FitzHugh [**1975**], Bell and Cook [**1978**], [**1979**], Carpenter and Knapp [**1978**], Connor and Stevens [**1971**], Goldman and Schauf [**1973**], Gulrajani and Roberge [**1978**], Hoyt [**1968**], Hunter, McNaughton, and Noble [**1975**], Jakobsson [**1973**], [**1978a**], Jakobsson and Scudiero [**1975**], Moore and Jakobsson [**1971**], Noble [**1966**], Plant [**1976**], [**1979**], Plant and Kim [**1975**], [**1976**], Ramón, Joyner and Moore [**1975**]). Two main concerns emerge repeatedly. First, limitations of time and technique prevent a complete HH-type experimental analysis of all but a few types of nerve cells. While many individual differences occur, the robust electrical and ionic properties seem to be universal. The goal of §3 will be to abstract from the HH model those mathematical properties which are fundamental from those which are artifacts of the curve-fitting procedure. Second, the Hodgkin-Huxley model fails to predict some subtle experimental paradigms (Hoyt and Adelman [**1970**], Goldman and Schauf [**1972**]). (See, however, contradictory evidence in Gillespie and Meves [**1980**].) This has led to the many revised models in the literature. However, claims about HH's failures are usually made in the absence of a complete mathematical analysis of solutions. Prior to Carpenter [**1979**], for example, all models of burst patterns assumed the necessity of more ionic processes than HH. A geometric analysis reveals that a large class of generalized HH models, with exactly the same variables as HH, admit solutions with burst patterns. This result suggests that many cells are capable of generating burst patterns, but that these cells are normally held in check by inhibitory processes such as feedback or other network connections. Indirect evidence for this conjecture is provided by the fact that in epileptic cortex, where bursts are often found (Figure 29A), there is evidence of a breakdown in inhibitory feedback (Anderson and Rutledge [**1979**], Schwartzkroin and Prince [**1980**]).

With the criteria for burst solutions established, numerical methods may be used to verify whether or not they are satisfied by a particular model. A major limitation of computer analysis, used by itself, is that only one particular example may be considered at a time. Thus, the investigator's choice of parameters may lead to false conclusions about the limitations of a class of models. The techniques developed in §4 will allow us to consider, at once, a large class of nerve impulse models, to understand their robust mathematical properties, and to see how variations in the basic model lead to variations in signaling properties. The main idea is to construct "singular solutions", which are relatively easy to analyze. General theorems (Carpenter [**1974**], [**1977a**], [**1977b**]) imply that each singular solution corresponds to a nearby solution of the full system.

In §5, the simplest version of the generalized HH model is examined and predictions made. "Prediction" here means a property of solutions of the model in question. If the prediction fails to hold in a particular case, then the observed effect must be due to a mechanism not embodied in the model. For example, the effect could be due to additional ionic currents, slow parameter shifts, or network interactions. Examples will be given in §6 in which the predictions of §5 fail, forcing modifications in the model for those cases. With the properties of the core model established it is relatively easy to see how changes in the model change the solution patterns. §6 also contains a discussion of the simplified FitzHugh-Nagumo model as seen from the point of view of the present analysis.

Some analytic studies of the FitzHugh-Nagumo equations include Casten, Cohen and Lagerstrom [**1975**], Conley [**1975**], Hastings [**1974**] and [**1976b**], McKean [**1970**], Rinzel [**1975**], Rinzel and Keller [**1973**], Schonbek [**1978**]. Studies of the Hodgkin-Huxley equations include Evans [**1972, 1975**], Evans and Feroe [**1978**], Hastings [**1976a**], Rinzel and Miller [**1980**], Troy [**1978**]. Many of the results outlined in the present article are in Carpenter [**1974**], [**1976**], [**1977a**], [**1977b**], [**1979**], which contain details of the proofs. The principal new results are contained in the list of predictions in §5. Although mathematical analysis is necessary for a complete understanding of these predications, most of them can be stated and used independently. The predictions are introduced here both to motivate the analysis in later sections and to assist the reader who wishes to apply the results without studying details of the method. Each prediction is a property of solutions of a generalized Hodgkin-Huxley system ((3.1), (4.1), or (4.2)) whose ionic current consists of fast and slow sodium components, a slow potassium component, and a small leakage current. That is, the model embodies the basic elements of nerve membrane mechanisms, and the predictions describe properties of signals which are characteristic of these basic elements. One surprising result is that the list includes many fairly complex phenomena previously ascribed to more intricate membrane mechanisms or to network interactions. Some common signal patterns, such as parabolic and paroxysmal bursts (Figure 45), are *not* seen as products of the most basic elements. However, they are found as solutions of augmented models, as discussed in §6.

DEFINITIONS. A few definitions are given to clarify terms used in the list of predictions. Each definition is made mathematically precise later in the article.

V (mV) denotes the voltage difference across the nerve membrane (Figure 1).

A *spike*, or *action potential*, consists of a large shift of V away from its resting value and a subsequent return to near rest (Figure 2).

A cell is *depolarized* when V is above its rest value; and *hyperpolarized* when V is below its rest value.

A *regular periodic* solution of the model corresponds to a sequence of evenly-spaced spikes. The analysis reveals the existence of two distinct types of regular periodic solutions: Ω-periodic spike trains (Figure 22D) and *bursts with one spike per burst*, or 1-*bursts* (Figure 22E). These two solution types have characteristic properties, as described in Prediction 9. Most important, they would respond very differently to drugs or other perturbations. An example *in vivo* of 1-bursts and Ω-periodic spike trains is given in Figure 37.

A burst with N spikes per burst, or N-burst, is a cluster of N spikes followed by a *quiet spell* (Q), during which V is near rest. A *periodic burst* is a repeated sequence of bursts (Figures 22A, 29, 37A, and 41).

A *finite wave train* is a single, isolated burst (Figure 22B).

A *single spike* is an isolated spike (Figure 22C).

Predictions.

Prediction 1. All generalized HH nerve models (4.1) and (4.2) admit Ω-periodic solutions.

Prediction 2. "Half" of these models also admit periodic burst and finite wave train solutions. (Henceforth we will call periodic burst solutions of (4.1) and (4.2) *HH bursts*.)

Fine structure of HH bursts.

Prediction 3. Within an HH burst or finite wave train, spiking frequency becomes nearly constant after the first few spikes. Early in the burst, the frequency tends to increase, giving rise to a "long first interval" (Figures 28 and 29).

Prediction 4. If two HH bursts have the same (temporal) period, the one with more spikes/burst travels faster (Figure 30A).

Prediction 5. With wave speed held constant, the length of the quiet spell (Q) increases with the number (N) of spikes in the previous burst (Figure 32). As N becomes large, Q approaches a finite upper bound. N is "large" when the spiking frequency becomes approximately constant. As seen in Figure 28, this may happen either early or late in the burst.

Prediction 6. The speed of a finite wave train increases with the number of spikes in the previous burst. The same is true of HH bursts with long quiet spells (asymptotically as $Q \to \infty$) (Figure 33A).

Prediction 7. Within an HH burst, V increases during the interspike interval. The longer the interval, the flatter the graph of V. At the end of a burst, there may be an afterdepolarization (arrow in Figure 34). During the quiet spell, V is relatively flat (Figures 34 and 29).

Prediction 8. During an HH burst sequence, the baseline of activity (V) is relatively flat (Figures 35A and 29). Large shifts in this baseline indicate the influence of external depolarizing shifts (DPS) during the oscillation (Figures 35B and 36).

Regular (evenly-spaced) periodic patterns.

Prediction 9. There are two types of regular periodic solutions of (4.1) and (4.2): Ω-periodic and bursts with 1 spike per burst ("1-bursts"). These two solution types have different qualitative properties. In particular, in a cell which bursts, Ω-periodics are far from equilibrium while bursts with one spike are near equilibrium. Thus, parametric shifts may easily extinguish a 1-burst while an Ω-periodic would first move through a burst phase to a 1-burst phase and then to equilibrium (Figure 37).

Distinguishing characteristics of Ω-periodic and 1-burst solutions are listed below.

Ω-periodic solutions.

(Ω: A) V increases during the interspike interval (Figures 37 and 38).

(Ω: B) In Cases 1, 2, and 4 of Theorem 1, §5, the spiking frequency of Ω-periodic solutions ranges from moderate to very high (Figures 27 and 37A).

In Case 3 of Theorem 1 (when there are no bursts) spiking frequency of Ω-periodic solutions ranges from low to very high (Figure 37B).

(Ω: C) The speed of an Ω-periodic solution increases as the frequency decreases (i.e., as the period increases). The wave speed ranges from low to moderate (Cases 1 and 3) or from low to high (Cases 1, 2, and 4).

(Ω: D) Various properties of Ω-periodic solutions are correlated. For example, the generalized HH model (4.2) predicts that if a parametric shift changes the frequency in a certain way then other properties such as amplitude must also change in a well-defined way at the same time. These correlated shifts would be a good test of the model; many of them are seen in Figure 41. Table 3 and Figure 40 summarize the properties.

Bursts with 1 *spike per burst.*

(1-burst: A) During the interspike interval, V first increases, but then has a long, relatively flat graph during the quiet spell, which may be preceded by an afterdepolarization (Figure 42). If there is an afterdepolarization at low frequency, it becomes smaller and then disappears as the frequency increases.

(1-burst: B) The frequency of 1-bursts ranges from very low to moderate (Figure 37A).

(1-burst: C) The wave speed is moderate and approximately constant over large changes in the period. It increases as frequency increases.

(1-burst: D) Spike amplitude is large (about the same as the large-amplitude Ω-periodics) and approximately constant over large changes in the period. Postspike hyperpolarization also stays approximately constant.

Figure 43 depicts a continuous family of regular periodic solutions, from the single spike (A) to 1-bursts (B and C) to the transitional point (D) to Ω-periodics (E and F).

Bursts in a noisy system. The HH bursts described so far are perfectly stereotyped: each burst is identical to the ones before and after. The presence of noise in the system could alter the stereotyped pattern so that, for instance, the number of spikes per burst varies. Variations in burst patterns may be explained by coupling (4.1) and (4.2) with noise *provided* that Predictions 10–12 are usually satisfied. Otherwise, other explanations for the variations must be sought.

Prediction 10. In a noisy system, HH bursts with many spikes per burst exhibit more variation in spike number than do bursts with few spikes per burst.

Prediction 11. "Variable first interval": Noise may produce large variations in the length of the first one or two interspike intervals (ISI's). Later in the burst, the frequency stays about constant, even if the total number of spikes per burst is varying.

Prediction 12. In a system with variable first interspike intervals, the longer first ISI's tend to correlate with longer quiet spells preceding the burst (Figure 44).

2. The ionic hypothesis and the Hodgkin-Huxley equations. The fluid inside the nerve axon contains a high concentration of potassium ions (K^+) and a low concentration of sodium ions (Na^+) relative to the outside. When an above-threshold stimulus is received, membrane permeability to Na^+ first increases, allowing a small number of sodium ions to enter the axon; then, permeability to Na^+ decreases and permeability to K^+ increases, so potassium ions leave the axon. The resulting shifts in transmembrane voltage constitute the nerve impulse, or action potential, which is propagated along the axons at an approximately constant wave speed.

The Hodgkin-Huxley equations (Table 1) summarize this picture of nerve impulse propagation, with functions and constants chosen for the special case of the squid giant axon. Stevens **[1966]** contains an excellent brief derivation of these equations.

Traveling wave solutions of HH correspond to waves propagated at fixed speeds. Consider the wave depicted in Figure 2 at times $t = 0$ and $t = T$. If we know $V(x, 0)$ for all x and wish to compute $V(x, t)$ at any later time T, we need only look (speed) $\cdot\, T$ cm to the right to obtain

$$V(x, T) = V(x + (\text{speed}) \cdot T, 0).$$

Thus, we consider solutions which depend on the single variable $s = x + (\text{speed}) \cdot t$. Henceforth, when we speak of a solution where $V = V(s)$, we shall mean that at position x at time t, $V(x, t) = V(s)$, where $s = x + (\text{speed}) \cdot t$.

According to the chain rule $\partial/\partial x = (d/ds) \cdot ds/dx = d/ds$ and $\partial/\partial t = (d/ds) \cdot ds/dt = (\text{speed}) \cdot d/ds$. Thus HH becomes

$$\begin{aligned} \frac{d^2V}{ds^2} &= \frac{2RC}{a} \cdot (\text{speed}) \cdot \frac{dV}{ds} + g(V, m, n, h), \\ (\text{speed}) \cdot dm/ds &= \gamma_m(V)(m_\infty(V) - m), \\ (\text{speed}) \cdot dn/ds &= \gamma_n(V)(n_\infty(V) - n), \\ (\text{speed}) \cdot dh/ds &= \gamma_h(V)(h_\infty(V) - h), \end{aligned}$$

where

$$g(V, m, n, h) = (2R/a) \cdot \left(\bar{g}_{\text{Na}} m^3 h(V - V_{\text{Na}}) + \bar{g}_{\text{K}} n^4 (V - V_{\text{K}}) + \bar{g}_L (V - V_L)\right).$$

If $dV/ds \equiv W$ and $\theta \equiv (2RC/a) \cdot (\text{speed})$, then we obtain the first order system

$$\begin{aligned} dV/ds &= W, \\ dW/ds &= \theta W + g(V, m, n, h), \\ (\text{speed}) \cdot dm/ds &= \gamma_m(V)(m_\infty(V) - m), \\ (\text{speed}) \cdot dn/ds &= \gamma_n(V)(n_\infty(V) - n), \\ (\text{speed}) \cdot dh/ds &= \gamma_h(V)(h_\infty(V) - h). \end{aligned} \tag{2.1}$$

Solutions of (2.1) are traveling wave solutions of the Hodgkin-Huxley equations.

TABLE 1. THE HODGKIN-HUXLEY EQUATIONS.

$$I = \frac{a}{2R}\frac{\partial^2 V}{\partial x^2} = C\frac{\partial V}{\partial t} + \bar{g}_{\mathrm{Na}} m^3 h(V - V_{\mathrm{Na}}) + \bar{g}_{\mathrm{K}} n^4 (V - V_{\mathrm{K}}) + \bar{g}_L (V - V_L),$$

$$\frac{\partial m}{\partial t} = \gamma_m(V)(m_\infty(V) - m), \frac{\partial n}{\partial t} = \gamma_n(V)(n_\infty(V) - n), \frac{\partial h}{\partial t} = \gamma_h(V)(h_\infty(V) - h).$$

variables and constants		units and approximate values
x	distance along the axon	cm
t	time	sec
I	total membrane current density (outward positive)	$\frac{\mathrm{mA}}{\mathrm{cm}^2} = \frac{\mathrm{mV}}{\Omega \cdot \mathrm{cm}^2}$
V	transmembrane voltage	mV; $V_{\mathrm{K}} < V < V_{\mathrm{Na}}$
m	measure of fast Na^+ activation	dimensionless; $0 < m < 1$
n	measure of slow K^+ inactivation	dimensionless; $0 < n < 1$
h	measure of slow Na^+ inactivation	dimensionless; $0 < h < 1$
a	axon radius	.0238cm
R	specific resistance of axoplasm	35.4 Ω-cm
C	membrane capacity/unit area	10^{-6} farad/cm^2 = 10^{-6} sec/(Ω-cm^2)
$C\partial V/\partial t$	capacitance current density	
$\bar{g}_{\mathrm{Na}} m^3 h(V - V_{\mathrm{Na}})$	(inward) sodium current density	negative
$\bar{g}_{\mathrm{K}} n^4 (V - V_{\mathrm{K}})$	(outward) potassium current density	positive
$\bar{g}_L (V - V_L)$	"leakage" current density	
$\bar{g}_{\mathrm{Na}}$	maximal conductances/unit area	.120(1/Ω · cm^2)
$\bar{g}_{\mathrm{K}}$	maximal conductances/unit area	.036(1/Ω · cm^2)
$\bar{g}_L$	maximal conductances/unit area	.0003(1/Ω · cm^2)
V_{Na}	sodium, potassium, and leakage potentials	115mV
V_{K}	sodium, potassium, and leakage potentials	−12mV
V_L	sodium, potassium, and leakage potentials	10.613mV
wave speed		cm/sec
$\gamma_m, \gamma_n, \gamma_h$	rate constants; $\gamma_m \gg \gamma_n, \gamma_h$	sec^{-1}; $\gamma_m, \gamma_n, \gamma_h > 0$
m_∞	increasing from 0 to 1	dimensionless
n_∞	increasing from 0 to 1	dimensionless
h_∞	decreasing from 1 to 0	dimensionless

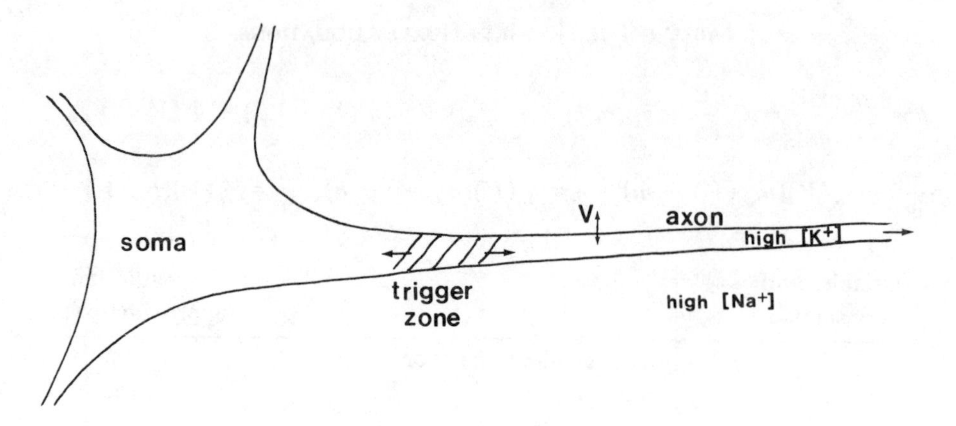

FIGURE 1. Schematic view of a portion of the nerve cell.

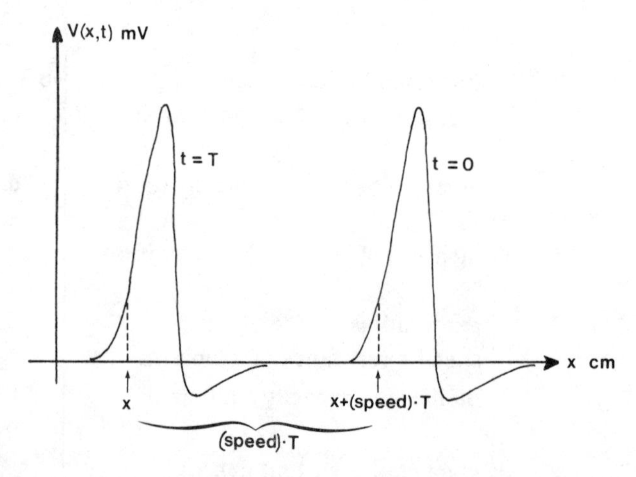

FIGURE 2. A wave traveling leftward with constant speed. The wave travels (speed) $\cdot$ T cm in T sec, and $V(x, T) = V(x + \text{(speed)} \cdot T, 0)$.

3. Generalized Hodgkin-Huxley systems: a simple case. In this section we will study a system of differential equations which look exactly like (2.1); the difference is that the empirical functions and parameters of HH will be replaced by hypotheses on those functions and parameters. These qualitative, rather than quantitative, hypotheses are appropriate for the qualitative, or geometric, analysis to be carried out in later sections.

A geometric approach to this problem is useful for several reasons. First, we wish to consider a very large number of examples of nerve impulse models, which may vary in particulars but share some fundamental properties. For example, we are able to study, all together, systems where $m_\infty(V)$ and $n_\infty(V)$ are increasing functions of V and $h_\infty(V)$ is a decreasing function of V. Second, even if we were able to obtain experimentally all parameters in the model, solutions lie in a high dimensional space, and a numerical analysis might overlook some interesting solution types. Also, even if completely successful, this analysis would apply only to the model in question; it would have to be repeated for each new model, and could not be extended to the many preparations where parameters cannot be easily measured. Third, any model is imprecise, but its value is greatly enhanced if we know which robust properties persist across many similar models and which ones depend critically on particular parameters or hypotheses. Finally, the most important features of nerve impulse patterns are generally their qualitative properties, such as spike frequency and amplitude, rather than, say, the precise value of V at any given moment. The presence of noise in all systems reinforces this last point.

The system.

$$\begin{aligned} \dot{V} &= W, \\ \dot{W} &= \theta W + g(V, m, n, h), \\ (\text{speed}) \cdot \dot{m} &= \delta^{-1}\gamma_m(V)(m_\infty(V) - m), \\ (\text{speed}) \cdot \dot{n} &= \varepsilon\gamma_n(V)(n_\infty(V) - n),4 \\ (\text{speed}) \cdot \dot{h} &= \varepsilon\gamma_h(V)(h_\infty(V) - h). \end{aligned} \tag{3.1}$$

The parameters δ and ε have been introduced for notational convenience. δ^{-1} and ε represent the scales of $\gamma_m(V)$ and of $\gamma_n(V)$ and $\gamma_h(V)$. δ and ε will be assumed to be small, which is equivalent to assuming that m represents a relatively fast process (Na^+ activation) and that n and h represent relatively slow processes (K^+ activation and Na^+ inactivation). In general, it is not necessary to assume that m be particularly fast; in the computations carried out in Carpenter [**1980**], $\gamma_m(V)$ is the same one used in HH and δ is set equal to 1. However, some hypothesis on $\gamma_m(V)$ is necessary: if K^+ left the nerve just as quickly as Na^+ entered, there would be no net signal. The simplest version of the hypothesis is that δ be small, so that version is used here.

The hypotheses. Except for part (G), all parts of Hypothesis 1 are easily verified for the original Hodgkin-Huxley equations, with (E) and (F) requiring numerical verification. (For HH, $V_{\text{rest}} = 0$, $n_0 = 0.315$ and $h_0 = 0.608$.) (G) can also be verified numerically, but requires many more computations than (A)–(F).

Hypothesis 1. Let $G(V, n, h) \equiv g(V, m_\infty(V), n, h)$.

(A) $g, \gamma_m, \gamma_n, \gamma_h, m_\infty, h_\infty$, and n_∞ are twice continuously differentiable.

(B) $\theta, \varepsilon, \delta, \gamma_m, \gamma_n, \gamma_h$, and the wave speed are all positive; and $0 < m_\infty, n_\infty, h_\infty < 1$.

(C) (Excitatory m, inhibitory n and h) $\partial g/\partial m < 0$, $m'_\infty > 0$, $\partial g/\partial h < 0$, $h'_\infty < 0$, $\partial g/\partial n > 0$, and $n'_\infty > 0$.

(D) (Unique rest point) There is some $V = V_{\text{rest}}$ such that $g(V, m_\infty(V), n_\infty(V), h_\infty(V)) = 0$ if and only if $V = V_{\text{rest}}$.

REMARK. Let $n_0 \equiv n_\infty(V_{\text{rest}})$ and $h_0 \equiv h_\infty(V_{\text{rest}})$. (D) implies that $G(V_{\text{rest}}, n_0, h_0) = 0$.

(E) (Maximal and minimal values of V). There exist $V_K < V_{rest} < V_{Na}$ such that $G(V_K, n, h) < 0 < G(V_{Na}, n, h)$ for all $n, h \in (0, 1)$.

(F) (Threshold of excitation) $G(V, n_0, h_0) = 0$ for exactly three values of V:

$$V_0(n_0, h_0) = V_{rest} < V_2(n_0, h_0) < V_1(n_0, h_0).$$

Assume that, when $V = V_0(n_0, h_0)$ or $V = V_1(n_0, h_0)$, $(\partial G/\partial V)(V, n_0, h_0) > 0$, and that

$$\int_{V_0(n_0,h_0)}^{V_1(n_0,h_0)} G(V, n_0, h_0)\, dV < 0 \quad \text{(Figure 3).}$$

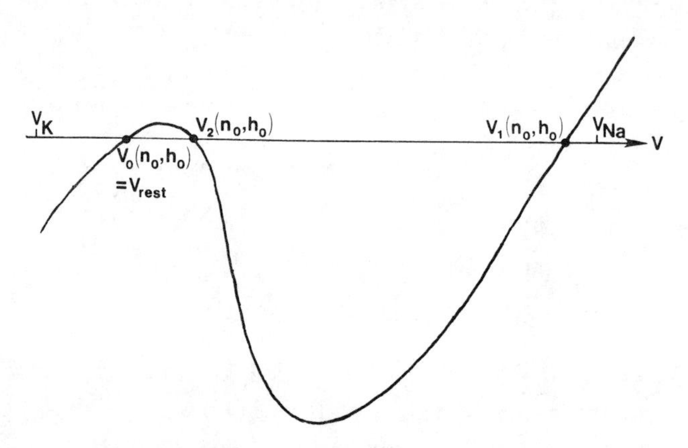

FIGURE 3. A typical graph of $G(V, n_0, h_0)$.

REMARK. Since, in HH, $g(V, m, n, h)$ is linear in V, the nonlinearity in $G(V, n_0, h_0) = g(V, m_\infty(V), n_0, h_0)$ depends on $m_\infty(V)$. The last inequality in (F) is equivalent to the statement that the threshold for excitation canot be too large.

(G) (V_0, V_1, and V_2: the left, right, and middle zeros of $G(V, n, h)$) (see Figures 4 and 5).

There exist three continuous functions and two open sets $\Pi_0, \Pi_1 \subseteq (0, 1) \times (0, 1)$ such that if $\partial\Pi_i$ is equal to the boundary of Π_i intersected with $(0, 1) \times (0, 1)$, then $V_0(n, h)$ is defined in $\Pi_0 \cup \partial\Pi_0$, $V_1(n, h)$ is defined in $\Pi_1 \cup \partial\Pi_i$, and $V_2(n, h)$ is defined in $[\Pi_0 \cup \partial\Pi_0] \cap [\Pi_1 \cup \partial\Pi_1]$. Moreover, for all $\langle n, h\rangle \in (0, 1) \times (0, 1)$:

(i) $G(V, n, h) = 0$ if and only if $V = V_i(n, h)$ for some $i = 0, 1,$ or 2;

(ii) $V_0(n, h) < V_2(n, h) < V_1(n, h)$ for $\langle n, h\rangle \in \Pi_0 \cap \Pi_1$;

(iii) $(\partial G/\partial V)(V_0(n, h), n, h) = 0$ if and only if $\langle n, h\rangle \in \partial\Pi_0$; in this case, $V_0(n, h) = V_2(n, h) < V_1(n, h)$ and $(\partial^2 G/\partial V^2)(V_0(n, h), n, h) < 0$; and

(iv) $(\partial G/\partial V)(V_1(n, h), n, h) = 0$ if and only if $\langle n, h\rangle \in \partial\Pi_1$; in this case, $V_0(n, h) < V_2(n, h) = V_1(n, h)$ and $(\partial^2 G/\partial V^2)(V_1(n, h), n, h) > 0$.

REMARKS. (F) implies that $\langle n_0, h_0\rangle \in \Pi_0 \cap \Pi_1$. (E) implies that $(\partial G/\partial V)(V_0(n, h), n, h) > 0$ for $\langle n, h\rangle \in \Pi_0$ and $(\partial G/\partial V)(V_1(n, h), n, h) > 0$ for $\langle n, h\rangle \in \Pi_1$. (E) also implies that $\Pi_0 \cup \Pi_1 = (0, 1) \times (0, 1)$. (C) implies that $\partial\Pi_0$ and $\partial\Pi_1$ are the graphs of increasing functions of n.

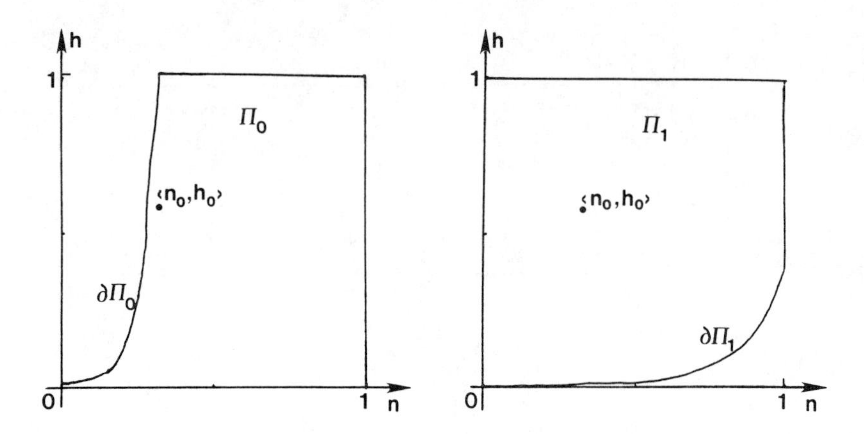

FIGURE 4. Π_0 and Π_1 for HH.

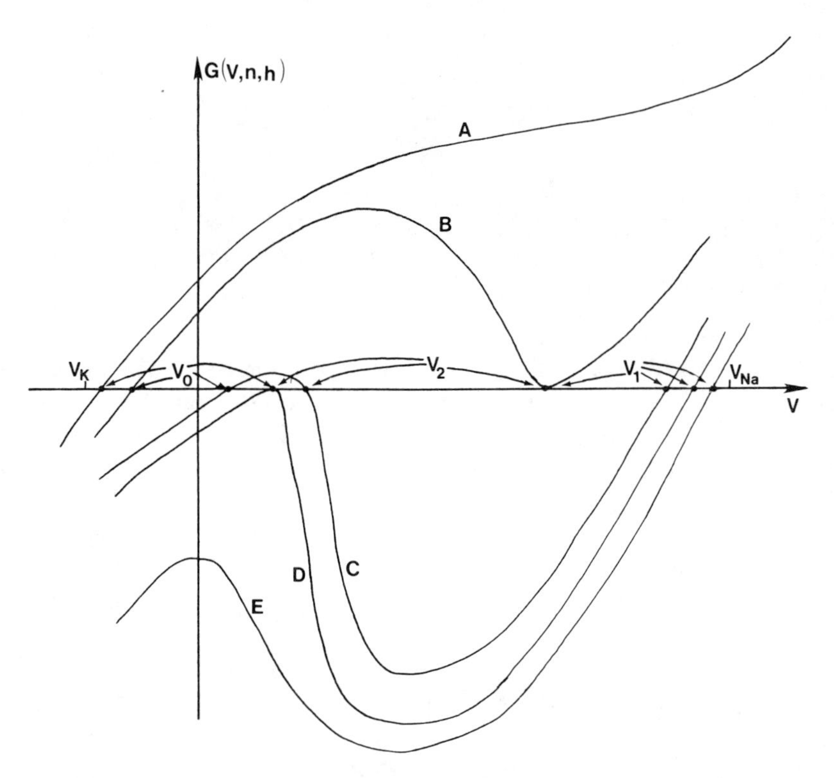

FIGURE 5. Typical graphs of $G(V, n, h)$ for (A) $\langle n, h \rangle \in \Pi_0$; (B) $\langle n, h \rangle \in \Pi_0 \cap \partial\Pi_1$; (C) $\langle n, h \rangle \in \Pi_0 \cap \Pi_1$; (D) $\langle n, h \rangle \in \Pi_1 \cap \partial\Pi_0$; and (E) $\langle n, h \rangle \in \Pi_1$.

EXAMPLE. Although (G) must, in general, be verified numerically, it is made plausible by examination of the following special case, for which $V_0(n, h)$, $V_1(n, h)$, $V_2(n, h)$, Π_0, and Π_1 can be computed directly.

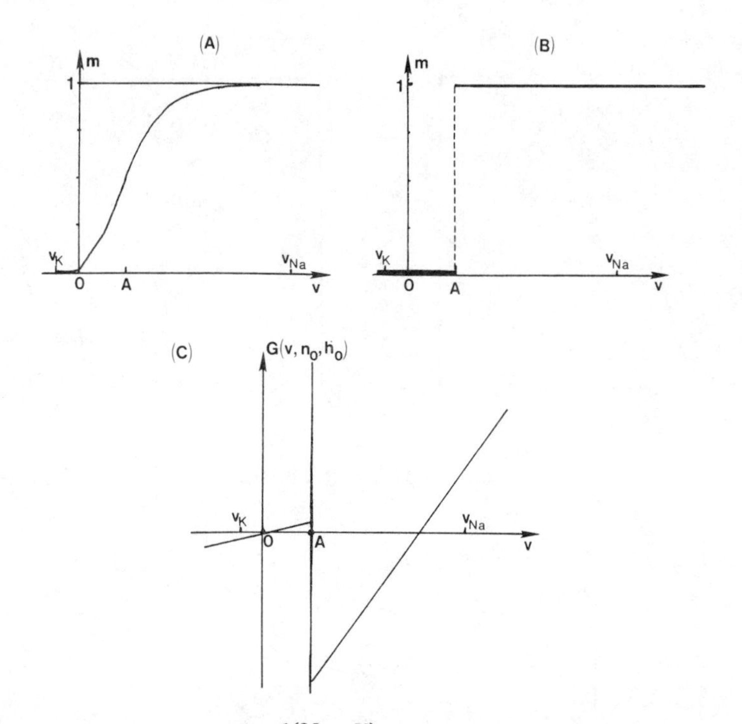

FIGURE 6. (A) $m_\infty(V) = \dfrac{.1(25 - V)}{.1(25 - V) + 4e^{-V/18}[e^{(25-V)/10} - 1]}$, from Hodgkin and Huxley [1952, pp. 515–516]. (B) $m_\infty(V) = 0$ if $V < A$ and $m_\infty(V) = 1$ if $V > A$. (C) $G(V, n_0, h_0)$, with $m_\infty(V)$ as in (B). $\bar{g}_L$ is set equal to $-\bar{g}_K n_0^4 V_K / V_L$ so that $V_{\text{rest}} = 0$.

In HH, $m_\infty(V)$ (Figure 6A) is a smooth s-shaped function, reflecting the fact that the threshold for sodium excitation is not precisely the same at all sites, but rather exists as a distribution of thresholds about some mean value A. If, instead, the threshold were always exactly at the same point $V = A$, then $m_\infty(V)$ would be the step function depicted in Figure 6B. For this $m_\infty(V)$,

$$G(V, n, h) = \bar{g}_{\text{Na}} m_\infty^3(V) h (V - V_{\text{Na}}) + \bar{g}_K n^4 (V - V_K) + \bar{g}_L (V - V_L)$$
$$= \begin{cases} \bar{g}_K n^4 (V - V_K) + \bar{g}_L (V - V_L) & \text{if } V < A, \\ \bar{g}_{\text{Na}} h (V - V_{\text{Na}}) + \bar{g}_K n^4 (V - V_K) + \bar{g}_L (V - V_L) & \text{if } V > A \end{cases}$$

is piecewise linear in V (Figure 6C). In this extreme case,

$$V_K < V_0(n, h) = \frac{\bar{g}_K n^4 V_K + \bar{g}_L V_L}{\bar{g}_{\text{Na}} h + \bar{g}_K n^4 + \bar{g}_L} < A$$
$$< V_1(n, h) = \frac{\bar{g}_{\text{Na}} h V_{\text{Na}} + \bar{g}_K n^4 V_K + \bar{g}_L V_L}{\bar{g}_{\text{Na}} h + \bar{g}_K n^4 + \bar{g}_L} < V_{\text{Na}}$$

and $V_2(n, h) = A$. On $\partial \Pi_0$,

$$n^4 = -\bar{g}_L (A - V_L) / \bar{g}_K (A - V_K).$$

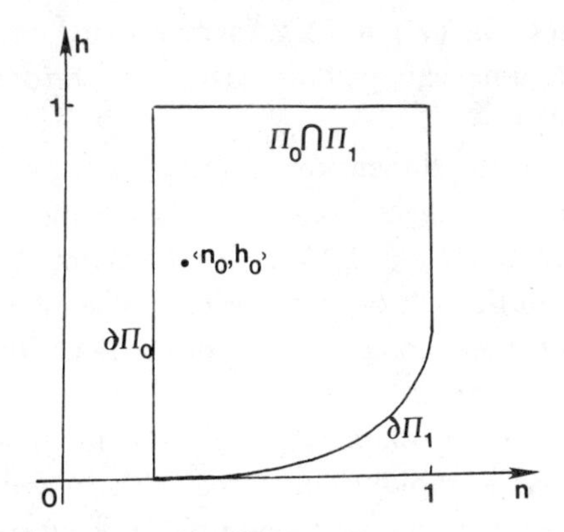

FIGURE 7. $\Pi_0 \cap \Pi_1$ when $m_\infty(V)$ is the step function illustrated in Figure 6B and $0 < A < V_L$. Note the similarity to $\Pi_0 \cap \Pi_1$ of Figure 4.

Thus $\partial\Pi_0$ is either the empty set (if $A > V_L$) or a vertical line segment (if $0 < A < V_L$) (Figure 7). Similarly, on $\partial\Pi_1$, $\bar{g}_{\mathrm{Na}}h(A - V_{\mathrm{Na}}) + \bar{g}_{\mathrm{K}}n^4(A - V_{\mathrm{K}}) + \bar{g}_L(A - V_L) = 0$, i.e.,

$$h = \frac{\bar{g}_{\mathrm{K}}n^4(A - V_{\mathrm{K}}) + \bar{g}_L(A - V_L)}{\bar{g}_{\mathrm{Na}}(V_{\mathrm{Na}} - A)}.$$

This example indicates that certain assumptions can be weakened: here, $m_\infty(V)$ is discontinuous at $V = A$, but the analysis still carries through.

4. Singular solutions. The singular solution is the central geometric idea of the present analysis. Once the singular geometry of a system is clearly in mind, one can easily derive most of the properties of solutions discussed in §§5 and 6. A few examples will illustrate the main points.

Consider a generalized Hodgkin-Huxley system

$$\begin{aligned} \dot{V} &= W, \\ \dot{W} &= \theta W + g(V, m, n, h), \\ (\text{speed}) \cdot \dot{m} &= \delta^{-1}\gamma_m(V)(m_\infty(V) - m), \\ (\text{speed}) \cdot \dot{n} &= \varepsilon\gamma_n(V)(n_\infty(V) - n), \\ (\text{speed}) \cdot \dot{h} &= \varepsilon\gamma_h(V)(h_\infty(V) - h). \end{aligned} \qquad (4.1;\ \theta, \varepsilon, \delta)$$

If δ is small, m rapidly approaches $m_\infty(V)$. Every solution (mentioned in this paper) with δ set equal to 0 corresponds to a similar, nearby solution with δ small and positive. That is, the transition from $\delta = 0$ to $\delta > 0$ is a regular perturbation. It is thus natural to set $\delta = 0$, so $m \equiv m_\infty(V)$, and to consider the reduced system

$$\begin{aligned} \dot{V} &= W, \\ \dot{W} &= \theta W + G(V, n, h), \\ (\text{speed}) \cdot \dot{n} &= \varepsilon\gamma_n(V)(n_\infty(V) - n), \\ (\text{speed}) \cdot \dot{h} &= \varepsilon\gamma_h(V)(h_\infty(V) - h), \end{aligned} \qquad (4.2;\ \theta, \varepsilon)$$

where $G(V, n, h) = g(V, m_\infty(V), n, h)$. (As mentioned in §3, setting $\delta = 0$ is unnecessary. Computations can just as easily be performed directly on the system (4.1; $\theta, \varepsilon, \delta = 1$).)

We have now reduced the dimension of the system by one. The presence of another small parameter, ε, suggests that we try the same trick to further reduce the dimension. Suppose we set $\varepsilon = 0$. Consider the claim: "If ε is small, solutions of (4.2; θ, ε) are near solutions of (4.2; $\theta, \varepsilon = 0$)." When $\varepsilon = 0$, $\dot{n} \equiv 0$ and $\dot{h} \equiv 0$, so n and h remain constant along each solution. Thus, the claim is true only when n and h vary slowly compared to V and W. This is, in fact, the case when ε is small *except* near the set of points where $\dot{V} = 0$ and $\dot{W} = 0$. If $\dot{V}$ and $\dot{W}$ are both zero, n and h change rapidly compared with V and W, no matter how small ε is. Consider, then, the set ("slow manifold") $\mathbb{S}$, where

$$\begin{aligned}\mathbb{S} &= \{\langle V, W, n, h\rangle : W = G(V, n, h) = 0\}\\ &= \{\langle V, W, n, h\rangle : W = 0 \text{ and } V = V_i(n, h) \text{ for some } i = 0, 1 \text{ or } 2\}.\end{aligned}$$

$\mathbb{S}$ is the set of rest points of (4.2; $\theta, \varepsilon = 0$) and contains $\langle V_{\text{rest}}, 0, n_0, h_0\rangle$, the unique rest point of (4.2; $\theta, \varepsilon > 0$).

The system (4.2) induces a natural flow on $\mathbb{S}$

$$\begin{aligned}V &= V_i(n, h),\\ W &= 0,\\ (\text{speed}) \cdot \dot{n} &= \varepsilon\gamma_n(V_i(n, h))(n_\infty(V_i(n, h)) - n),\\ (\text{speed}) \cdot \dot{h} &= \varepsilon\gamma_h(V_i(n, h))(h_\infty(V_i(n, h)) - h),\end{aligned} \tag{4.3; i}$$

where $i = 0, 1$, or 2. This system governs the dynamics of a solution of (4.2) when that solution is near the set $\mathbb{S}$. A *singular solution* consists of a sequence of solution segments of (4.2; $\theta, \varepsilon = 0$) as illustrated below. Every singular solution corresponds to a nearby family of solutions of (4.2; θ, ε). It will be convenient to sketch singular solutions (ss) in a variety of ways. The most complete would be a picture of the ss in $\mathbf{R}^4$, if such were possible. One typical ss is sketched in Figure 9. In Figure 14A the ss is represented by just one fixed point. This reduction is useful when we consider whole families of ss's, which are represented by curves of fixed points (Figures 14B, 24, and 25).

We will need one main fact about (4.2; $\theta, \varepsilon = 0$):

For each $\bar{\theta} \geqslant 0$ there is a one-dimensional set $\mathrm{UP}(\bar{\theta}) \subseteq (\Pi_0 \cap \Pi_1) \cup \partial\Pi_0$ such that for each $\langle \bar{n}, \bar{h}\rangle \in \mathrm{UP}(\bar{\theta})$ there is a solution of (4.2; $\theta = \bar{\theta}, \varepsilon = 0$) from $\langle V_0(\bar{n}, \bar{h}), 0, \bar{n}, \bar{h}\rangle$ to $\langle V_1(\bar{n}, \bar{h}), 0, \bar{n}, \bar{h}\rangle$. Also, there is a one-dimensional set $\mathrm{DOWN}(\bar{\theta}) \subseteq (\Pi_0 \cap \Pi_1) \cup \partial\Pi_1$ such that for each $\langle \bar{n}, \bar{h}\rangle \in \mathrm{DOWN}(\bar{\theta})$ there is a solution of (4.2; $\theta = \bar{\theta}, \varepsilon = 0$) from $\langle V_1(\bar{n}, \bar{h}), 0, \bar{n}, \bar{h}\rangle$ to $\langle V_0(\bar{n}, \bar{h}), 0, \bar{n}, \bar{h}\rangle$. $\mathrm{UP}(\bar{\theta})$ and $\mathrm{DOWN}(\bar{\theta})$ are graphs of increasing functions of $\bar{\theta}$.

$\langle n, h\rangle \in \mathrm{UP}(0) = \mathrm{DOWN}(0)$ if and only if $\int_{V_0(n,h)}^{V_1(n,h)} G(V, h, h)\, dV = 0$.
$\langle n, h\rangle \in \mathrm{UP}(\theta)$ for some $\theta > 0$ if and only if $\int_{V_0(n,h)}^{V_1(n,h)} G(V, n, h)\, dV < 0$.
$\langle n, h\rangle \in \mathrm{DOWN}(\theta)$ for some $\theta > 0$ if and only if $\int_{V_0(n,h)}^{V_1(n,h)} G(V, n, h)\, dV > 0$.

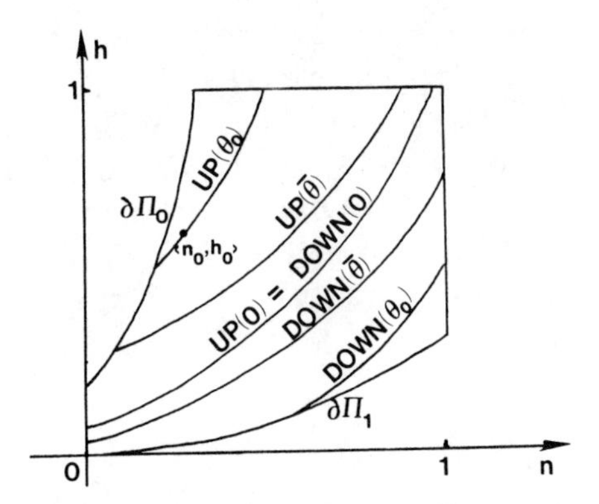

FIGURE 8. Typical sets UP(θ) and DOWN(θ). Here $0 < \bar{\theta} < \theta_0$. UP($\theta_0$) and UP($\bar{\theta}$) run along $\partial\Pi_0$ up to the left edge, and DOWN(θ_0) runs along $\partial\Pi_1$ to the left edge.

Let θ_0 be the value of θ for which $\langle n_0, h_0\rangle \in$ UP(θ_0) (Figure 8). In other words, there is a solution of (4.2; $\theta = \theta_0$, $\varepsilon = 0$) from $\langle V_0(n_0, h_0), 0, n_0, h_0\rangle = \langle V_{\text{rest}}, 0, n_0, h_0\rangle$ to $\langle V_1(n_0, h_0), 0, n_0, h_0\rangle$. Such a value of θ exists since Hypothesis 1(G) implies that $\int_{V_0(n_0,h_0)}^{V_1(n_0,h_0)} G(V, n_0, h_0)\, dV < 0$.

Representations of a singular solution of (4.2; $\theta = \bar{\theta}$, ε).

Representations 1–5 give, by example, various ways of depicting a periodic singular solution. The example is one of the Ω-periodic solutions discussed in §5.

Representation 1. *Projection into* $\mathbf{R}^3$ (V-n-h) *space.*

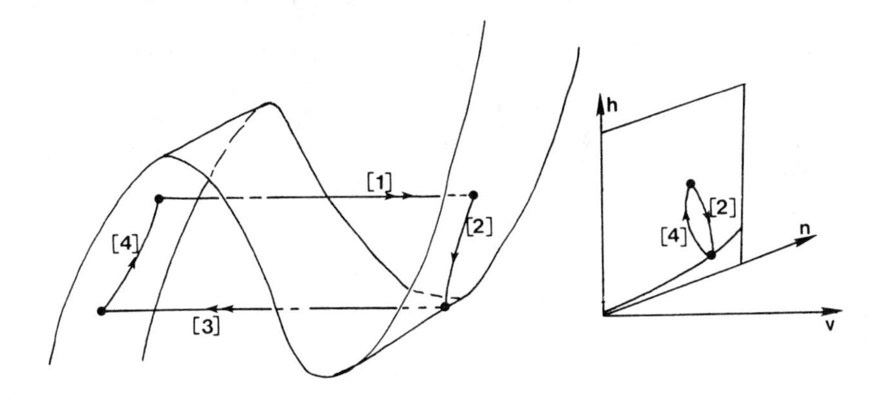

FIGURE 9. Representation 1. A periodic ss projected into $\mathbf{R}^3$ (V-n-h space). The coordinate axes are indicated on the right, where the ss is shown projected into the n-h plane.

In Figure 9, the W-axis has been suppressed. In $\mathbf{R}^4$,

$$A \equiv \langle V_0(\bar{n}, \bar{h}), 0, \bar{n}, \bar{h}\rangle, \qquad B \equiv \langle V_1(\bar{n}, \bar{h}), 0, \bar{n}, \bar{h}\rangle,$$

$$C \equiv \langle V_1(\tilde{n}, \tilde{h}), 0, \tilde{n}, \tilde{h}\rangle, \qquad D \equiv \langle V_0(\tilde{n}, \tilde{h}), 0, \tilde{n}, \tilde{h}\rangle.$$

The ss consists of:

[1]–a solution of (4.2; $\theta = \bar{\theta}$, $\varepsilon = 0$) from A to B;

[2]–a solution segment of (4.3; $i = 1$) from B to C;

[3]–a solution of (4.2; $\theta = \bar{\theta}$, $\varepsilon = 0$) from C to D; and

[4]–a solution segment of (4.3; $i = 0$) from D to A.

In this example, $\langle \tilde{n}, \tilde{h} \rangle \in \Pi_0 \cap \partial\Pi_1$ and $\langle \bar{n}, \bar{h} \rangle \in \Pi_0 \cap \Pi_1$.

On the middle section of $\mathbb{S}$, where $V = V_2(n, h)$, $\partial G/\partial V(V, n, h) < 0$. This implies that no solutions of (4.2; θ, $\varepsilon = 0$) approach that section, so all ss's run between the left and right sections, where $V = V_0(n, h)$ and $V = V_1(n, h)$.

Representation 2. *Graph of* $V(s)$.

We can use Representation 1 to sketch $V(s)$ for $-\infty < s < \infty$ (Figure 10).

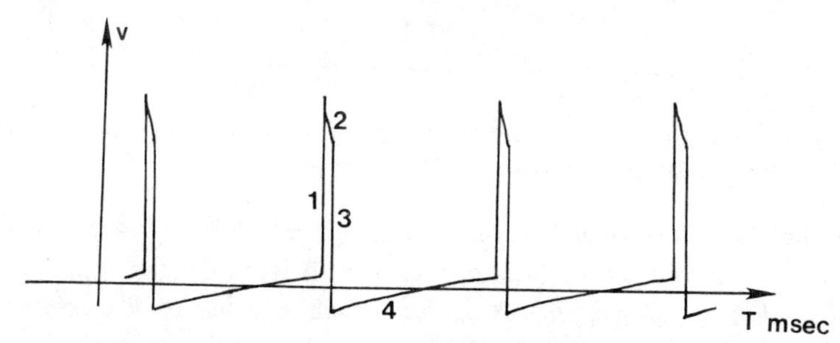

FIGURE 10. Representation 2. A graph of $V(s)$, rescaled to a new time scale of $T = \varepsilon \cdot s/\text{speed}$ msec. In the singular limit as $\varepsilon \to 0^+$, [1] and [3] become vertical lines, as shown here.

By implicit differentiation of the identity $G(V_i(n, h), n, h) = 0$ we obtain

$$\frac{\partial V_i}{\partial n} = \frac{-\partial G/\partial n}{\partial G/\partial V} < 0 \quad \text{and} \quad \frac{\partial V_i}{\partial h} = \frac{-\partial G/\partial h}{\partial G/\partial V} > 0.$$

Thus, $V = V_1(n, h)$ decreases along [2] and $V = V_0(n, h)$ increases along [4].

Representation 3. *Projections into* V-W *and* n-h *space.*

Each of the solution segments [1], [2], [3], [4] is contained in a two-dimensional subspace of $\mathbf{R}^4$. In this representation, each segment is projected into the V-W or n-h plane.

In Figure 11, [1] is projected to a solution of

$$\dot{V} = W, \qquad \dot{W} = \bar{\theta} W + G(V, \bar{n}, \bar{h})$$

from $\langle V_0(\bar{n}, \bar{h}), 0 \rangle$ to $\langle V_1(\bar{n}, \bar{h}), 0 \rangle$.

[2] is projected to a solution segment in Π_1 from $\langle \bar{n}, \bar{h} \rangle$ to $\langle \tilde{n}, \tilde{h} \rangle$ with

$$n' = \gamma_n(V_1(n, h))(n_\infty(V_1(n, h)) - n),$$
$$h' = \gamma_h(V_1(n, h))(h_\infty(V_1(n, h)) - h). \qquad (4.4;\ i = 1)$$

Here, $T = \varepsilon \cdot s/\text{speed}$ msec and $' = d/dT$.

[3] is projected to a solution of

$$\dot{V} = W, \qquad \dot{W} = \bar{\theta} W + G(V, \tilde{n}, \tilde{h})$$

from $\langle V_1(\tilde{n}, \tilde{h}), 0 \rangle$ to $\langle V_0(\tilde{n}, \tilde{h}), 0 \rangle$.

[4] is projected to a solution segment from $\langle \tilde{n}, \tilde{h} \rangle$ to $\langle \bar{n}, \bar{h} \rangle$ with

$$n' = \gamma_n(V_0(n, h))(n_\infty(V_0(n, h)) - n),$$
$$h' = \gamma_h(V_0(n, h))(h_\infty(V_0(n, h)) - h). \qquad (4.4;\ i = 0)$$

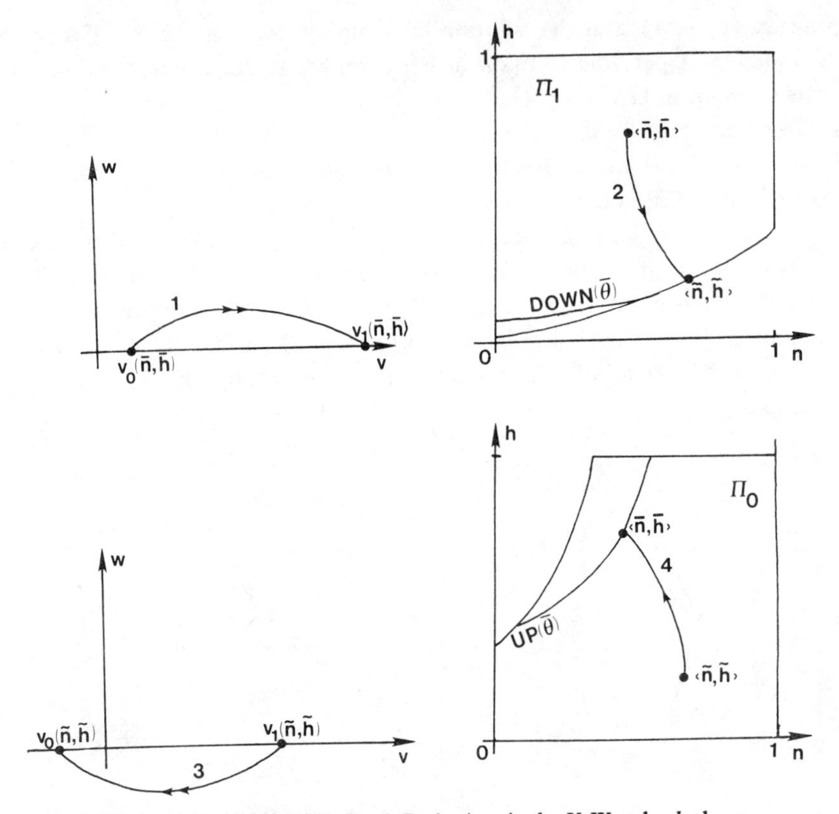

FIGURE 11. Representation 3. Projections in the V-W and n-h planes.

Representation 4. *Projection into* $\Pi_0 \cap \Pi_1$.

Representation 3 has the advantage that solutions are projected into the plane ($\mathbf{R}^2$), where we can easily sketch and study their properties. It has the disadvantage of not looking very much like a periodic solution. We next combine the four pictures of Figure 11 into the one picture of Figure 12. There, we need to imagine the solution [1]

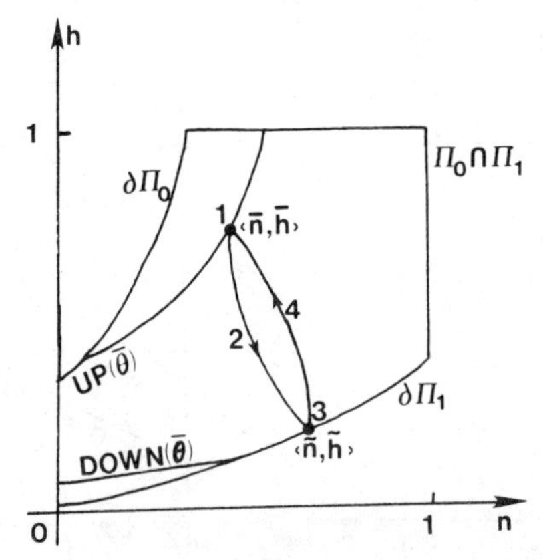

FIGURE 12. Representation 4. The singular solution projected into $\Pi_0 \cap \Pi_1$.

jumping up at $\langle \bar{n}, \bar{h} \rangle$ and the solution [3] jumping down at $\langle \tilde{n}, \tilde{h} \rangle$. Otherwise, the ss resembles a periodic solution in $\mathbf{R}^2$, except that the segment in Π_0 could cross the segment in Π_1.

Representation 5. *A fixed point.*

The point $\langle \bar{n}, \bar{h} \rangle$ is a fixed point of a natural map from $\mathrm{UP}(\bar{\theta})$ into itself. This map is defined as follows.

For each $\langle n, h \rangle \in \mathrm{UP}(\bar{\theta})$, trace the solution of (4.4; $i = 1$) in Π_1 from $\langle n, h \rangle$ to the first point in $\mathrm{DOWN}(\bar{\theta})$ on that solution. Call this point $F_1(n, h)$. Similarly, for each point $\langle n, h \rangle \in \mathrm{DOWN}(\bar{\theta})$, trace the solution of (4.4; $i = 0$) in Π_0 from $\langle n, h \rangle$ to the first point (if any) in $\mathrm{UP}(\bar{\theta})$ on that solution. Call this point $F_0(n, h)$. If, instead of ever crossing $\mathrm{UP}(\bar{\theta})$, the solution in Π_0 beginning at $\langle n, h \rangle$ approaches $\langle n_0, h_0 \rangle$ as $T \to \infty$, let $F_0(n, h) \equiv \langle n_0, h_0 \rangle$.

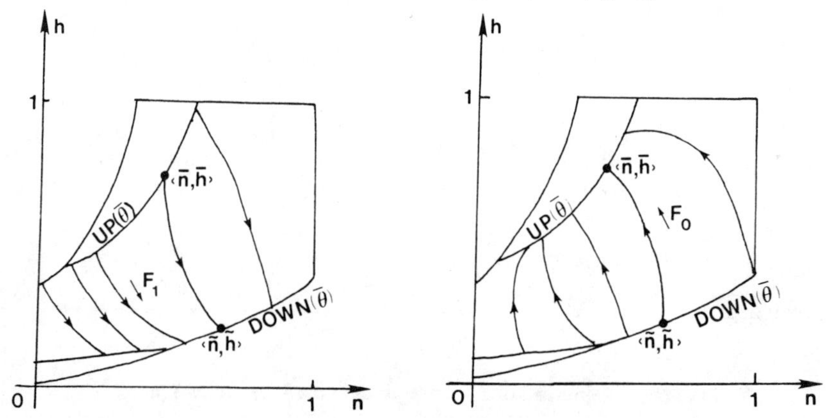

FIGURE 13. The maps F_0 and F_1. $\langle \tilde{n}, \tilde{h} \rangle = F_1(\bar{n}, \bar{h})$ and $\langle \bar{n}, \bar{h} \rangle = F_0(\tilde{n}, \tilde{h})$, so $\langle \bar{n}, \bar{h} \rangle = F_0 \circ F_1(\bar{n}, \bar{h})$.

In Figure 13, $\langle \bar{n}, \bar{h} \rangle$ is a fixed point of $F_0 \circ F_1$. In Figure 14A, the periodic ss is represented by the single point $\langle \bar{n}, \bar{h} \rangle$. Implicit in this representation is the statement that $\langle \bar{n}, \bar{h} \rangle$ is a fixed point of the map $F_0 \circ F_1$, where F_0 and F_1 are extended in the obvious way to any point in $\mathrm{DOWN}(\theta)$ or $\mathrm{UP}(\theta)$.

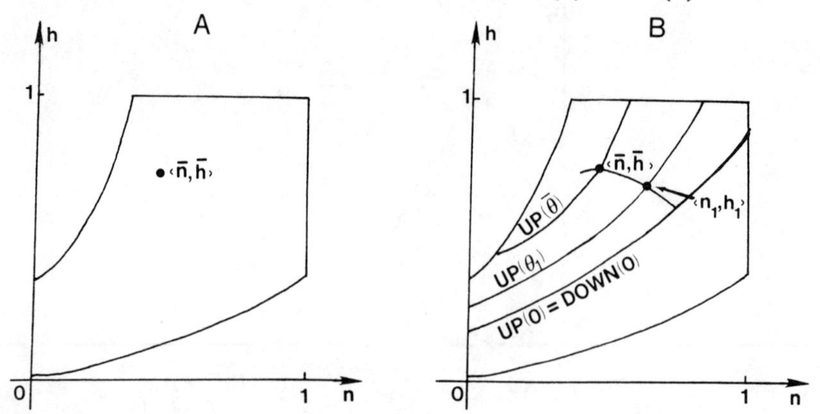

FIGURE 14. Representation 5. (A) A ss as a fixed point, $\langle \bar{n}, \bar{h} \rangle$, of the map $F_0 \circ F_1$. (B) An arc of fixed points of $F_0 \circ F_1$, representing a family of ss's.

Using the fixed-point representation we can speak of a family of ss's by referring to a curve of fixed points of $F_0 \circ F_1$. $\langle \bar{n}, \bar{h} \rangle$ is contained in an arc (Figure 14B), each point of which is a fixed point of $F_0 \circ F_1$. In Figure 14B, $0 < \theta_1 < \bar{\theta}$ and $\langle n_1, h_1 \rangle$ is a ss of (4.2; $\theta = \theta_1$, ε). The family of ss's forms a two-dimensional surface in $\mathbf{R}^4$.

5. Analysis of the generalized Hodgkin-Huxley model, with predictions. Hypothesis 1 implies that the phase portraits of the flows on Π_0 and Π_1 have the qualitative properties depicted in Figure 15. Although nonlinear, each flow resembles part of a two-dimensional linear system with one stable node.

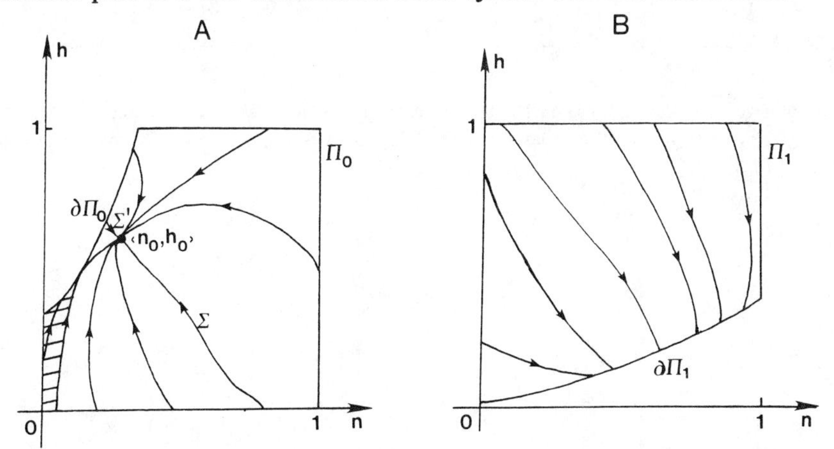

FIGURE 15. (A) A typical flow on Π_0. The solutions Σ and Σ' separate solutions which approach $\langle n_0, h_0 \rangle$ from above from those which approach $\langle n_0, h_0 \rangle$ from below. (B) A typical flow on Π_1. All solutions go to $\partial\Pi_1$.

In Π_0, all solutions either approach $\langle n_0, h_0 \rangle$ as $T \to \infty$ or go to $\partial\Pi_0$ in finite time (shaded region in Figure 15A). The separating solutions, Σ and Σ', approach $\langle n_0, h_0 \rangle$ along a line with negative slope. All other solutions which approach $\langle n_0, h_0 \rangle$ do so along a line with positive slope. Roughly speaking, points in the flow move more slowly the closer they are to $\langle n_0, h_0 \rangle$.

The flow on Π_1 is very simple: all solutions run to $\partial\Pi_1$ in finite time. In particular, for any $\theta > 0$, if $\langle n, h \rangle \in \mathrm{UP}(\bar{\theta})$ then the solution in Π_1 beginning at $\langle n, h \rangle$ crosses $\mathrm{DOWN}(\bar{\theta})$ in finite time.

One observation about the flow on Π_0 is crucial to the study of burst solutions. As illustrated in Figure 16, Σ splits $\mathrm{DOWN}(\theta_0)$ into two pieces, one piece above Σ, the other below. Every point in one or the other of these pieces crosses $\mathrm{UP}(\theta_0)$ in finite time.

Consider, now, a system in which the solution in Π_1 beginning at $\langle n_0, h_0 \rangle$ crosses $\mathrm{DOWN}(\theta_0)$ in one of the shaded regions (Figure 17). Then $B_1 \equiv F_0 \circ F_1(n_0, h_0) \neq \langle n_0, h_0 \rangle$. Therefore (since solutions cannot cross one another) $F_1(B_1) \equiv A_2$ is also in the shaded region, so $F_0(A_2) \equiv B_2 \neq \langle n_0, h_0 \rangle$, and so on. The points B_i approach $B \in \mathrm{UP}(\theta_0)$ and the points A_i approach $A \in \mathrm{DOWN}(\theta_0)$. Moreover, $F_0 \circ F_1(B) = B$, so B is a fixed point of $F_0 \circ F_1$. (Compare Figure 14.) The ss through A and B is here called *Ω-periodic*, since the

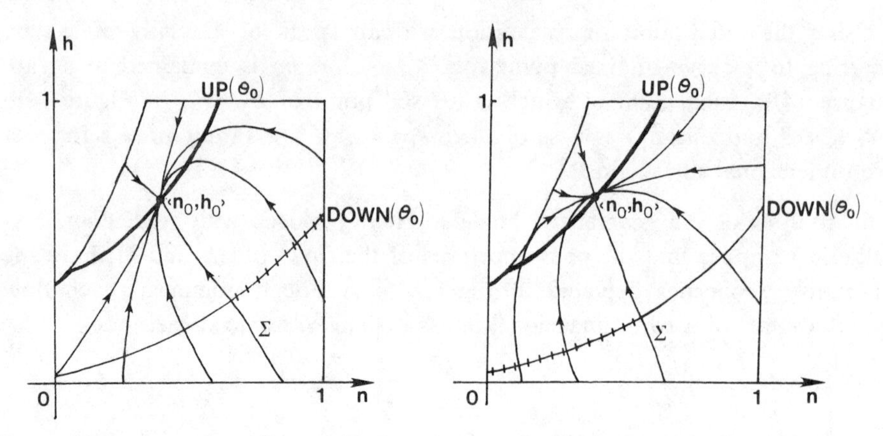

FIGURE 16. Two examples of flows on Π_0. In each case, points in the shaded portions of DOWN(θ_0) cross UP(θ_0) in finite time. Other points in DOWN(θ_0) approach $\langle n_0, h_0 \rangle$ as $T \to \infty$ without first crossing UP(θ_0).

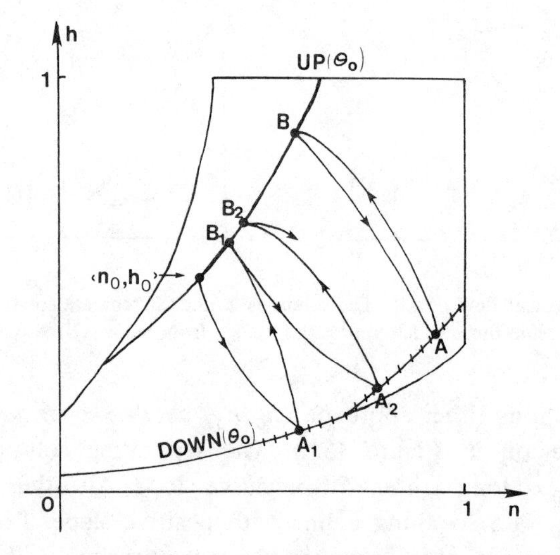

FIGURE 17. Points in the shaded part of DOWN(θ_0) all cross UP(θ_0) in finite time. $A_1 = F_1(n_0, h_0)$, $B_1 = F_0 \circ F_1(n_0, h_0) \neq \langle n_0, h_0 \rangle$, $A_2 = F_1 \circ (F_0 \circ F_1)(n_0, h_0)$, $B_2 = (F_0 \circ F_1)^2(n_0, h_0)$,etc.

solution segment from A to B is the omega-limit of the solution segments from A_i to B_i, as $i \to \infty$; and the solution segment from B to A is the omega-limit of the segments from B_i to A_i.

These observations imply that one of the following four cases holds (Figures 18–21).

Let

$$A_i = F_1 \circ (F_0 \circ F_1)^{i-1}(n_0, h_0) \in DOWN(\theta_0),$$

$$B_i = (F_0 \circ F_1)^i(n_0, h_0) \in UP(\theta_0),$$

$$A = \lim_{i \to \infty} A_i,$$

$$B = \lim_{i \to \infty} B_i.$$

Case 1. The solution in Π_0 beginning at A_1 crosses UP(θ_0) in finite time, and the solution in Π_0 beginning at B goes to $\langle n_0, h_0 \rangle$ as $T \to \infty$ (Figure 18).

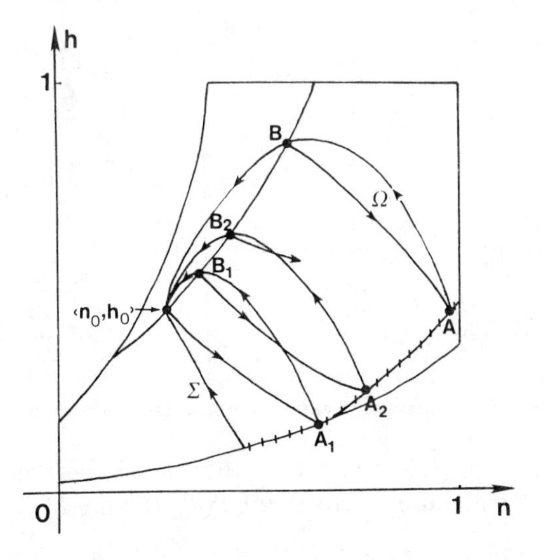

FIGURE 18. Case 1. The figure contains three ss's: (i) the ss from $\langle n_0, h_0\rangle$ to A_1 to B_1 to $\langle n_0, h_0\rangle$ has one spike; (ii) the ss from $\langle n_0, h_0\rangle$ to A_1 to B_1 to A_2 to B_2 to $\langle n_0, h_0\rangle$ has two spikes; and (iii) the ss through A and B is Ω-periodic.

Case 2. The solution in Π_0 beginning at A_1 crosses UP(θ_0) in finite time; for some $I \geqslant 2$ the solution in Π_0 beginning at B_I goes to $\partial\Pi_0$; and for all $i < I$, the solution in Π_0 beginning at B_i goes to $\langle n_0, h_0\rangle$ as $T \to \infty$ (Figure 19).

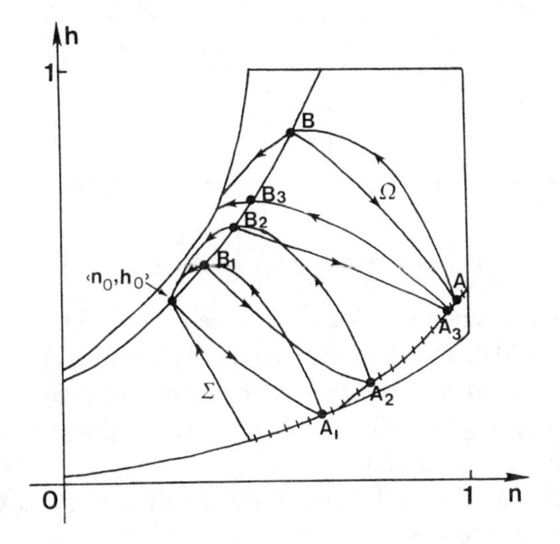

FIGURE 19. Case 2. $I = 3$. Pictured are the same ss's as in Figure 18. The solution segment beginning at A_3 goes to $\partial\Pi_0$.

Case 3. The solution in Π_0 beginning at A_1 goes to $\partial\Pi_0$ (Figure 20).

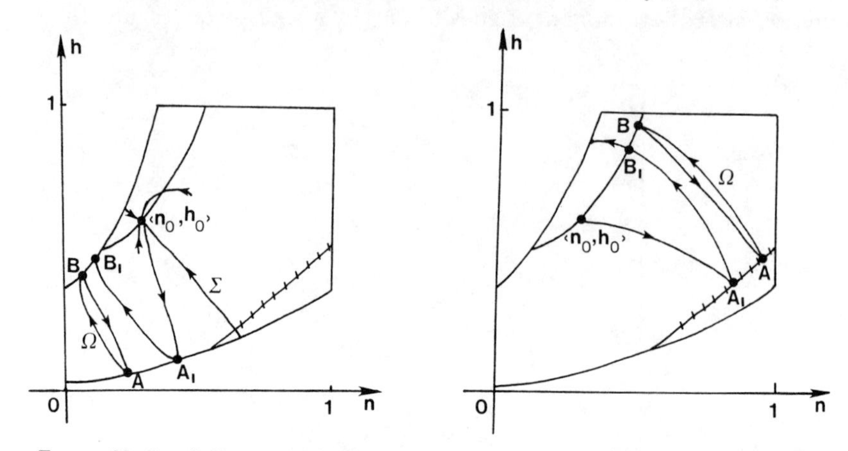

FIGURE 20. Case 3. Two examples. In each case, the solution beginning at A_1 goes to $\partial\Pi_0$.

Case 4. $F_0(A_1) = \langle n_0, h_0\rangle$, i.e., the solution in Π_0 beginning at A_1 goes to $\langle n_0, h_0\rangle$ as $T \to \infty$ without ever crossing UP(θ_0) (Figure 21).

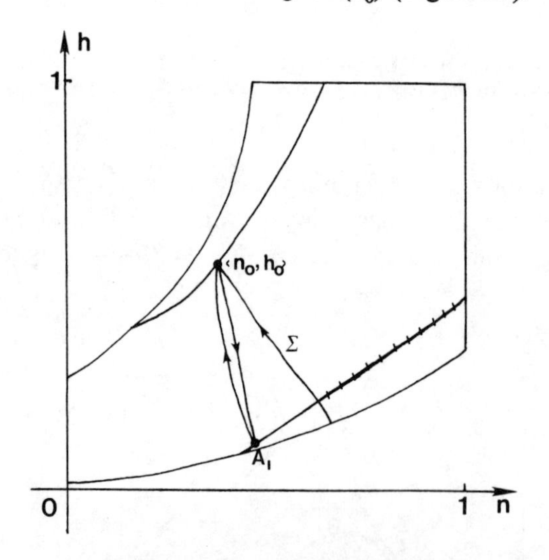

FIGURE 21. Case 4. $F_0(A_1) = \langle n_0, h_0\rangle$.

These four cases correspond to the four cases of Theorem 1. (4.2) has burst solutions in Cases 1 and 2, and Ω-periodic solutions in all cases.

Technical assumption. In each sketch, we have implicitly assumed that:

(i) in Π_0, all solutions cross DOWN(θ_0) in the same direction and the flow is not tangent to UP(θ_0) at any B_i; and

(ii) in Π_1, all solutions cross UP(θ_0) in the same direction and the flow is not tangent to DOWN(θ_0) at any A_i.

If (i) and (ii) hold, we will say that the system (4.2) is *admissible*. This assumption can be weakened, but simplifies the proof.

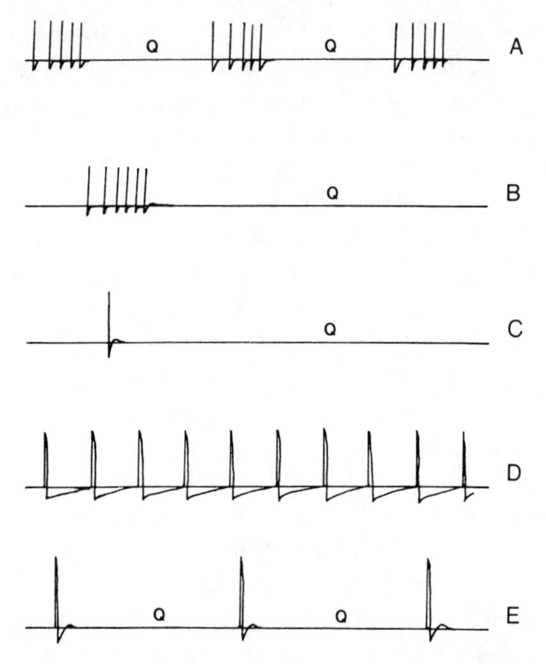

FIGURE 22. Examples of (A) burst-periodic, (B) finite wave train, (C) single spike, (D) Ω-periodic, and (E) burst-with-one-spike solutions. Q indicates the location of the quiet spells, which are infinite in (B) and (C).

THEOREM 1. *Half of all generalized Hodgkin-Huxley models have burst solutions.*

Assume that the generalized HH system (4.2; θ, ε) is admissible and satisfies Hypothesis 1. Then (4.2) has a family of Ω-periodic solutions with evenly-spaced spikes. In addition, one of the following four cases holds.

TABLE 2

	Case 1	Case 2	Case 3	Case 4
A family of Ω-periodic solutions, bounded away from equilibrium	X	X	X	
A family of Ω-periodic solutions not bounded away from equilibrium				X
Burst-periodic and finite wave train solutions with N spikes	X ($N \geqslant 1$)	X ($1 \leqslant N < I$)		
Single pulse solutions	X	X		X

Case 1. For all $N \geqslant 1$ there is a family of burst-periodic solutions with N spikes per burst. The family of bursts with N spikes converges to a finite wave train with N spikes as the length of the quiet spell goes to infinity. The family of Ω-periodic solutions is bounded away from infinity.

Case 2. For all N with $1 \leqslant N < I$ there is a family of burst-periodic solutions with N spikes per burst. The family of bursts with N spikes converges to a finite wave train with N spikes as the length of the quiet spell goes to infinity. The family of Ω-periodic solutions is bounded away from equilibrium.

Case 3. The family of Ω-periodic solutions is bounded away from equilibrium. For small ε and δ, there are no single pulse, finite wave train, or burst solutions.

Case 4. The family of Ω-periodic solutions converges to a single pulse solution as the interspike interval goes to infinity. For small ε and δ there are no finite wave train or burst solutions.

PROOF. The proof of the existence of burst-periodic and Ω-periodic solutions uses topological degree theory, along with the notion of an "n-dimensional singular solution" (Carpenter **[1977b]**, **[1979]**). Here $n = 2$, the dimension of the slow manifold. However, the intuitive idea of the proof may be seen simply by following the various 1-dimensional singular solutions sketched in Figures 18–21. Figures 18 and 19 contain singular solutions with one and two spikes. Each of these ss's is a finite wave train since the quiet spell is infinite. However, each singular finite wave train solution with N spikes bounds a family of burst solutions with N spikes per burst. As the length of the quiet spell goes to infinity, the family of burst solutions approaches the finite wave train solution.

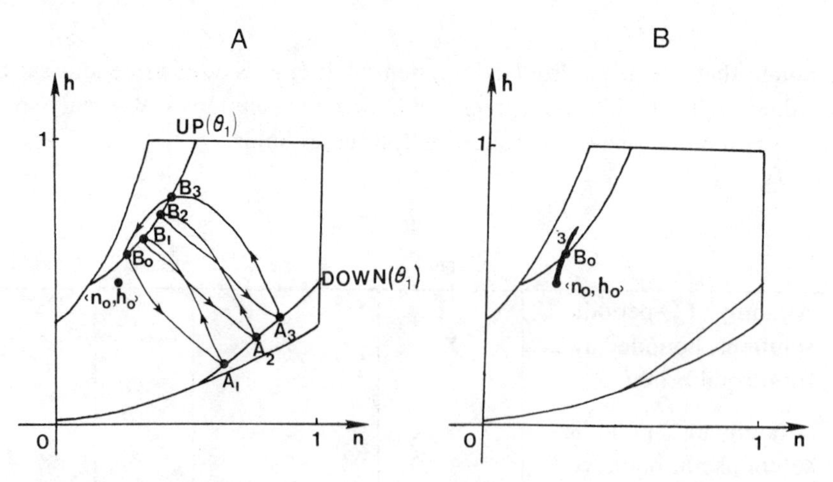

FIGURE 23. (A) Representation 4. A singular burst solution with 3 spikes per burst. $\theta_1 > \theta_0$. (B) Representation 5. $B_0 = \phi \circ (F_0 \circ F_1)^3(B_0)$. The arc represents an entire family of fixed points of $\phi \circ (F_0 \circ F_1)^3$, and hence a family of burst solutions with 3 spikes per bust.

In Figure 23A we see a burst-periodic ss with 3 spikes per burst. For this ss, $\theta = \theta_1 > \theta_0$. The quiet spell occurs between B_3 and B_0. If B_0 is near $\langle n_0, h_0\rangle$, the quiet spell is long. In Figure 23, $B_3 = (F_0 \circ F_1)^3(B_0)$ and $B_0 = \phi(B_3)$, where, if $\langle n, h\rangle \in \mathrm{UP}(\theta)$, then $\phi(n, h)$ is the *last* point (if any) in $\mathrm{UP}(\theta)$ and on the solution in Π_0 beginning at $\langle n, h\rangle$. Thus

$$B_0 = \phi \circ (F_0 \circ F_1)^3(B_0)$$

i.e., B_0 is a fixed point of the map $\phi \circ (F_0 \circ F_1)^3$. Similarly, each burst-periodic ss with N spikes (Representation 4) contains a fixed point of the map $\phi \circ (F_0 \circ F_1)^N$. Using Representation 5, we can summarize the ss of Figure 23A by locating the single point B_0. Moreover, in Figure 23B we can represent the entire family of burst solutions of which the ss of Figure 23A is one member. One end of the arc is the point $\langle n_0, h_0\rangle$, where the length of the quiet spell goes to infinity. At the other end, either the arc "3" joins the arc "Ω" of fixed points of $F_0 \circ F_1$ (Figure 24A) and $Q \to 0$; or the solution beginning at $(F_0 \circ F_1)^3$ touches $\partial\Pi_0$ (Figure 24B), and the family of bursts terminates with $Q > 0$.

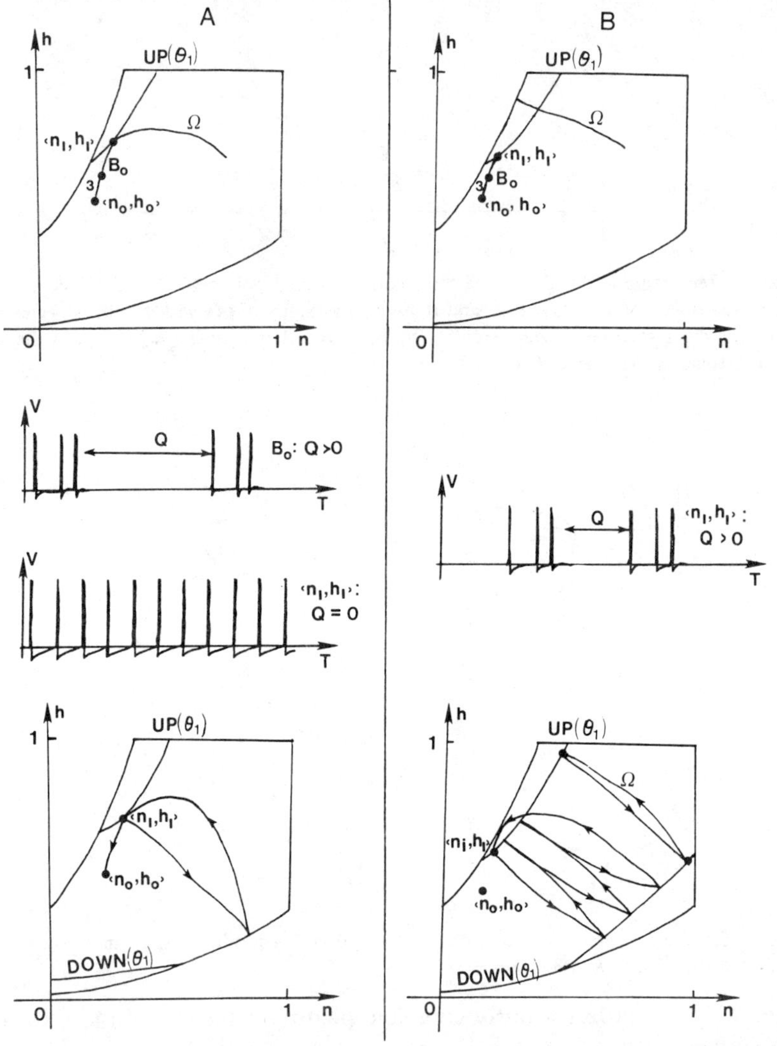

FIGURE 24. (A) At $\langle n_1, h_1\rangle$, $Q = 0$ and all spikes are evenly-spaced. The solution in Π_0 is tangent to UP(θ_1) at $\langle n_1, h_1\rangle$. (B) At $\langle n_1, h_1\rangle$, $Q > 0$ but the solution segment touches $\partial\Pi_1$ during the quiet spell. $\langle n_1, h_1\rangle$ is the last point on the curve "3" of fixed points of $\phi \circ (F_0 \circ F_1)^3$.

Figure 25 completes the picture of Figure 24A (Case 1 of Theorem 1). There is a curve of fixed points of $\phi \circ (F_0 \circ F_1)^N$ from $\langle n_0, h_0 \rangle$ to $\langle n_1, h_1 \rangle$ for each $N \geqslant 1$. When N is large, most of the spikes of the burst ($\theta = \bar{\theta}$) are near the Ω-periodic solution with $\theta = \bar{\theta}$, as illustrated schematically in Figure 26.

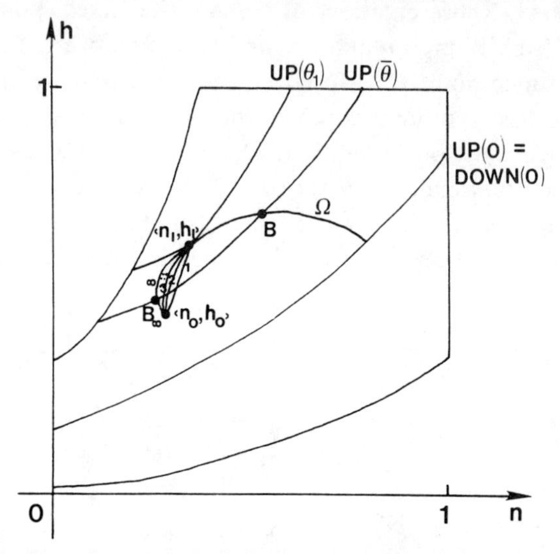

FIGURE 25. Representation 5. The curve "Ω" contains fixed points of $F_0 \circ F_1$. Each curve "N" contains fixed points of $\phi \circ (F_0 \circ F_1)^N$ and represents a family of ss's with N spikes per burst. The length of the quiet spell goes from zero (at $\langle n_1, h_1 \rangle$) to infinity (at $\langle n_0, h_0 \rangle$). B_∞, in the limiting curve "∞", equals $\phi(B)$, where B is in "Ω".

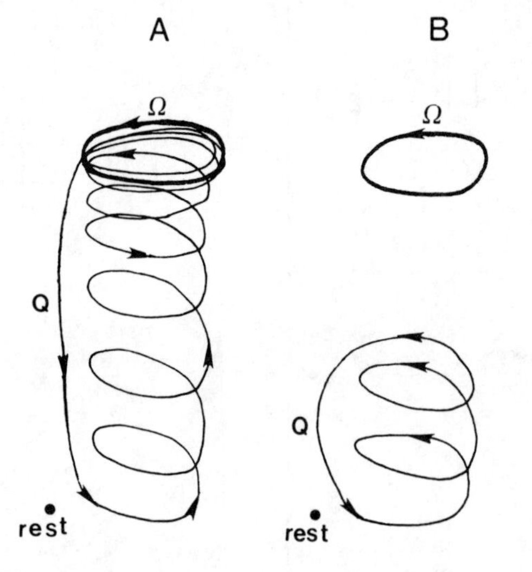

FIGURE 26. Schematic representation, in phase space, of an Ω-periodic solution and (A) a burst with N spikes and (B) a burst with 3 spikes.

Figure 25 summarizes a fairly complete picture of the ss's of (4.2), Case 1. It also contains information leading to predictions about the parameters and qualitative properties of solutions. Define the *period of the singular solution* through the point $B_0 = \phi \circ (F_0 \circ F_1)^N(B_0)$ as the total time the solution takes to go from B_0 to A_1 to B_1 to A_2, etc., and back to B_0. Then Figure 27 indicates the relationship between θ (and hence the wave speed) and the period of the ss, which is approximately equal to the period of the true solution of (4.2; θ, ε, δ).

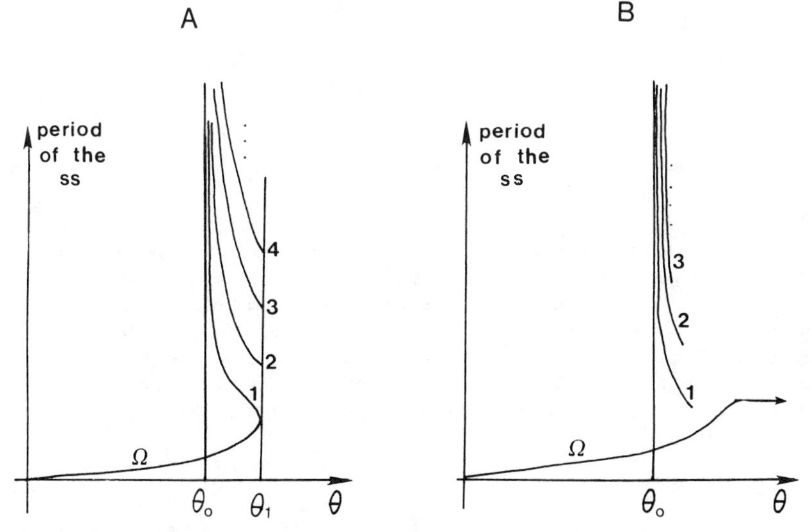

FIGURE 27. (A) Period of the ss as a function of θ for the case depicted in Figures 24A and 25. (B) Period of the ss as a function of θ for the case depicted in Figure 24B.

Predictions.

Prediction 1. All generalized HH nerve models (4.2) admit Ω-periodic solutions.

Prediction 2. "Half" of these models also admit periodic burst and finite wave train solutions. (Henceforth we will call periodic burst solutions of (4.2) *HH bursts*.)

Fine structure of HH bursts.

Prediction 3. Within an HH burst or finite wave train, spiking frequency becomes nearly constant after the first few spikes. Early in the burst, the frequency tends to increase, giving rise to a "long first interval" (Figures 28 and 29).

Prediction 4. If two HH bursts have the same (temporal) period, the one with more spikes/burst travels faster (Figures 27 and 30).

Prediction 5. With wave speed held constant, the length of the quiet spell (Q) increases with the number (N) of spikes in the previous burst (Figures 31 and 32).
As N becomes large, Q approaches a finite upper bound. N is "large" when the spiking frequency becomes approximately constant. As seen in Figure 28, this may happen either early or late in the burst.

Prediction 6. The speed of a finite wave train increases with the number of spikes in the previous burst. The same is true of HH bursts with long quiet spells (asymptotically as $Q \to \infty$) (Figure 33).

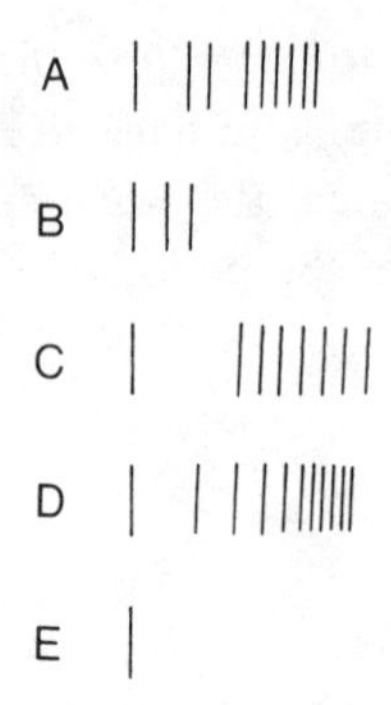

FIGURE 28. Examples of typical HH burst patterns, as described in Prediction 3. In each case (A)–(E), the burst pictured would be repeated over and over, with a quiet spell after each one. (A) As a whole, the frequency increases within the burst, but some variation is possible. (B) With few spikes in the burst, no conclusions about the increase or decease of frequency can be drawn. (C) A distinct long first interval, followed by a volley of spikes with approximately constant frequency. In this case, $F_0 \circ F_1(n_0, h_0)$ is near $\langle n_0, h_0 \rangle$, but $(F_0 \circ F_1)^N(n_0, h_0)$ is near Ω when $N > 2$. (D) Frequency increases gradually. (E) A burst may have only one spike, and hence is part of a regular periodic solution. Its qualitative properties are distinct from those of a regular Ω-periodic solution, as discussed in Prediction 9.

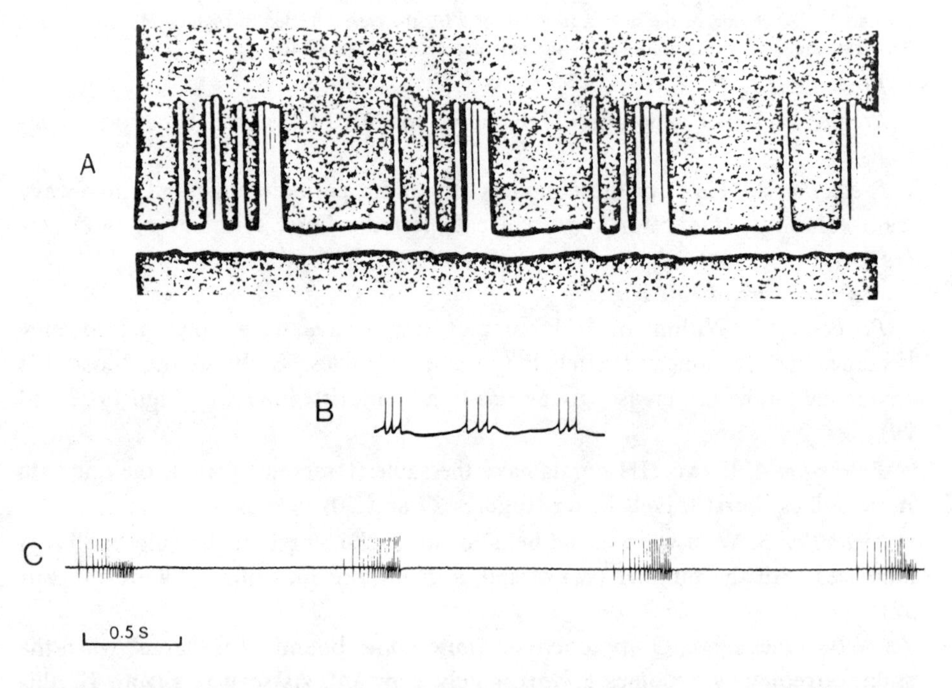

FIGURE 29. Prediction 3. Typical HH burst patterns in intracellular recordings. (A) Spontaneous action potentials in monkey epileptic cortex (Atkinson and Ward [1964, p. 291]). (B) HH bursts in the lobster stomatogastric ganglion (Russell and Hartline [1978, p. 454]). (C) Bursts in the motor neuron controlling expiration in the dragonfly (Mill [1977, p. 193]). Note the increasing frequency within the burst and the very flat interburst interval.

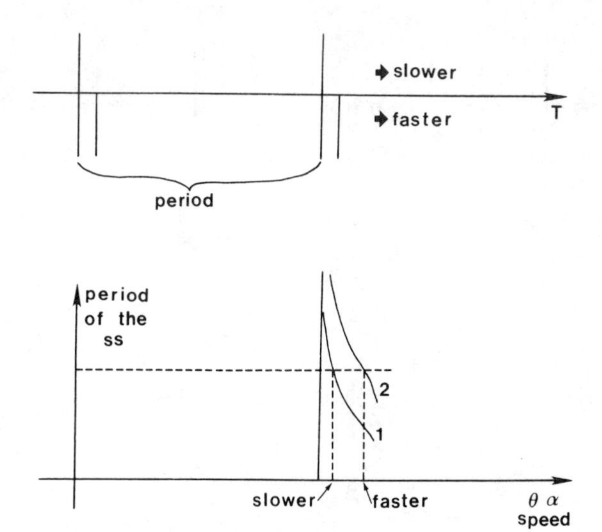

FIGURE 30. Prediction 4. θ is proportional to the wave speed.

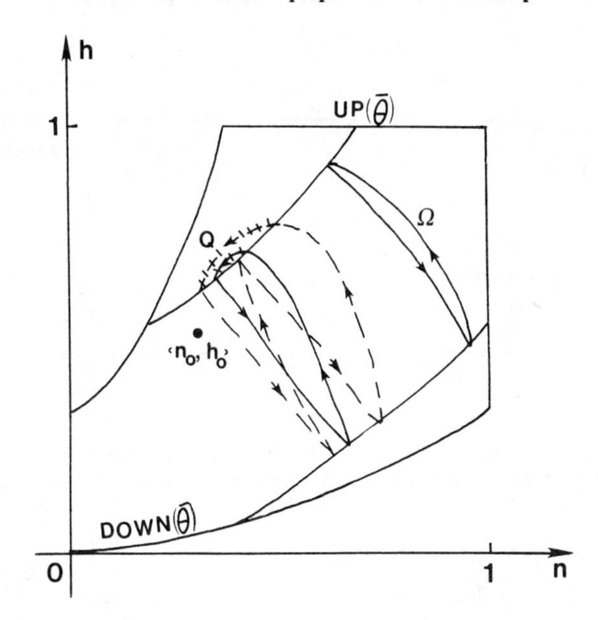

FIGURE 31. Prediction 5. HH bursts with 1 (solid line) and 2 (dotted line) spikes per burst. The quiet spells are shaded. For each, $\theta = \bar{\theta}$.

This prediction is a result of the proof of the existence of finite wave train solutions (Carpenter [**1976**], [**1977a**]). It is one of the few predictions which cannot be easily obtained by examining the ss. In the singular limit, all finite wave train solutions have $\theta = \theta_0$.

Prediction 7. Within an HH burst, V increases during the interspike interval. The longer the interval, the flatter the graph of V. At the end of a burst, there may be an afterdepolarization (arrow in Figure 34).

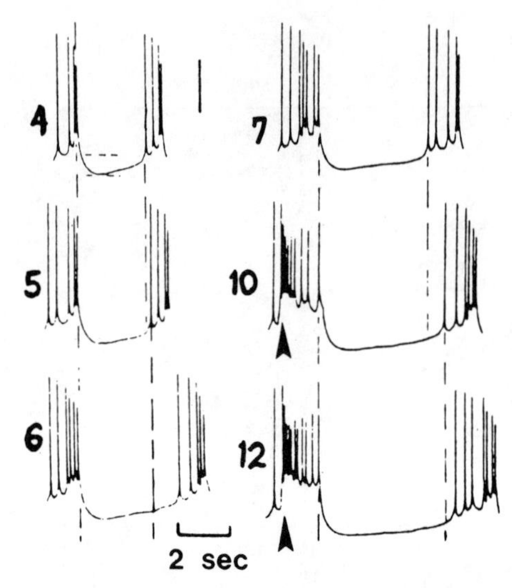

FIGURE 32. Prediction 5. The length of the quiet spell increases with the number of spikes in the previous burst: snail (*Lymnae stagnalis*) yellow cells (Benjamin [1978, p. 209]).

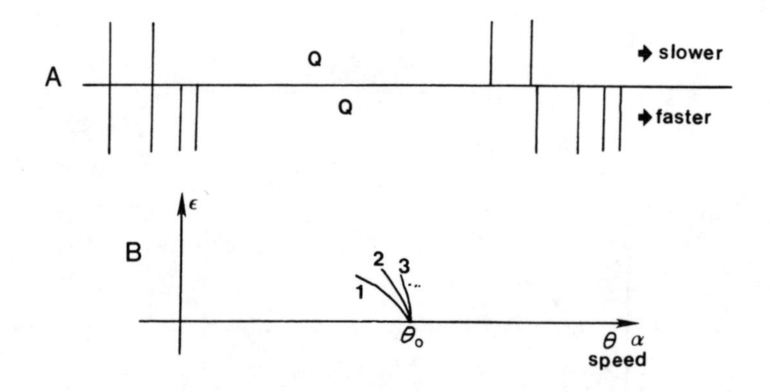

FIGURE 33. Prediction 6. (A) Two burst solutions with long quiet spells. (B) 1, 2, 3, . . . represent parameters in the θ-ε plane for which there exist finite weave trains with 1, 2, 3, . . . spikes.

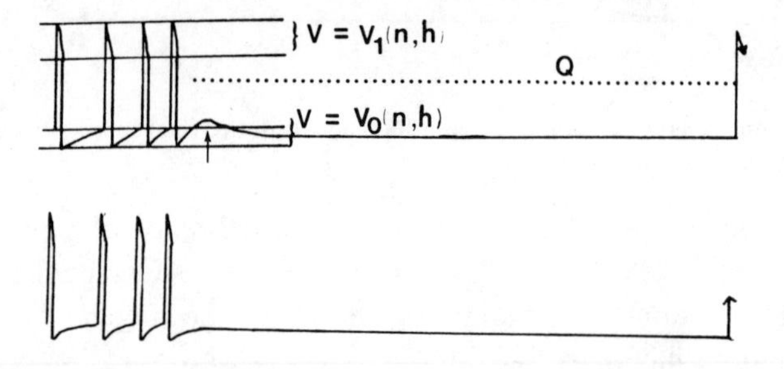

FIGURE 34. Prediction 7. Two typical wave forms within an HH burst. The arrow marks the afterdepolarization, which may or may not occur, depending on whether or not $V_0(n, h)$ is monotonic during the quiet spell.

Prediction 8. During an HH burst sequence, the baseline of activity (V) is relatively flat (Figures 35A and 29). Large shifts in this baseline indicate the influence of external depolarizing shifts (DPS) during the oscillation (Figure 35B and 36).

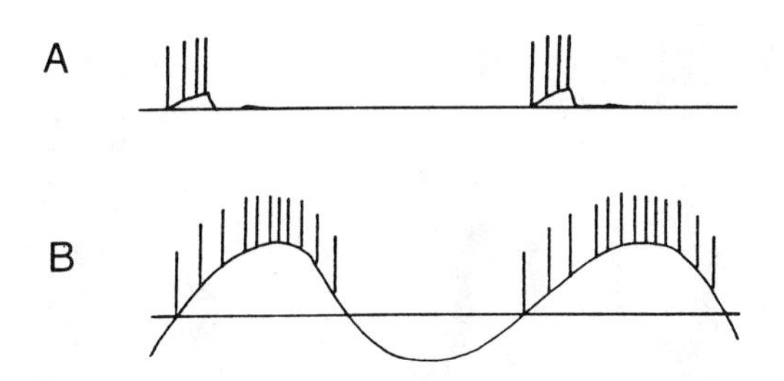

FIGURE 35. Prediction 8. (A) The baseline of activity is relatively flat. (B) The baseline oscillates substantially, indicating the influence of an external driving force.

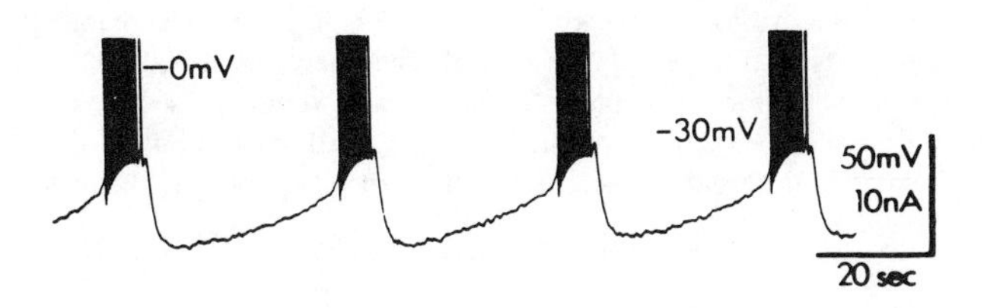

FIGURE 36. A burst riding the crest of a slow potential wave (*Otala lactea*, cell 11) (Barker and Smith [1978, p. 380]). The large oscillations in the baseline of potential are not seen in HH bursts, although small oscillations may be present.

Regular (evenly-spaced) periodic patterns.

Prediction 9. There are two types of regular periodic solutions of (4.2): Ω-periodic and bursts with 1 spike per burst ("1-bursts"). These two solution types have different qualitative properties. In particular, in a cell which bursts, Ω-periodics are far from equilibrium while bursts with one spike are near equilibrium. Thus, parametric shifts may easily extinguish a 1-burst while an Ω-periodic would first move through a burst phase to a 1-burst phase and then to equilibrium (Figure 37).

Distinguishing characteristics of Ω-periodic and 1-burst solutions are listed below.

Ω-periodic solutions (fixed points of $F_0 \circ F_1$).

(Ω:A) V increases during the interspike interval (Figures 37 and 38).

(Ω:B) In Cases 1, 2, and 4 of Theorem 1, the spiking frequency of Ω-periodic solutions ranges from moderate to very high (Figures 27 and 37A). In Case 3 of Theorem 1 (when there are no bursts) spiking frequency of Ω-periodic solutions ranges from low to very high (Figures 37B and 39). Note that the frequency is high when the period of the ss is near zero.

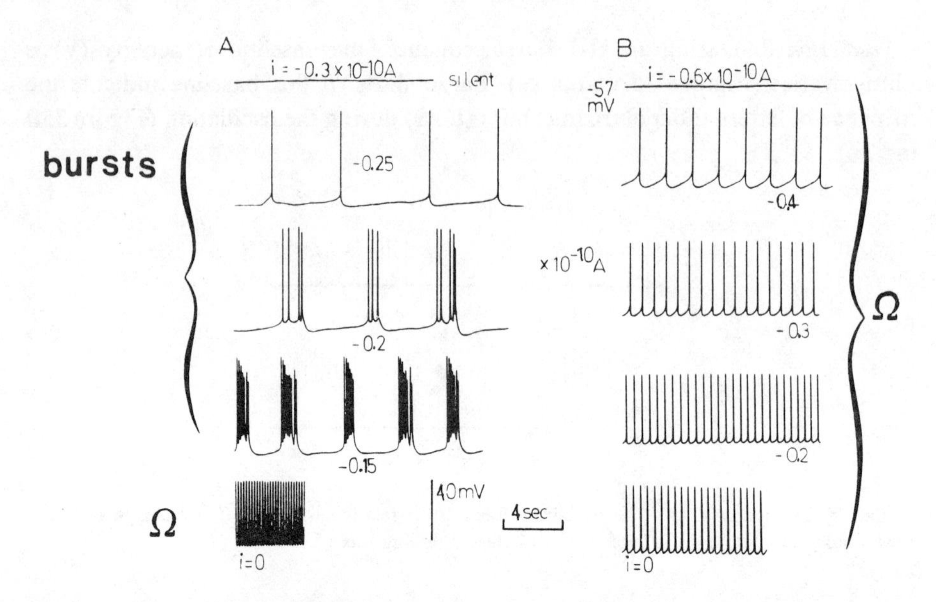

FIGURE 37. Prediction 9. Recordings from two similar snail yellow cells (Benjamin [1978, p. 208]). From bottom to top, cells are hyperpolarized until they become silent. (A) High frequency Ω-periodic spikes ($i = 0$) pass through a phase of bursts with many spikes, then bursts with few spikes, then 1-bursts, then silent: exactly as predicted. (B) In a cell without bursts (the other "half" of Prediction 2) high frequency Ω-periodics become low frequency Ω-periodics, then silent.

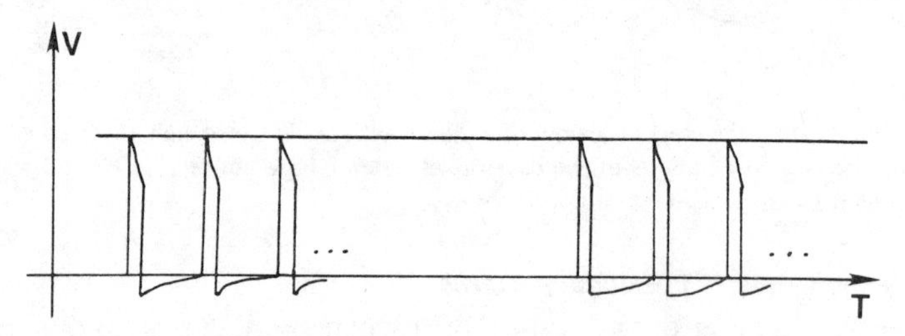

FIGURE 38. Typical wave forms of Ω-periodic spikes.

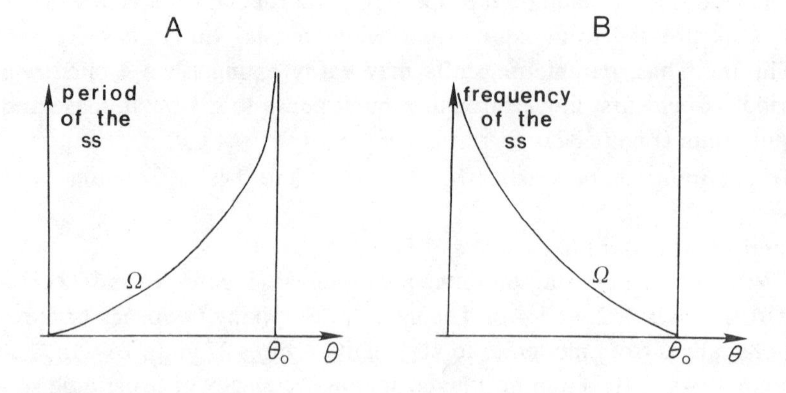

FIGURE 39. (A) Period of an Ω-periodic ss for Case 3 of Theorem 1, when there are no burst solutions. (B) Frequency of the same ss.

(Ω:C) The speed of an Ω-periodic solution increases as the frequency decreases (i.e., as the period increases) (Figure 27 and 39). The wave speed ranges from low to moderate (Cases 1 and 3) or from low to high (Cases 1, 2, and 4).

(Ω:D) Various properties of Ω-periodic solutions are correlated. For example, the generalized HH model (4.2) predicts that if a parametric shift changes the frequency in a certain way then other properties such as amplitude must also change in a well-defined way at the same time. These correlated shifts would be a good test of the model; many of them are seen in Figure 41. Table 3 and Figure 40 summarize the properties.

TABLE 3. CORRELATED PROPERTIES OF Ω-PERIODIC SOLUTIONS.

higher frequency	lower frequency
lower amplitude	higher amplitude
smaller post-spike hyperpolarization	larger post-spike hyperpolarization
low speed	moderate or high speed
lower threshold	higher threshold

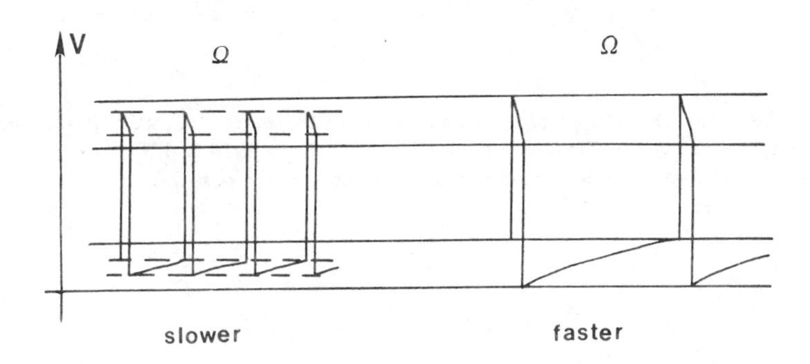

FIGURE 40. Correlated properties of Ω-periodic solutions.

Bursts with 1 *spike per burst* (*fixed points of* $\phi \circ F_0 \circ F_1$).

(1-burst:A) During the interspike interval, V first increases, but then has a long, relatively flat graph during the quiet spell, which may be preceded by an afterdepolarization (Figure 42). If there is an afterdepolarization at low frequency, it becomes smaller and then disappears as the frequency increases.

(1-burst:B) The frequency of 1-bursts ranges from very low to moderate (Figure 37A).

(1-burst:C) The wave speed is moderate and approximately constant over large changes in the period. θ is larger than θ_0, and the wave speed increases as frequency increases (curve "1" in Figures 25, 27, and 30).

(1-burst:D) Spike amplitude is large (about the same as the large-amplitude Ω-periodics) and approximately constant over large changes in the period. Postspike hyperpolarization also stays approximately constant.

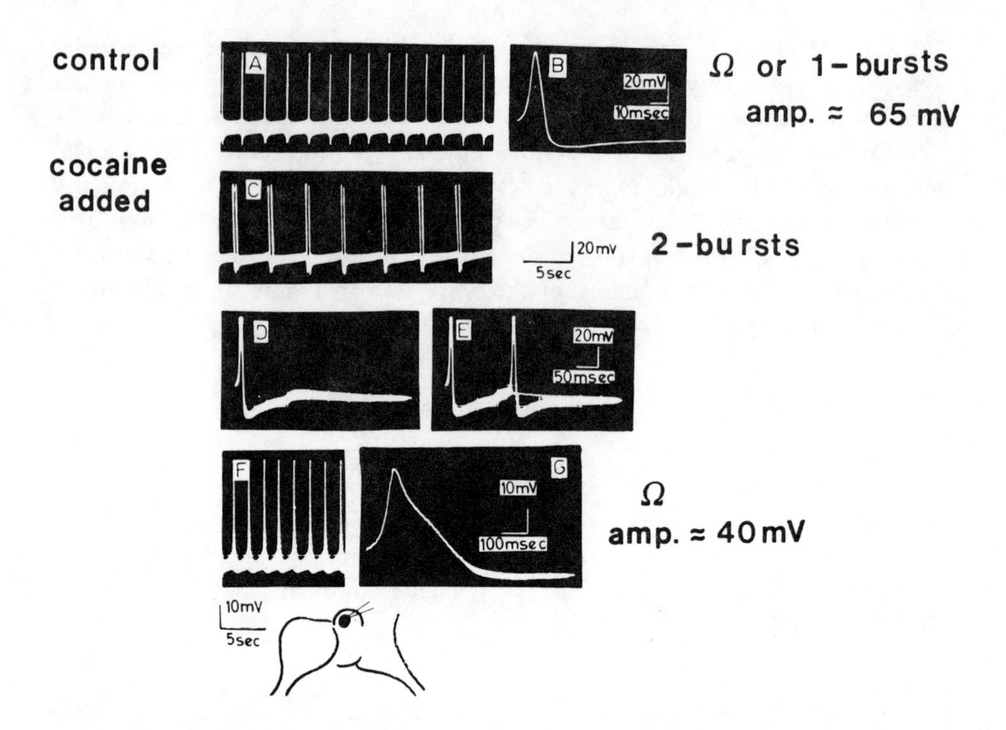

FIGURE 41. Recordings from snail (*Helix*) neuron. (A)–(B): control; (C)–(E): 10–25 min after administration of cocaine; (F)–(G) after 30 min (Lábos and Láng [1978, p. 179]). Compare (A) with (F): in (A) the frequency is lower; the amplitude is higher; and the post-spike hyperpolarization is larger–all as predicted in Table 3.

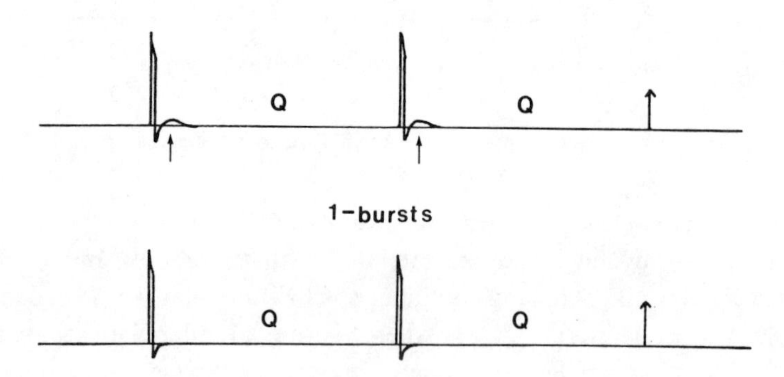

FIGURE 42. Typical graphs of bursts with one spike. The arrows indicate afterdepolarization, which may or may not occur.

Figure 43 depicts a continuous family of regular periodic solutions, from the single spike (A) to 1-bursts (B) and (C) to the transitional point (D) to Ω-periodics (E) and (F). The curves "1" and "Ω" in (G) are part of Figure 25 and in (H) are part of Figure 27.

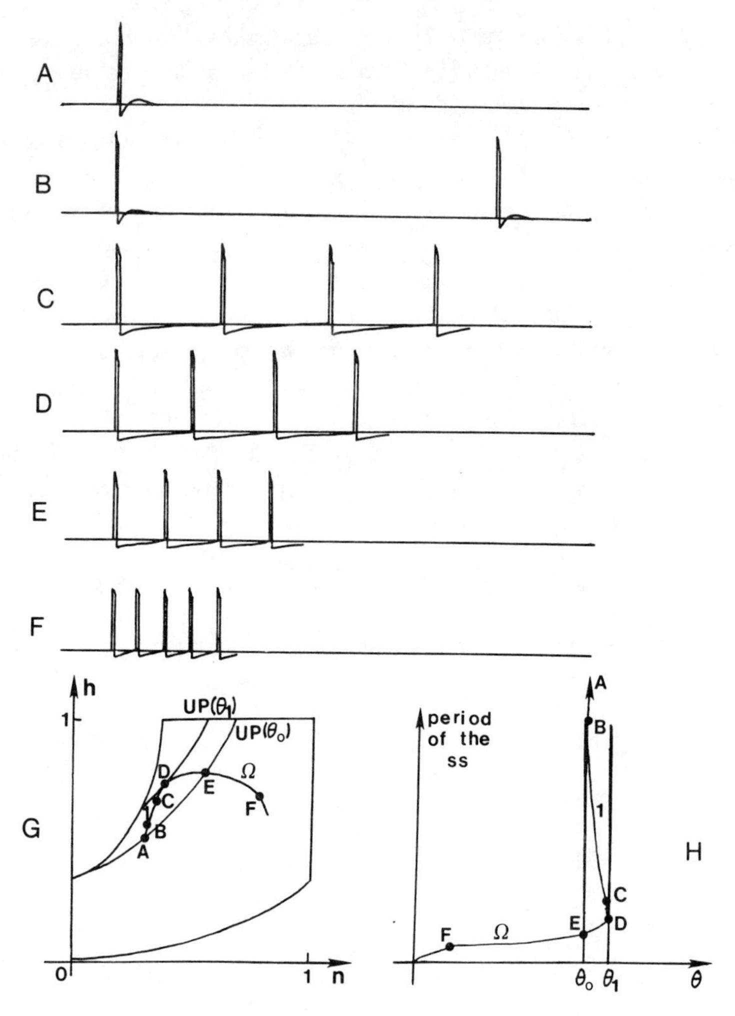

FIGURE 43. A typical family of regular periodic solutions of a generalized HH model. The quiet spell goes from ∞ at the single pulse (A) to 0 at the dividing point (D). (A) A single pulse, with $Q = \infty$ and $\theta = \theta_0$ in the singular limit. (B) A 1-burst with low frequency (large Q). (C) A 1-burst with moderate frequency (small Q). (D) The singular solution dividing Ω-periodic and 1-burst solutions. $Q = 0$ and $\theta = \theta_1$. (E) An Ω-periodic solution with moderate frequency. $\theta = \theta_0$. (F) An Ω-periodic solution with high frequency. θ is small. (G) Representation 5. Points A, B, and C are fixed points of $\phi \circ F_0 \circ F_1$. D, E, and F are fixed points of $F_0 \circ F_1$. (H) Period of the singular solutions (A)–(F).

Bursts in a noisy system. The HH bursts described so far are perfectly stereotyped: each burst is identical to the ones before and after. The presence of noise in the system could alter the stereotyped pattern so that, for instance, the number of spikes per burst varies. Variations in burst patterns may be explained by coupling (4.2) with noise *provided* that Predictions 10–12 are usually satisfied. Otherwise, other explanations for the variations must be sought.

Prediction 10. In a noisy system, HH bursts with many spikes per burst exhibit more variation in spike number than do bursts with few spikes per burst.

Reason. Burst solutions of (4.2) with many spikes per burst are near one another in phase space (Figure 25). Thus, small disturbances could easily move a point from one burst solution to another.

Prediction 11. "Variable first interval": Noise may produce large variations in the length of the first one or two interspike intervals (ISI's). Later in the burst, the frequency stays about constant, even if the total number of spikes per burst is varying.

Reason. The first one or two solution segments in Π_0 are close to $\langle n_0, h_0 \rangle$. Small variations in the location of these segments may move them nearer to $\langle n_0, h_0 \rangle$ (increasing the ISI) or farther from $\langle n_0, h_0 \rangle$ (decreasing the ISI). After several spikes, the burst solution approaches an Ω-periodic solution and spike frequency, for bursts of all lengths, approaches the frequency of the Ω solution.

Prediction 12. In a system with variable first interspike intervals, the longer first ISI's tend to correlate with longer quiet spells preceding the burst (Figure 44).

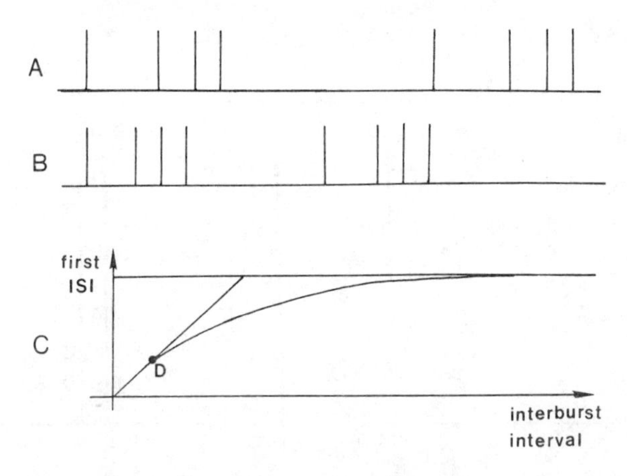

FIGURE 44. (A) Longer Q and longer first ISI. (B) Shorter Q and shorter first ISI. (C) Length of the preceding interburst interval vs. the length of the first interspike interval. Point D corresponds to the point (Figure 43) where the burst solutions meet the Ω-periodic solutions and all spikes are evenly spaced.

Reason. In Π_0, if a point approaches $\langle n_0, h_0 \rangle$ during a (long) quiet spell, the first segment in Π_0 of the next burst will also be comparatively near $\langle n_0, h_0 \rangle$.

6. Other types of bursts. A survey of the experimental literature reveals that some burst patterns resemble the HH bursts of §5 (Figure 29); some are determined by slow depolarization shifts (Figures 35 and 36); and some resemble neither of these. Two of the most common burst patterns are parabolic bursts (Figure 45A) and paroxysmal bursts (Figure 45B). A *parabolic burst* is characterized by the property that frequency first increases and then decreases. During a *paroxysmal burst* a sharp decrease in spike amplitude is followed by small oscillations about a plateau and a sudden

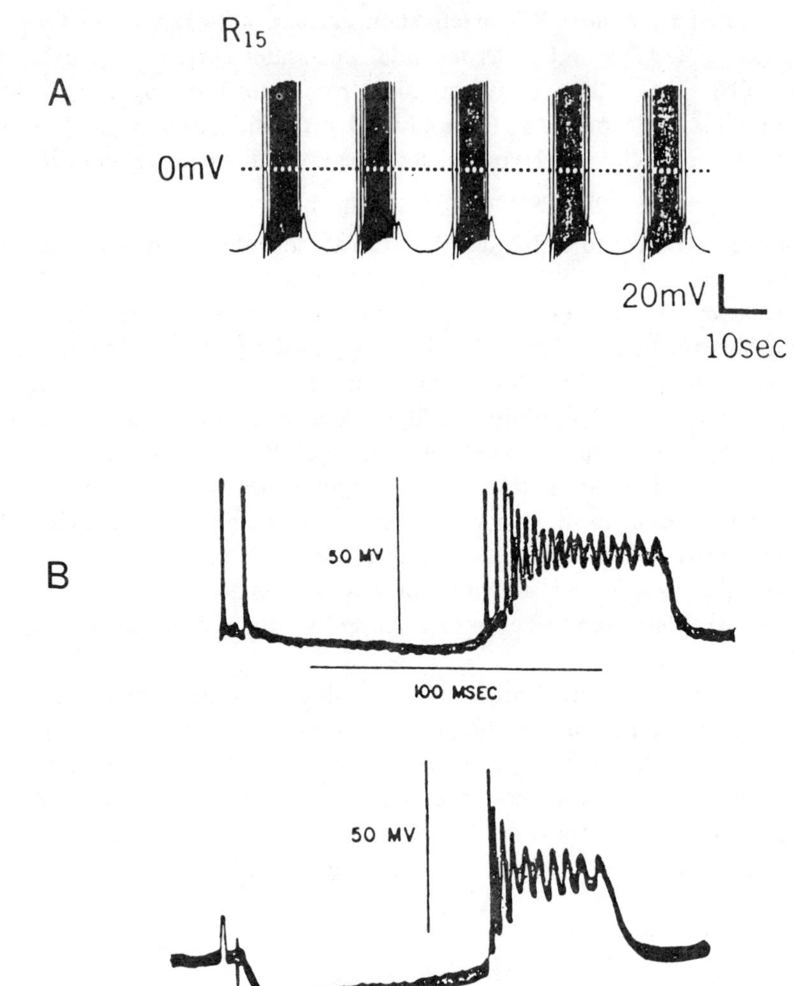

FIGURE 45. (A) Parabolic bursts in *Aplysia* abdominal ganglion (Roberge et al. [**1978**, p. 392]). (B) Paroxysmal bursts in the cat hippocampus (Kandel and Spencer [**1961**, p. 245]).

return to rest. Neither of these burst types has occurred in the analysis of the basic HH model (4.2). However, modifications of the model provide intuitive explanations, as seen below.

One striking feature of the parabolic burst is that it at first looks just like an HH burst. It is thus plausible that the usual sodium and potassium currents induce the same basic wave form as before, and spikes approach an Ω-periodic solution. Then, another process, acting on the time scale of the burst (sec) rather than on the scale of the single spike (msec) gradually shifts the cell out of the burst range and into the range where no bursts occur. The slow shift could be caused, for example, by K^+ accumulation, which would change V_K. It could also be caused by a slow K^+ inactivation current. An example in Carpenter [**1979**] takes $\bar{g}_K(np)^4(V - V_K)$ as the total potassium current, where $\partial p/\partial t = \gamma_p(V)(p_\infty(V) - p)$, with $p_\infty(V)$ decreasing from 1 to 0 and $\gamma_p \ll \gamma_n, \gamma_h \ll \gamma_m$. Thus, $p_\infty(V)$ is analogous to $h_\infty(V)$ and $n_\infty(V)$ is analogous to $m_\infty(V)$. Another model (Faber and Klee [**1972**]) postulates an extra additive potassium current, I_A

$$\text{ionic current} = I_{Na} + I_K + I_L + I_A.$$

It is an open problem to distinguish between these and various other models of parabolic bursts.

No versions of the present analysis have yet shown the existence of paroxysmal bursts solutions. This has led to the conjecture that paroxysmal bursts are never solutions of the basic or modified HH model. How, then, do they arise? Recordings in Chalazonitis [**1978**] show a paroxysmal burst in the soma (Figure 1) occurring simultaneously with a burst of spikes in the axon of the same cell. This observation leads to an amplification of the conjecture: paroxysmal bursts occur in the soma (or other region with synaptic inputs), where antidromic spikes feed back to the dendrites which in turn depolarize the cell and raise the baseline of activity, forming the plateau. In particular, the conjecture postulates the presence of a current source not included in the HH model.

Shifts from one rhythmic firing mode to another are often seen when external factors, such as temperature or drug concentrations, change membrane parameters. In Figure 41, for example, regular periodic spikes shift to spike doublets, or 2-bursts, in the presence of cocaine. Note that, in that example, the average spiking frequency is identical in (A) and (C), so any change in output is elicited by the change in pattern. Other examples are given in Figures 46 and 47.

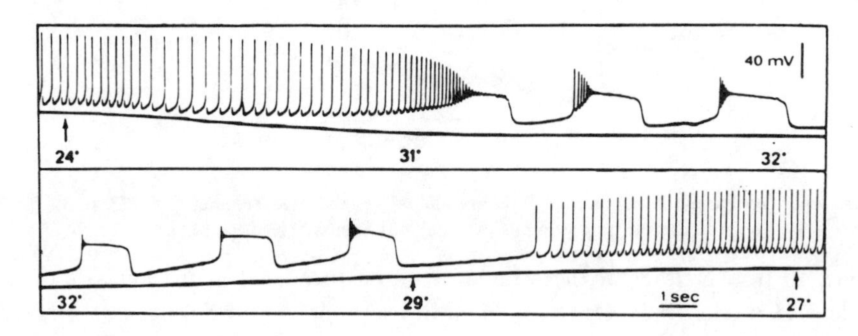

FIGURE 46. Regular periodic spikes are converted to paroxysmal bursts when the cell's temperature is increased (*Aplysia*) (Chalazonitis [**1978**, p. 121]).

The goal of Carpenter [**1980**] is to examine which parametric shifts induce which changes in signal pattern. This is accomplished by a computer program that traces the steps of the analysis outlined in §5.

The assumptions of Hypothesis 1 are natural and generally valid over a wide variety of preparations. If, in addition, some special hypotheses are satisfied by a

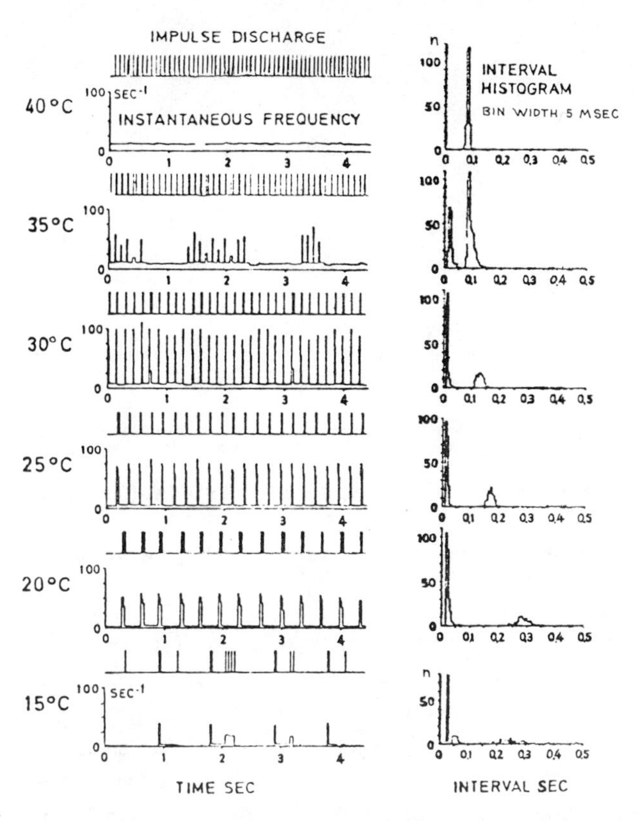

FIGURE 47. Regular periodic spikes are converted to periodic bursts when temperature of the cat lingual nerve is decreased (Bade, Braun and Hensel [**1979**, p. 2]).

generalized HH model (4.2), then types of signal patterns not already described may be possible, even for the basic *m-n-h* model. For example, a flow on Π_0 may be constructed (Carpenter [**1979**]) such that for *each* sequence of positive integers $N_1, N_2, N_3, \ldots$ there is a solution of (4.2) with N_1 spikes in the first burst, N_2 spikes in the second burst, and so on. Moreover, these solutions are arranged in lexicographic order, parametrized by wave speed. That is, if two solutions with burst sequences $N_1, N_2, N_3, \ldots, N_I, \ldots$ and $M_1, M_2, M_3, \ldots, M_I, \ldots$ have speeds θ_N and θ_M, respectively; if the two sequences are identical up to the Ith term; and if $N_I < M_I$; then $\theta_N < \theta_M$. Thus the single constant θ parametrizes an infinite-dimensional signal code. Moreover, the code is stable since two sequences agree over a large initial segment if and only if their two wave speeds are close. Other special examples are constructed in the same article.

In the opposite direction, a simplifying assumption

$$n + h \equiv \text{constant}$$

is at the heart of the FitzHugh-Nagumo equations (FitzHugh [**1961**] and Nagumo, Arimoto and Yoshizawa [**1962**]). In the present analysis, $n + h \equiv$ constant implies, in particular, that in Π_0 all solutions approach $\langle n_0, h_0 \rangle$ along

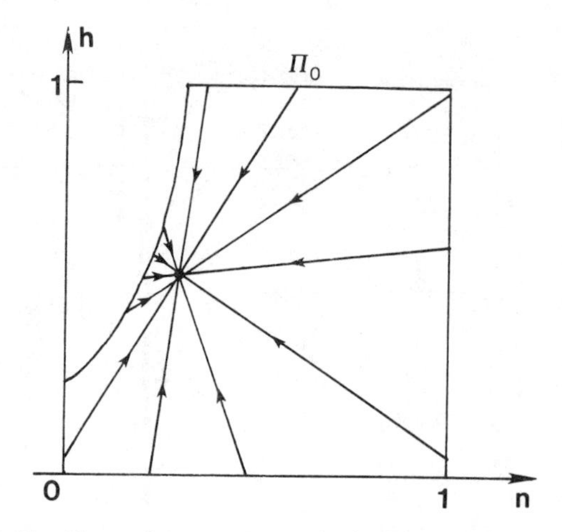

FIGURE 48. A flow on Π_0 with $n + h \equiv$ constant, as in the FitzHugh-Nagumo equations.

straight lines (Figure 48). Also, UP(θ_0) is the straight line whose equation is

$$n + h \equiv n_0 + h_0.$$

Thus, no solution in Π_0 starting at DOWN(θ_0) ever crosses UP(θ_0). There exist, therefore, Ω-periodic and single pulse solutions, but no burst or finite wave train solutions for small ε and δ. Recent work by Evans, Fenichel, and Feroe **[1980]** implies that if ε is sufficiently *large*, so that two of the eigenvalues at the rest point become complex conjugate, and if certain other hypotheses are satisfied, then there exist finite wave train solutions of the FitzHugh-Nagumo equations. Computations by Feroe **[1980]** for a piecewise-linear model and by Hastings **[1980]** for a model with cubic G-function indicate that the necessary hypotheses may be satisfied in certain parameter ranges. These Evans-Fenichel-Feroe wave trains have qualitative properties different from those discussed previously: the spikes are widely- and nearly evenly-spaced, with each spike an approximate copy of a single spike. They resemble a finite set of low-frequency Ω-periodic spikes.

REFERENCES

1. W. J. Adelman and R. FitzHugh, *Solutions of the Hodgkin-Huxley equations modified for potassium accumulation in a periaxonal space*, Fed. Proc. **34** (1975), 1322–1329.

2. T. E. Anderson and L. T. Rutledge, *Inhibition in penicillin-induced epileptic foci*, Electroenceph. Clin. Neurophysiol. **46** (1979), 498–509.

3. J. Atkinson and A. Ward, *Intracellular studies of cortical neurons in chronic epileptogenic foci in the monkey*, Experimental Neurol. **10** (1964), 285–295.

4. H. Bade, H. A. Braun and H. Hensel, *Parameters of the static burst discharge of lingual cold receptors in the cat*, Pflügers Arch. **382** (1979), 1–5.

5. J. L. Barker and T. G. Smith, Jr., *Electrophysiological studies of molluscan neurons generating bursting pacemaker potential activity*, Abnormal Neuronal Discharges, Raven Press, New York, 1978, pp. 359–387.

6. J. Bell and L. P. Cook, *On the solutions of a nerve conduction equation*, SIAM J. Appl. Math. **35** (1978), 678–688.

7. ______, *A model of the nerve action potential*, Math. Biosci. **46** (1979), 11–36.

8. P. R. Benjamin, *Endogenous and synaptic factors affecting the bursting of double spiking molluscan neurosecretory neurons (Yellow Cells of Lymnae Stagnalis)*, Abnormal Neuronal Discharges, Raven Press, New York, 1978, pp. 205–216.

9. M. Boisson and N. Chalazonitis, eds., *Abnormal neuronal discharges*, Raven Press, New York, 1978.

10. G. A. Carpenter, *Traveling wave solutions of nerve impulse equations*, Ph. D. thesis, Univ. of Wisconsin, 1974.

11. ______, *Nerve impulse equations*, Structural Stability, the Theory of Catastrophes, and Applications in the Sciences (P. Hilton, ed.), Lecture Notes in Math., vol. 525, Springer-Verlag, Berlin and New York, 1976, pp. 58–76.

12. ______, *A geometric approach to singular perturbation problems with applications to nerve impulse equations*, J. Differential Equations **23** (1977a), 335–367.

13. ______, *Periodic solutions of nerve impulse equations*, J. Math. Anal. Appl. **58** (1977b), 152–173.

14. ______, *Bursting phenomena in excitable membranes*, SIAM J. Appl. Math. **36** (1979), 334–372.

15. ______, *Parametric studies of nerve impulse models* (in preparation).

16. G. A. Carpenter and V. Knapp, *Analysis of the mammalian ventricular action potential*, J. Math. Biol. **6** (1978), 305–316.

17. R. Casten, H. Cohen and P. Lagerstrom, *Perturbation analysis of an approximation to Hodgkin-Huxley theory*, Quart. Appl. Math. **32** (1975), 365–402.

18. N. Chalazonitis, *Some intrinsic and synaptic properties of abnormal oscillators*, Abnormal Neuronal Discharges, Raven Press, New York, 1978, pp. 115–132.

19. C. Conley, *On traveling wave solutions of non-linear diffusion equation*, MRC Tech. Summary Report #1492, 1975.

20. J. A. Connor and C. F. Stevens, *Prediction of repetitive firing behaviour from voltage clamp data on an isolated neurone soma*, J. Physiol. **213** (1971), 31–53.

21. J. W. Evans, *Nerve axon equations*. I, II, III, IV, Indiana Univ. Math. J. **21** (1972), 877–885; **22** (1972), 75–90, 577–593; **24** (1975), 1169–1190.

22. J. W. Evans, N. Fenichel and J. Feroe, *Double impulse solutions in nerve axon equations*, manuscript, 1980.

23. J. W. Evans and J. Feroe, *Local stability of the nerve impulse*, Math. Biosci. **37** (1978), 23–50.

24. D. Faber and M. Klee, *Membrane characteristics of bursting pacemaker neurons in Aplysia*, Nature, New Biol. **240** (1972), 29–31.

24A. J. A. Feroe, *Existence and stability of multiple impulse solutions of a nerve equation*, manuscript, 1980.

25. R. FitzHugh, *Impulses and physiological states in theoretical models of nerve membrane*, Biophys. J. **1** (1961), 445–466.

26. J. I. Gillespie and H. Meves, *The time course of sodium inactivation in squid giant axons*, J. Physiol. **299** (1980), 289–307.

27. L. Goldman and C. L. Schauf, *Inactivation of the sodium current in Myxicola giant axons*, J. Gen. Physiol. **59** (1972), 659–675.

28. ______, *Quantitative description of sodium and potassium currents and computed action potentials in Myxicola giant axons*, J. Gen. Physiol. **61** (1973), 361–384.

29. R. M. Gulrajani and F. A. Roberge, *Possible mechanisms underlying bursting pacemaker discharges in invertebrate neurons*, Fed. Proc. **37** (1978), 2146–2152.

30. S. P. Hastings, *The existence of periodic solutions to Nagumo's equation*, Quart. J. Math. (Oxford) **25** (1974), 369–378.

31. ______, *On travelling wave solutions of the Hodgkin-Huxley equations*, Arch. Rational Mech. Anal. **60** (1976a), 229–257.

31A. ______, *On the existence of homoclinic and periodic orbits for the FitzHugh-Nagumo equations*, Quart. J. Math. (Oxford) **27** (1976b), 123–134.

31B. ______, *Single and multiple pulse waves for the FitzHugh-Nagumo equations*, manuscript, 1980.

32. A. L. Hodgkin and A. F. Huxley, *A quantitative description of membrane current and its application to conduction and excitation in nerve*, J. Physiol. **117** (1952), 500–544.

33. G. Hoyle, ed., *Identified neurons and behavior of arthropods*, Plenum Press, New York, 1977.

34. R. C. Hoyt, *Sodium inactivation in nerve fibers*, Biophys. J. **8** (1968), 1074–1097.

35. R. C. Hoyt and W. J. Adelman, Jr., *Sodium inactivation: experimental test of two models*, Biophys. J. **10** (1970), 610–617.

36. J. P. Hunter, P. A. McNaughton and D. Noble, *Analytical models of propagation in excitable cells*, Prog. Biophys. Molec. Biol. **30** (1975), 99–144.

37. E. Jakobsson, *The physical interpretation of mathematical models for sodium permeability changes in excitable membranes*, Biophys. J. **13** (1973), 1200–1211.

38. ______, *A fully coupled transient excited state model for the sodium channel*, I. *Conductance in the voltage clamped case*, J. Math. Biol. **5** (1978a), 121–142; II. *Implications for action potential generation, threshold, repetitive firing, and accommodation* **6** (1978b), 235–248.

39. E. Jakobsson and C. Scudiero, *A transient excited state model for sodium permeability changes in excitable membranes*, Biophys. J. **15** (1975), 577–590.

40. E. Kandel and W. Spencer, *Electrophysiology of hippocampal neurons* II. *After-potentials and repetitive firing*, J. Neurophysiol. **24** (1961), 243–259.

41. E. Lábos and E. Láng, *On the behavior of snail* (*Helix pomatia*) *neurons in the presence of cocaine*, Abnormal Neuronal Discharges, Raven Press, New York, 1978, pp. 177–188.

42. H. P. McKean, Jr., *Nagumo's equation*, Adv. in Math. **4** (1970), 209–223.

43. P. J. Mill, *Ventilation motor mechanisms in the dragonfly and other insects*, Identified Neurons and Behavior of Arthropods, Plenum Press, New York, 1977, pp. 187–208.

44. L. E. Moore and E. Jakobsson, *Interpretation of the sodium permeability changes of myelinated nerve in terms of linear relaxation theory*, J. Theoret. Biol. **33** (1971), 77–89.

45. J. Nagumo, S. Arimoto and S. Yoshizawa, *An active pulse transmission line simulating nerve axon*, Proc. IRE **50** (1962), 2061–2070.

46. D. Noble, *Applications of Hodgkin-Huxley equations to excitable tissue*, Physiol. Rev. **46** (1966), 1–50.

47. R. E. Plant, *The effects of* Ca^{++} *on bursting neurons: a modelling study*, Biophys. J. **16** (1976), 227–244.

48. ______, *Bifurcation and resonance in a model for bursting nerve cells*, manuscript, 1979.

49. R. E. Plant and M. Kim, *On the mechanism underlying bursting in the Aplysia abdominal ganglion* R15 *cell*, Math. Biosci. **26** (1975), 357–375.

50. ______, *Mathematical description of a bursting pacemaker neuron by a modification of the Hodgkin-Huxley equations*, Biophys. J. **16** (1976), 227–244.

51. F. Ramón, R. W. Joyner and J. W. Moore, *Propagation of action potentials in inhomogeneous axon regions*, Fed. Proc. **34** (1975), 1357–1363.

52. J. Rauch and J. Smoller, *Qualitative theory of the FitzHugh-Nagumo equations*, Adv. in Math. **27** (1978), 12–44.

53. J. Rinzel, *Spatial stability of traveling wave solutions of a nerve conduction equation*, Biophys. J. **15** (1975), 975–988.

54. J. Rinzel and J. B. Keller, *Traveling wave solutions of a nerve conduction equation*, Biophys. J. **13** (1973), 1313–1337.

55. J. Rinzel and R. N. Miller, *Numerical calculation of stable and unstable periodic solutions to the Hodgkin-Huxley equations*, Math. Biosci. **49** (1980), 27–59.

56. F. A. Roberge, R. M. Gulrajani, H. H. Jasper and P. A. Mathieu, *Ionic mechanisms for rhythmic activity and bursting in nerve cells*, Abnormal Neuronal Discharges, Raven Press, New York, 1978, pp. 389–405.

57. D. F. Russell and D. K. Hartline, *Bursting neural networks: a reexamination*, Science **200** (1978), 453–456.

58. M. E. Schonbek, *Boundary value problems for the FitzHugh-Nagumo equations*, J. Differential Equations **30** (1978), 119–147.

59. P. A. Schwartzkroin and D. A. Prince, *Changes in excitatory and inhibitory synaptic potentials leading to epileptogenic activity*, Brain Res. **183** (1980), 61–76.

60. C. Stevens, *Neurophysiology: a primer*, Wiley, New York, 1966.

61. W. C. Troy, *Bifurcation of periodic solutions in the Hodgkin-Huxley equations*, Quart. Appl. Math. **36** (1978), 73–83.

Department of Mathematics, Northeastern University, Boston, Massachusetts 02115

SIAM-AMS Proceedings
Volume **13**
1981

The Law of Large Numbers in Neural Modelling

STUART GEMAN[1]

Put loosely, the law of large numbers (LLN) says that the average of a large number of independent, or nearly independent, random variables is usually close to its mean. For some of the mathematics that typically arise in neural modelling, this simple principle has a natural and rewarding application. In one version of this application, equations for the development of long term memory traces (usually modelled as changes in "synaptic efficacies") are well approximated by more elementary equations, and from these the performance of the model can be more easily anticipated. In a second version, a large system of equations modelling the individual activities of interconnected homogeneous populations of neurons is replaced by a small number of prototype equations which accurately describe the macroscopic dynamics of the network. Models of this latter type might be relevant, for example, to the generation of phrenic nerve activity by the brainstem respiratory centers.

What I mean to present is more a point of view than a strict mathematical technique. It is another, more simple, way of looking at models which may be very complex, or even intractable, in their first formulation. For this purpose, I feel that a presentation completely by example will be most effective. A reader interested in a more formal and rigorous development, and a more general context, is referred to **[9]**, **[10]**, **[12]**, and **[14]**, and the references therein to other authors.

Time averaging: the behavior of models for the development of long term memory. In the three examples of this section, the LLN takes the form of a stochastic "method of averaging" for differential equations, through which method the behavior of a complex neural or cognitive model can often be anticipated with surprising ease. The method applies to differential equations in which the solution is slowly varying relative to the other time dependent terms

1980 *Mathematics Subject Classification.* Primary 60F99.
[1]Supported by the National Science Foundation under grant MCS76-80762.

of the equation. Equations modelling the development of a long term memory are typically of this form: the dependent variable represents some component of the long term trace, and it is slowly varying relative to the stimuli which effect changes in that trace. It may be that a particular model has been formalized using a system of integral or difference equations, but such models typically have a natural reformalization using differential equations. And, conversely, a method of averaging can be formulated for these other settings as well.

As both an introduction and a good demonstration of the method's utility, I will first apply it to a neural network memory model proposed by Uttley (in **[22]**–**[24]**). (My discussion of averaging in Uttley's model is, by and large, a repeat of what was said in **[8]**.) The model consists of a network of units called "informons". These are adaptive neuron-like elements which can learn to signal whether or not a vector input belongs to a particular classification. The dynamics of the informon and the rule by which it self-organizes are defined by the following equations

$$F(Y) = \sum_{i=1}^{n} F(X_i)\gamma_i + F(Z)\gamma_z, \qquad \Delta\gamma_i = -bF(X_i)F(Y) \tag{i}$$

where

$F(Y)$ is the output (firing rate) of the informon (neuron) labelled "Y";

$F(X_i)$, $i = 1, 2, \ldots, n$, are outputs from other units in the network, and comprise the input to the unit Y;

$F(Z)$ is a binary classifying signal which indicates, during training, which inputs $(F(X_1), \ldots, F(X_n))$ belong to a particular category;

γ_i, $i = 1, \ldots, n$, are modifiable conductivities, which determine the extent to which the signals $F(X_i)$, $i = 1, \ldots, n$, contribute to the output at Y;

γ_z is a fixed and negative conductivity transmitting the classifying signal to Y;

$\Delta\gamma_i$ is the change in the conductance γ_i due to a simultaneous appearance of an input $F(X_i)$ with an output $F(Y)$;

b is a positive constant determining the rate at which this latter modification proceeds.

The connection between this example and those which follow will be more transparent if I use, as nearly as is possible, a unified notation. For this purpose, let $y(t)$ be the output at time t of the unit designated "Y" in the Uttley model (replacing "$F(Y)$"). $x_1(t), \ldots, x_n(t)$ will replace $F(X_1), \ldots, F(X_n)$ as input signals to Y, and $\gamma_1(t), \ldots, \gamma_n(t)$ will, again, denote the corresponding conductivities. Since γ_z is fixed and negative, it will be convenient to denote the net classifying signal, $\gamma_z F(Z)$, simply by $-z(t)$. Finally, writing ε in place of b, a continuous time formulation for (i) is

$$y(t) = \sum_{i=1}^{n} \gamma_i(t)x_i(t) - z(t),$$

$$\frac{d}{d\tau}\gamma_i(t) = -\varepsilon x_i(t)y(t) = \varepsilon x_i(t)\left\{ z(t) - \sum_{j=1}^{n} \gamma_j(t)x_j(t) \right\}. \tag{ii}$$

Uttley does not analyze this system directly, but instead replaces it by a new system which is intended to be a more tractable approximation. The analysis of this approximating system, together with simulation results, indicates that the informon and networks of interconnected informons have properties suggestive of classical and operant conditioning as well as a capability for pattern classification. We will see that the method of averaging yields some additional insights, and a more direct and precise analysis of the behavior of (ii).

This is a long term memory model, and Uttley assumes that $\gamma_i(t)$ reflects only the long term behavior of $x_i(t)$ and $y(t)$, not their most immediate fluctuations. The translation, for (ii), is the assumption that ε is small; changes in $\gamma_i(t)$ are slow relative to those in $x_i(t)$ and $y(t)$. This assumption is an important one, because it means that, essentially, $d\gamma_i(t)/dt$ "sees only the average" of the right-hand side of the differential equation in (ii). To make this a little more precise and to see why it should be true, consider the change in $\gamma_i(t)$ over a period of time $\Delta t = \delta/\varepsilon$, where δ is small, but not nearly as small as ε. In this period of time, $\gamma_i(t)$ will not appreciably change (its derivative being of order ε), but this is a considerable interval relative to the time course of $x_i(t)$ and of $y(t)$. From (ii)

$$\gamma_i\left(t + \frac{\delta}{\varepsilon}\right) - \gamma_i(t) = -\varepsilon \int_t^{t+\delta/\varepsilon} x_i(s)y(s)\, ds$$

i.e.

$$\gamma_i(t + \Delta t) - \gamma_i(t) = \frac{\delta}{\Delta t}\int_t^{t+\Delta t} x_i(s)\left\{ z(s) - \sum_{j=1}^{n} \gamma_j(s)x_j(s)\right\} ds$$

$$\approx \frac{\delta}{\Delta t}\int_t^{t+\Delta t} x_i(s)z(s)\, ds - \sum_{j=1}^{n} \gamma_j(t)\frac{\delta}{\Delta t}\int_t^{t+\Delta t} x_i(s)x_j(s)\, ds. \qquad \text{(iii)}$$

The latter (very rough) approximation is because $\gamma_j(s)$ is nearly constant over the interval $[t, t + \Delta t]$.

Now, let us take the point of view (deferring discussion on this) that $(x_1(t), \ldots, x_n(t), z(t))$ is a random process, and that over large periods of time it is essentially "independent of itself" (current observations of the process tell us very little about its distant future). Then, $(\Delta t)^{-1}\int_t^{t+\Delta t} x_i(s)z(s)\, ds$ and $(\Delta t)^{-1}\int_t^{t+\Delta t} x_i(s)x_j(s)\, ds$ are long run averages (since $\Delta t = \delta/\varepsilon$ is large) of nearly independent random variables and should be well approximated by means (i.e. expected values)

$$\frac{1}{\Delta t}\int_t^{t+\Delta t} x_i(s)z(s)\, ds \approx \frac{1}{\Delta t}\int_t^{t+\Delta t} E\left[x_i(s)z(s)\right] ds,$$

$$\frac{1}{\Delta t}\int_t^{t+\Delta t} x_i(s)x_j(s)\, ds \approx \frac{1}{\Delta t}\int_t^{t+\Delta t} E\left[x_i(s)x_j(s)\right] ds.$$

Finally, put this back in (iii) and take derivatives

$$\frac{d}{dt}\gamma_i(t) \approx \varepsilon E\left[x_i(t)z(t)\right] - \varepsilon \sum_{j=1}^{n} \gamma_j(t)E\left[x_i(t)x_j(t)\right], \qquad \text{(iv)}$$

and this is what I meant when I said that $d\gamma_i(t)/dt$ sees only the average of the right-hand side in (ii). The point is this: (iv) relates the "memory trace", $\gamma_i(t)$, to the statistical structure of the environment (as revealed by the operator, E). When the solution to (iv) (with $\approx$ replaced by $=$) is close to that in (ii), we will be able to infer from (iv) the most important features of the model's behavior.

In fact, under very general conditions the solution to (ii) is well approximated by the solution to (iv). The smaller ε, the better the approximation, and, in particular, the error goes to zero with ε. Details about the conditions, as well as a precise statement of the sense in which the approximation holds, can be found in **[9]** or **[10]**.

The next step, then, is to develop the consequences of (iv). But before this, we should briefly examine in a nontechnical manner the two most important assumptions implicit in this approximation procedure. Above all, the reader may question the use of a random process model for $x_i(t)$, $i = 1, 2, \ldots, n$, and $z(t)$, the "environment" of the conductivity, $\gamma_i(t)$. In fact, we did not have to take this approach at all, since the "method of averaging" is, originally, a technique for approximating *deterministic* equations (see, for example, Mitropolsky **[21]**). Thus, a deterministic model would lead us to a version of (iv) in which a certain time average plays the role of expectation, E, and we could then proceed to analyze instead this analogue to (iv). But I believe that the probabilistic point of view has something special to offer, and the further discussion of this example together with the examples below should convincingly support this position. Whether the environment is in some sense truly random is of no importance; the probability model offers a convenient framework in which to describe characteristics of that environment. It does not in any sense suggest that the environment is *unstructured*. Indeed, a determinstic model is merely a special case.

There has also been made a "mixing" assumption: that the past and future are, asymptotically, independent. Mixing is an ergodic-like property that, practically speaking, puts very little constraint on potential models of the environment. (Any deterministic model, for example, is mixing. But then (iv) is (ii), and the method offers no simplification.) For example, a wide variety of Markov processes, and in particular those which would be most appropriate in representing a model's environment (i.e. bounded and obeying some mild regularity conditions), are mixing in a way suitable for application of the method of averaging. In the pattern recognition literature, a much stronger assumption is typical: successive scenes or patterns are statistically independent.

In short, (iv) will approximate (ii) under assumptions which are natural for the system being modelled. What, then, can (iv) tell us about the behavior of Uttley's model? A good place to start is with the asymptotic behavior: How does the model perform after a theoretically infinite period of time? If there is to be an "asymptotic behavior", then we must first assume that something like an equilibrium for $\gamma_i(t)$ exists, and this amounts to making an assumption of stationarity. Or, at the least, an assumption that the expectations appearing in

(iv) do not depend on time (but averaging is appropriate whether or not this is the case). Really, this is not much of an additional assumption, since there would be no point in a long term memory if the environment did not possess some degree or stationarity. Let us assume, then, that $E[x_i(t)z(t)] = E[x_i z]$ and $E[x_i(t)x_j(t)] = E[x_i x_j]$ do not depend on t (certainly *not*, however, that $x_i(t)$ or $z(t)$ are constant). Then, the equilibrium for (iv), and therefore the approximate equilibrium for (ii), is immediately available. Simply set the derivative in (iv) equal to 0 and solve for γ_i,

$$\sum_{j=1}^{n} E[x_i x_j]\gamma_j = E[x_i z], \qquad i = 1, 2, \ldots, n. \tag{v}$$

Define $n \times 1$ column vectors $X(t)$ and $\Gamma(t)$ by $X(t) = (x_1(t), \ldots, x_n(t))^T$ and $\Gamma(t) = (\gamma_1(t), \ldots, \gamma_n(t))^T$ (using T to denote transpose). In vector-matrix notation, (v) is $E[XX^T]\Gamma = E[Xz]$. And therefore, assuming that $E[XX^T]$ is nonsingular,[2]

$$\Gamma = E[XX^T]^{-1}E[Xz]. \tag{vi}$$

Since (ii) behaves like (iv), the conclusion is that $\Gamma(t)$ will approach and remain close to $E[XX^T]^{-1}E[Xz]$.

The reader familiar with multivariate analysis will recognize (vi) as the solution to the linear regression problem: Choose $\gamma_1, \ldots, \gamma_n$ so as to minimize the mean square error in approximating $z(t)$ by the linear combination $\sum_{j=1}^n \gamma_j x_j(t)$, i.e. minimize

$$E\left|z - \sum_{j=1}^{n} \gamma_j x_j\right|^2 \tag{vii}$$

over all possible values of $\Gamma = (\gamma_1, \ldots, \gamma_n)^T$. In words, the conductivities of the informon modify in such a way that the output of Y in *absence* of the classifying signal (i.e. $\sum_{j=1}^n \gamma_j(t)x_j(t)$) approaches the best linear predictor of $z(t)$ (the classification) given $x_1(t), \ldots, x_n(t)$. Actually, we know much more. (iv) is an autonomous system of linear differential equations, and its exact solution is well known. Then, since (ii) stays close to (iv), we have available essentially the entire time course of $\Gamma(t)$. Roughly, $\Gamma(t)$ approaches $E[XX^T]^{-1}E[Xz]$ exponentially with rate determined by the eigenvalues of the positive definite matrix $E[XX^T]$.

The method of averaging, really an application of the LLN, gives us a virtually complete description of the dynamics of the informon. It reveals details about the unit's behavior not obviously apparent in (i) and not found in the system which Uttley offers as a more tractable alternative. Thus we know that the informon is asymptotically a nearly optimal classifier–at least among linear machines. In fact, $\sum_{j=1}^n \gamma_j x_j(t)$ will predict $z(t)$, in an approximately minimum

---

[2]Equivalent is the assumption that no component of $X(t)$, say $x_i(t)$, is a *deterministic* linear combination of the remaining components $\{x_j(t)\}, j \neq i$. Any such deterministic relation would be undone by "noise" in a real system.

mean square error sense, whether or not $z(t)$ is the binary signal assumed in the model. That is, asymptotically, the solution to (iv) minimizes (vii) whatever the nature (discrete or continuous) of the "classifying signal", $z(t)$. And, if $x_1(t), \ldots, x_n(t), z(t)$ jointly form a Gaussian process, then the best linear predictor of $z(t)$ is also the best unconstrained predictor.

The analysis, then, supports (ii) as an appropriate system for learning to predict a "classification", $z(t)$, from the information contained in the channels $x_1(t), \ldots, x_n(t)$. But, by making the connection to some well-studed areas of statistics and pattern recognition, the analysis also suggests some possibly unattractive features of the model. For example, unless $(x_1(t), \ldots, x_n(t), z(t))$ is a Gaussian process, the best linear predictor of $z(t)$ may be quite inferior to the overall best predictor. Although the inevitable noise present in neural activity is probably well approximated by a Gaussian model, I would doubt that the signals themselves are anything like a Gaussian process. If these signals are not Gaussian, would the nevous system employ a suboptimal solution? Also, there is reason to question the efficiency of the modification procedure defined in (ii). It is, essentially, a stochastic approximation procedure for finding the least mean square error linear predictor of $z(t)$ given $x_1(t), \ldots, x_n(t)$ (see, for example, Duda and Hart [**6**], or Wasan [**25**]). We must, then, ask why the nervous system would utilize this particular version of stochastic approximation when there are other versions known to perform more efficiently. Again, there is raised a question of optimality. There may, of course, be good answers for these questions, and it may be that the model is entirely appropriate. But, at the least, we have established a framework in which the model can be meaningfully compared to already existing theory.

Uttley points out that since $\gamma_i(t)$ may be positive or negative, its neural realization would require both excitatory and inhibitory synapses. Amari (in [**2**]) has proposed a model quite similar in spirit to Uttley's, but one which more explicitly addresses the problem of achieving a net conductivity which may be positive or negative, out of couplings which are individually constrained to be excitatory or inhibitory. In [**2**], Amari is already aware of the method of averaging and applies it, much as we did above, to determine the equilibrium behavior of his model. I will retrace some of Amari's analysis, and interpret the conclusions with special attention to the close relationship between the Uttley and the Amari theories.

The fundamental unit in Amari's model is a neuron-like device, which I will again call "Y", receiving inputs $x_1(t), \ldots, x_n(t)$, possibly from other units or possibly from an external source. Each of these inputs $x_i(t)$ influences Y through both an excitatory and an inhibitory coupling; let us denote the strengths of these couplings by $\gamma_i^+(t)$ and $\gamma_i^-(t)$ respectively. The unit learns in the sense that these coupling strengths are modified by its experience. The net input to Y through this variable pathway is $\sum_{j=1}^n \gamma_j^+(t)x_j(t) - \sum_{j=1}^n \gamma_j^-(t)x_j(t)$. Or, in terms of the corresponding vector quantities (for notation, refer back to the discussion of Uttley's model): $\Gamma^+(t)^T X(t) - \Gamma^-(t)^T X(t)$. There is also at Y an unmodifiable channel which receives a "teacher" input, $z(t)$. Learning is by modification of the γ connectivities, as is described (in its continuous time formulation) by the following equations:

$$\frac{d}{dt}\gamma_i^+(t) = \varepsilon(\alpha_1 z(t)x_i(t) - \alpha_2\gamma_i^+(t))$$

$$\frac{d}{dt}\gamma_i^-(t) = \varepsilon\left\{\alpha_3 x_i(t)\left(\sum_{j=1}^n \gamma_j^+(t)x_j(t) - \sum_{j=1}^n \gamma_j^-(t)x_j(t)\right) - \alpha_4\gamma_i^-(t)\right\}.$$

Here again ε is a small positive constant. α_1, α_2, α_3 and α_4 are for the time being arbitrary positive constants. Writing $\Gamma(t)$ for $\Gamma^+(t) - \Gamma^-(t)$, the system is rewritten in more convenient vector-matrix notation as

$$\frac{d}{dt}\Gamma^+(t) = \varepsilon(\alpha_1 z(t)X(t) - \alpha_2\Gamma^+(t)),$$

$$\frac{d}{dt}\Gamma^-(t) = \varepsilon(\alpha_3 X(t)X(t)^T\Gamma(t) - \alpha_4\Gamma^-(t)). \tag{viii}$$

Now let us apply the LLN. When ε is small, a good approximation to (viii) (making all the necessary assumptions, as discussed in the previous example) is:

$$\frac{d}{dt}\Gamma^+(t) = \varepsilon(\alpha_1 E[zX] - \alpha_2\Gamma^+(t)),$$

$$\frac{d}{dt}\Gamma^-(t) = \varepsilon(\alpha_3 E[XX^T]\Gamma(t) - \alpha_4\Gamma^-(t)). \tag{ix}$$

This is an autonomous linear system, and we could if we wished analyze it in complete detail. But the asymptotics (equilibrium) are the most revealing:

$$\Gamma^+(t) \to (\alpha_1/\alpha_2)E[zX], \qquad \Gamma^-(t) \to (\alpha_3/\alpha_4)E[XX^T]\Gamma(t).$$

Therefore, at equilibrium,

$$\Gamma = \Gamma^+ - \Gamma^- = \frac{\alpha_1}{\alpha_2}E[zX] - \frac{\alpha_3}{\alpha_4}E[XX^T]\Gamma$$

$$\Rightarrow \Gamma = \frac{\alpha_1\alpha_4}{\alpha_2\alpha_3}\left(\frac{\alpha_4}{\alpha_3}I + E[XX^T]\right)^{-1}E[zX] \tag{x}$$

where I is the $n \times n$ identity matrix.

Let us examine the information that Y receives after learning (i.e. with Γ given by (x)) and in the absence of the teacher signal, $z(t)$. Then,

$$\Gamma^{+T}X(t) - \Gamma^{-T}X(t) = X(t)^T\Gamma$$

$$= \frac{\alpha_1\alpha_4}{\alpha_2\alpha_3}X^T(t)\left(\frac{\alpha_4}{\alpha_3}I + E[XX^T]\right)^{-1}E[zX]. \tag{xi}$$

The constant, $\alpha_1\alpha_4/\alpha_2\alpha_3$, is obviously unimportant; the interpretation of (xi) will be clearest if we choose $\alpha_2/\alpha_1 = \alpha_4/\alpha_3 = \delta$ so that

$$X(t)^T\Gamma = X(t)^T(\delta I + E[XX^T])^{-1}E[zX]. \tag{xii}$$

Notice: *If* δ were 0 (it cannot be) this would be exactly the asymptotic output of Uttley's informon in the absence of the "classifying signal", $z(t)$. Thus (xii) approximates the minimum mean square error linear predictor of the classifying signal (here called the teacher signal), $z(t)$. The term δI, which may at first appear to be a nuisance, actually represents a potentially important improvement over the unmodified "optimal" solution. In fact, (xii) is a "ridge estimator" for $z(t)$, introduced by Hoerl and Kennard [**19**], and since then analyzed in some detail (see for example [**20**]). When $E[XX^T]$ is "nearly singular" (more precisely, "ill-conditioned"), the addition of δI stabilizes the inverse in (xii), making it more accurately computable in a real system.

To appreciate the relevance of this in the present context, consider again the Uttley system (ii), but when $E[XX^T]$ is nearly singular (as would be the case, for example, if two of the channels $x_i(t)$ and $x_{i'}(t)$ were essentially redundant). For fixed ε, the approximation of (ii) by (iv) (i.e. the method of averaging) is made less accurate as (iv) is brought closer to instability–which is just what happens when $E[XX^T]$ is brought closer to singularity. Although the solution to (iv) will still asymptotically approach the desired (optimal) equilibrium, the solution to (ii) the *real system* will behave erratically, wandering far from the course predicted for it by (iv). Hence the system is, under these circumstances, unreliable. In contrast, the relative stability of (ix) (the "averaged system" for the Amari theory) is essentially unaffected by an ill-conditioned matrix $E[XX^T]$. As a consequence, the method of averaging remains in force and the desired solution, (xii), is still realized to within a good approximation.

Amari and Uttley, in the papers reviewed here, have each proposed neural-like mechanisms capable of learning pattern classifications. Thus modelled neurons in these theories learn to predict a one dimensional "classifying signal" based on the evidence available in an n dimensional pattern. There is also the problem of postulating mechanisms by which the nervous system can commit to memory patterns themselves, both motor and sensory, as it is evidently capable of this task as well. Grossberg has proposed a theory for pattern learning in which individual neuron-like units learn to reproduce an entire pattern of activity (see, for example, [**17**] and [**18**]). Excitation of one of these units elicits an activity pattern in the "postsynaptic" units, and this pattern is identical (in the sense of relative, "figure to ground" activities) to a practice pattern arriving at these postsynasptic units during learning. Grossberg's analysis is deterministic and rather sophisticated. Although I will not add to the conclusions reached by that analysis, I will show how the essential properties of the learned behavior of the system can be anticipated by an application of the method of averaging.

All units (modelled neurons) of the network belong to one or both of two subpopulations: "I" represents the collection of subscripts belonging to those units in one of these subpopulations, and "J" represents the collection of subscripts associated with the other subpopulation. The units in the subpopulation "J" receive a pattern of input from outside the immediate network. Each unit in "I" contacts all units in "J", and, under appropriate conditions, will learn to reproduce the pattern seen at "J". What makes this model particularly complex is that I and J are not assumed to be disjoint; the intersection, $I \cap J$, is arbitrary. In other words, the "receptor cells" in "J" may themselves be "sampling cells" in "I", realizing a feedback, rather than feedforward, system.

Let us represent by $b_i(t)$, $i \in I$, the output of the i element of the "I" subpopulation at time t. The activity of the j element of the "J" subpopulation, call it $x_j(t)$, is determined by the outputs of the units in "I", and an exogenous input, $c_j(t)$. Formally,

$$\frac{d}{dt} x_j(t) = -a(t)x_j(t) + \sum_{i \in I} b_i(t)\gamma_{ij}(t) + c_j(t) \qquad \text{(xiii)}$$

for each $j \in J$, where $1/a(t)$ is an "instantaneous decay time" ($a(t) \geqslant 0$ for all t), and $\gamma_{ij}(t)$ is the synaptic or coupling strength for the $i \in I$ to $j \in J$ contact. The exogenous input to "J" is a pattern in the sense that it takes the form

$$c_j(t) = \psi(t)\theta_j, \quad \text{where } \sum_{j \in J} \theta_j = 1 \qquad \text{(xiv)}$$

and $\theta_j \geqslant 0$ for all $j \in J$. $\psi(t)$ is the input intensity at time t, and may vary arbitrarily during the learning period.

As in the previous examples, learning is by modification of the "synaptic weights", $\gamma_{ij}(t)$, $i \in I, j \in J$,

$$d\gamma_{ij}(t)/dt = -d_i(t)\gamma_{ij}(t) + e_i(t)x_j(t). \qquad \text{(xv)}$$

$e_i(t)$ plays a role analogous to $b_i(t)$, representing the signal from $i \in I$ available to effect change in $\gamma_{ij}(t)$. In the absence of a correlated $i \in I$ and $j \in J$ activity, i.e. when $e_i(t)x_j(t) = 0$, $\gamma_{ij}(t)$ decays towards 0 ($d_i(t) \geqslant 0$ for all t). Observe that (xv) is physically "realizable", in the sense that modification of $\gamma_{ij}(t)$ depends only on signals locally available, i.e. it depends only on pre- and postsynaptic activities. (Actually, the theory in **[17]** is developed for a system somewhat more general than (xiii) and (xv). The slightly specialized version here will serve for better illustration. The general system can be discussed in much the same way.)

What sort of results should we be looking for? Grossberg gives conditions under which the system demonstrates, asymptotically, the following learned behavior: With or without the exogenous pattern of input (xiv), activity in the "I" subpopulation leads to a reproduction of the learned pattern at J. In particular, after learning, "I" activity will produce a relative activity at x_j equal to θ_j

$$\frac{x_j(t)}{\Sigma_{k \in J}\, x_k(t)} \to \theta_j, \qquad \text{(xvi)}$$

the relative strength of the exogenous signal at j. (See **[17]** for a precise formulation of results.) Let us see how we might anticipate this behavior by taking a probabilistic point of view and applying an LLN.

$x_j(t)$ models the activity of the $j \in J$ neuron, and should be "fast" relative to the other time dependent terms appearing in the right-hand side of (xiii). As a first approximation then, it is not unreasonable to replace $x_j(t)$ by its "instantaneous equilibrium value", determined by setting $dx_j(t)/dt = 0$ in (xiii),

$$x_j(t) = \sum_{i \in I} \frac{b_i(t)}{a(t)} \gamma_{ij}(t) + \frac{c_j(t)}{a(t)}. \tag{xvii}$$

With this substitution, plus the one in (xiv), (xv) becomes

$$\frac{d}{dt}\gamma_{ij}(t) = -d_i(t)\gamma_{ij}(t) + \sum_{k \in I} \frac{e_i(t)b_k(t)}{a(t)} \gamma_{kj}(t) + \frac{e_i(t)\psi(t)}{a(t)}\theta_j. \tag{xviii}$$

Now let us again make the assumption that $\gamma_{ij}(t)$ is slowly varying, representing a long term memory trace. If, in equation (xv), we write $\varepsilon \tilde{d}_i(t)$ for $d_i(t)$ and $\varepsilon \tilde{e}_i(t)$ for $e_i(t)$, then with this assumption, we may take ε to be small while $\tilde{d}_i(t)$ and $\tilde{e}_i(t)$ are still of order 1. (xviii) then becomes

$$\frac{d}{dt}\gamma_{ij}(t) = \varepsilon\left(-\tilde{d}_i(t)\gamma_{ij}(t) + \sum_{k \in I} \frac{\tilde{e}_i(t)b_k(t)}{a(t)} \gamma_{kj}(t) + \frac{\tilde{e}_i(t)\psi(t)}{a(t)}\theta_j\right),$$

which should be well approximated by the "averaged equation"

$$\frac{d}{dt}\gamma_{ij}(t) = \varepsilon\Bigg(-E\left[\tilde{d}_i(t)\right]\gamma_{ij}(t) + \sum_{k \in I} E\left[\frac{\tilde{e}_i(t)b_k(t)}{a(t)}\right]\gamma_{kj}(t) + E\left[\frac{\tilde{e}_i(t)\psi(t)}{a(t)}\right]\theta_j\Bigg). \tag{xix}$$

Although (xix) is deterministic, it is not at all simple. In Grossberg's theory, $a(t)$, $b_i(t)$, $d_i(t)$, $e_i(t)$, and $\psi(t)$ may themselves depend on $\{x_i(t)\}$, $i \in I \cup J$, and $\{\gamma_{ij}(t)\}$, $i \in I$, $j \in J$, as long as the subscript conditions (e.g. $a(t)$ does not depend on j) are not violated. The possible dependence on the $\gamma_{ij}(t)$'s in particular prevents us from assuming that the expectations in (xix) are constant or that (xix) is linear. However, at any equilibrium point for (xix) the $\gamma_{ij}(t)$'s entering into these expectations are constant (although unknown), and in this case these expectations themselves may be assumed constant (just as in the previous two examples). At equilibrium, then, we may write

$$\gamma_{ij} = \sum_{k \in I} \frac{E[\tilde{e}_i b_k/a]}{E[\tilde{d}_i]} \gamma_{kj} + \frac{E[\tilde{e}_i \psi/a]}{E[\tilde{d}_i]}\theta_j, \tag{xx}$$

where the expectations, which may depend on $\{\gamma_{ij}\}$, $i \in I, j \in J$, do not depend on time.

Fix j. Then (xx) is a linear system of equations for γ_{kj}, $k \in I$, in which the only dependence on j appears in the inhomogeneous terms (due to θ_j). Hence, its solution (which I will assume exists) has the form

$$\gamma_{ij} = \sum_{k \in I} m_{ik} \frac{E[\tilde{e}_k \psi/a]}{E[\tilde{d}_k]}\theta_j,$$

where $M = \{m_{ik}\}$, $i, k \in I$, is a square matrix determined by the coefficients of γ_{kj} in (xx) and does not depend on j. The point is that, at an equilibrium, γ_{ij} (of the averaged equation) must have the form $\gamma_{ij} = \alpha_i\theta_j$, in which case (from (xvii))

$$x_j(t) = \left(\sum_{i \in I} \frac{b_i(t)}{a(t)} \alpha_i + \frac{\psi(t)}{a(t)} \right) \theta_j,$$

and this obviously implies that (xvi) holds. In other words, activity in "I" reproduces the practiced pattern at "J", even when that pattern is no longer present as an exogenous input (i.e. even when $\psi(t) = 0$). In fact, a direct (but much more involved) analysis of the *unapproximated* system, (xiii) and (xv), shows that under suitable conditions (xvi) holds there as well, and without a slowly varying assumption for $\gamma_{ij}(t)$ (see **[17]**).

Certain generalizations are immediately available, "free of charge". Since (xv) is well approximated by the averaged equation (xix), any substitution for the system (xiii) and (xv) that leaves (xix) unmodified will demonstrate essentially the same learned behavior. This includes, for example, allowing $a(t)$, $b_i(t)$, $d_i(t)$, and $e_i(t)$ to depend on j, provided that $E[d_{ij}]$, $E[e_{ij}b_{kj}/a_j]$, and $E[e_{ij}\psi/a_j]$ are still independent of j. Of course, this is as true in the previous two examples: the average equation represents a class of systems that must all exhibit approximately the same asymptotic behavior.

There are in the literature many other examples which can be, or already have been, treated in very much the same way. Although the three which I have discussed above should serve as a good introduction to this application of the method of averaging, I would also recommend (3), (4), (5), and (11), each of which contains an example of the explicit use of this technique in problems of neural or cognitive modelling.

Population averaging: stable oscillations in a large system of modelled neurons. In mathematical models of neural network activity, considerable use has been made of equations for the average activity of homogeneous collections of modelled neurons (some examples are in **[7]**, **[13]**, **[15]**, **[16]**, and **[26]**). The implicit assumption is that "macroscopic" (average) activity has a description which does not involve "microscopic" (individual neuronal) activities, in close analogy to the situation in statistical mechanics. Simulation experiments (see especially **[1]**) indicate that such a description is in fact broadly available, but there have been very few rigorous analytic results. This "population averaging" can again be viewed as an application of the LLN; here I will quote some analytic results (from **[12]**) which, in some instances, rigorously justify this application.

The discussion will be through a specific example. For this purpose, I will use essentially the model proposed by Miller and me (in **[13]**) for the generation of periodic phrenic nerve activity by the brainstem respiratory centers. Our analysis was based on the hypothesis that the LLN was operating in the proposed system. As I will indicate, for the equations discussed here the hypothesis is indeed correct, meaning that the behavior of the entire (very large) system can be accurately described by a small number of prototype (averaged) equations.

In [13], Miller and I argue for a respiratory model based on reciprocating activities of negatively coupled inspiratory and expiratory populations of neurons, each of which is capable of independent stable oscillation if (theoretically) isolated from the other. The model postulates that these populations are further divided into excitatory and inhibitory subpopulations, and that the interaction between these subpopulations is responsible for the inspiratory and expiratory oscillations. For the discussion here, it will be enough to examine just one population, let us say the inspiratory population of neurons. Suppose that there are n excitatory inspiratory neurons and m inhibitory inspiratory neurons. $x_i(t)$ will denote the cell body membrane potential of the ith excitatory neuron at time t, and $y_i(t)$ will denote this potential for the ith inhibitory neuron. The dynamics of the modelled inspiratory population are described by the following systems of equations

$$\frac{d}{dt}x_i(t) = -\alpha x_i(t) + \frac{1}{n}\sum_{j=1}^{n} \gamma_{ji}^{++} f(x_j(t)) - \frac{1}{m}\sum_{j=1}^{m} \gamma_{ji}^{-+} g(y_j(t)), \qquad 1 \leqslant i \leqslant n,$$

$$\frac{d}{dt}y_i(t) = -\beta y_i(t) + \frac{1}{n}\sum_{j=1}^{n} \gamma_{ji}^{+-} f(x_j(t)) - \frac{1}{m}\sum_{j=1}^{m} \gamma_{ji}^{--} g(y_j(t)), \qquad 1 \leqslant i \leqslant m. \tag{xxi}$$

Here,

α (β) is the inverse of the membrane decay time of excitatory (inhibitory) neurons;

$f(x_i(t))$ ($g(y_i(t))$) is the frequency of action potentials generated in the axon of the ith excitatory (inhibitory) neuron by a cell body membrane potential of $x_i(t)$ ($y_i(t)$). $f(x)$ and $g(x)$ are assumed to be bounded and increasing functions.

All γ_{ji}'s are nonnegative. $n^{-1}\gamma_{ji}^{++}$ is the coupling strength ("synaptic weight") from the jth excitatory to the ith excitatory neuron, $m^{-1}\gamma_{ji}^{-+}$ is the coupling strength from the jth inhibitory to the ith excitatory neuron, etc. When there is no synaptic connection between two neurons, the corresponding γ_{ji} is zero. "$1/n$" and "$1/m$" embody the assumption that the total synaptic contribution to a neuron's input is "order 1", regardless of the number of these synapses.

Let us suppose that the system (xxi) represents a reasonable first approximation of the dynamics of the inspiratory population of brainstem neurons. With n and m of order 10^5 (conservatively), there is the ontological problem of choosing 10^{10} or more parameters in (xxi) so as to achieve a stable oscillation with a specified period and wave form. If it were possible to specify that all synaptic weights of a given type of synapse (such as excitatory to excitatory) were identical, then (xxi) would be perfectly described by just two prototype equations, one for the excitatory and one for the inhibitory subpopulations. Then, the biological problem would be entirely manageable, requiring the appropriate specification of only a very small number of parameters. But it is not a tenable proposition that this level of precision is achieved in a developing nervous system.

Miller and I (in **[13]**) have argued that the ontological problem is limited to the specification of target values for each of the four types of connections in (xxi), and that random fluctuations about these target values (means) will not influence the dynamics of the network as a whole. We reasoned heuristically, as follows: For each type of connection, take the case of excitatory to excitatory, let us model the synaptic weights, γ_{ji}^{++}, $1 \leqslant j, i \leqslant n$, as independent and identically distributed random variables, chosen from a distribution in which the mean only is genetically specified. Let γ^{++} be the mean strength of an excitatory to excitatory synapse. When n is very large, the dependence between any two excitatory neurons, or between any excitatory neuron and any synaptic weight, should be very small. Then, the excitatory input to the ith excitatory cell,

$$\frac{1}{n}\sum_{j=1}^{n} \gamma_{ji}^{++} f(x_j(t)), \qquad \text{(xxii)}$$

should "look like" an average of independent random variables. The LLN would replace (xxii) by its mean

$$\frac{1}{n}\sum_{j=1}^{n} \gamma_{ji}^{++} f(x_j(x)) \approx \frac{1}{n}\sum_{j=1}^{n} E\left[\gamma_{ji}^{++} f(x_j(t))\right]$$

$$\approx \frac{1}{n}\sum_{j=1}^{n} E\left[\gamma_{ji}^{++}\right] E\left[f(x_j(t))\right]$$

(because γ_{ji}^{++} and $x_j(t)$ are "nearly independent")

$$= E\left[\gamma_{ii}^{++}\right] E\left[f(x_i(t))\right]$$

(because all of the γ_{ji}^{++}'s and all of the $x_j(t)$'s are identically distributed)

$$= \gamma^{++} E\left[f(x_i(t))\right].$$

If this makes sense, then it applies as well to the other three input terms in (xxi). For each i, then, we expect that

$$\frac{d}{dt}x_i(t) \approx -\alpha x_i(t) + \gamma^{++} E\left[f(x_i(t))\right] - \gamma^{-+} E\left[g(y_i(t))\right],$$

$$\frac{d}{dt}y_i(t) \approx -\beta y_i(t) + \gamma^{+-} E\left[f(x_i(t))\right] - \gamma^{--} E\left[g(x_i(t))\right]. \qquad \text{(xxiii)}$$

Since the right-hand side of (xxiii) is deterministic, $x_i(t)$ and $y_i(t)$ are nearly deterministic. In this case, $E[f(x_i(t))] \approx f(x_i(t))$ and $E[g(y_i(t))] \approx g(y_i(t))$. Put this back in (xxiii) and conclude that the behavior of the entire system (xxi) should be well described by the two dimensional prototype system

$$\frac{d}{dt}x(t) = -\alpha x(t) + \gamma^{++}f(x(t)) - \gamma^{-+}g(y(t)),$$
$$\frac{d}{dt}y(t) = -\beta y(t) + \gamma^{+-}f(x(t)) - \gamma^{--}g(y(t)). \qquad \text{(xxiv)}$$

Miller and I assumed that (xxiv) provided an adequate description for (xxi), and analyzed the dynamics of (xxiv) as a possible model for the generation of inspiratory neuronal activity. In **[12]** it has been shown that (xxiv) does in fact provide an arbitrarily good approximation for (xxi) as n and m go to ∞. More specifically, as $n \to \infty$ and $m \to \infty$ all excitatory activities $x_i(t)$, $1 \leqslant i \leqslant n$, and all inhibitory activities $y_i(t)$, $1 \leqslant i \leqslant m$, will remain, respectively, arbitrarily close to the trajectories of $x(t)$ and $y(t)$, as defined by the prototype equations in (xxiv) (see **[12]** for details). In other words, the LLN is in force in (xxi), and the consequence is that the behavior of the entire system is determined by the parameters in the two dimensional system, (xxiv).

Simulations of this averaging effect can be quite striking. For example, we may choose the functions f and g and the six parameters in (xxiv) so that $(x(t), y(t))$ has a globally stable limit cycle. Then, in a typical experiment, with the standard deviation of each γ larger than 50% of its mean, (xxi) already oscillates when $n = m = 7$. (For smaller n and m, all activities approach an equilibrium.) But, for this still small system, the oscillation is quite different from the limit cycle trajectory predicted by (xxiv). When n and m are 80, however, the $x_i(t)$ and $y_i(t)$, $1 \leqslant i \leqslant 80$, trajectories are virtually indistinguishable from the prototype $x(t)$ and $y(t)$ trajectories. See **[12]** for phase portraits from one such experiment.

If the output of the respiratory centers can be simulated by a four dimensional system (two subpopulations of inspiratory and two subpopulations of expiratory neurons), why commit large numbers of neurons to the generation of this activity? One obvious reason is reliability. It is widely appreciated that there is advantage to redundancy in the nervous system, especially when promoting vital functions such as breathing. There may also be a second purpose, as is suggested by the averaging effect discussed here. The dynamics of a low dimensional system will be more critically dependent on individual parameters. The relevance of this is that each component of a small system generating respiratory activity would need to be specified with extreme accuracy, if the system is precisely to achieve a desired output. The alternative is to reach this precision by averaging: when in force, the LLN guarantees arbitrary precision in arbitrarily large systems. Of course, nothing said here needs to be limited to models for the control of breathing. Averaging is an available mechanism for the reliable and precise generation of activity by a homogeneous collection of neurons, whatever the physiological application.

REFERENCES

1. S. I. Amari, *Characteristics of random nets of analog neuron-like elements*, IEEE Trans. on Systems, Man and Cybernetics, vol. SMC-2, 1972, pp. 643–657.

2. ______, *Neural theory of association and concept-formation*, Biol. Cybernet. **26** (1977), 175–185.

3. ______, *Topographic organization of nerve fields*, Bull. Math. Biol. **42** (1980), 339–364.

4. S. I. Amari and A. Takeuchi, *Mathematical theory on formation of category detecting nerve cells*, Biol. Cybernet. **29** (1978), 127–136.

5. E. L. Bienenstock, L. N. Cooper and P. Munro, *On the development of neuronal selectivity: orientation specificity and binocular interaction in visual cortex* (in preparation).

6. R. O. Duda and P. E. Hart, *Pattern classification and scene analysis*, Wiley, New York, 1973.

7. J. L. Feldman and J. D. Cowan, *Large-scale activity in neural nets*. II. *A model for the brainstem respiratory oscillator*, Biol. Cybernet. **17** (1975), 39–51.

8. S. Geman, *Application of stochastic averaging to learning systems*, Brain Theory Newsletter **3** (1978), 69–71.

9. ______, *Some averaging and stability results for random differential equations*, SIAM J. Appl. Math. **36** (1979), 86–105.

10. ______, *A method of averaging for random differential equations with applications to stability and stochastic approximations*, Approximate Solution of Random Equations (A. T. Bharucha-Reid, ed.), North-Holland Series in Probability and Appl. Math., North-Holland, Amsterdam, 1979, pp. 49–85.

11. ______, *Notes on a self-organizing machine*, Parallel Models of Associative Memory (G. Hinton and J. Anderson, eds.), Erlbaum Associates, Hillsdale, N. J., 1980.

12. ______, *Almost sure stable oscillations in a large system of randomly coupled equations*, Reports on Pattern Analysis, no. 97, Div. Appl. Math., Brown University, 1980 (submitted).

13. S. Geman and M. Miller, *Computer simulation of brainstem respiratory activity*, J. Appl. Physiol. **41** (1976), 931–938.

14. S. Geman and C. R. Hwang, *A chaos hypothesis for some large systems of random equations*, Reports on Pattern Analysis, no. 82, Div. Appl. Math., Brown University, 1980 (submitted).

15. S. Grossberg, *A neural theory of punishment and avoidance*. I. *Qualitative theory*. Math. Biosci. **15** (1972), 39–67.

16. ______, *A neural theory of punishment and avoidance*. II. *Quantitative theory*. Math. Biosci. **15** (1972), 253–285.

17. ______, *Pattern learning by functional-differential neural networks with arbitrary path weights*, Delay and Functional Differential Equations and Their Applications (K. Schmitt, ed.), Academic Press, New York, 1972, pp. 121–160.

18. ______, *Classical and instrumental learning by neural networks*, Progress in Theoret. Biol. **3** (1974), 51–141.

19. A. E. Hoerl and R. W. Kennard, *Ridge regression. Biased estimation for non-orthogonal problems*, Technometrics **12** (1970), 55–67.

20. K. Kadiyala, *Operational ridge regression estimator under the prediction goal*, Comm. Statist. A-Theory Methods **8** (1979), 1377–1391.

21. Iu. A. Mitropolsky, *Averaging method in non-linear mechanics*, Internat. J. Non-Linear Mech. **2** (1967), 69–96.

22. A. M. Uttley, *A two-pathway informon theory of conditioning and adaptive pattern recognition*, Brain Res. **102** (1976), 23–35.

23. ______, *Simulation studies of learning in an informon network*, Brain Res. **102** (1976), 37–53.

24. ______, *Neurophysiological predictions of a two-pathway informon theory of neural conditioning*, Brain Res. **102** (1976), 55–70.

25. M. T. Wasan, *Stochastic approximation*, Cambridge Univ. Press, Cambridge, 1969.

26. H. R. Wilson and J. D. Cowan, *A mathematical theory of the functional dynamics of cortical and thalamic nervous tissue*, Kybernetik **13** (1973), 55–80.

DIVISION OF APPLIED MATHEMATICS, BROWN UNIVERSITY, PROVIDENCE, RHODE ISLAND 02912

SIAM-AMS Proceedings
Volume **13**
1981

Adaptive Resonance in Development, Perception and Cognition

STEPHEN GROSSBERG[1]

1. Introduction. Several of the best physicists of the last half of the nineteenth century were also distinguished psychologists or physiologists. The contributions of Helmholtz, Maxwell, and Mach to perception are notable examples of this productive interdisciplinary activity (Boring [**1950**], Koenigsberger [**1906**], Campbell and Garnett [**1882**], Ratliff [**1965**]). Why then did not interdisciplinary studies, based on these inspiring successes, flourish at the beginning of the twentieth century?

One reason lies in the nature of psychophysiological phenomena. The mathematics of traditional physics has been centered in linear and stationary concepts, whereas the data of psychophysiology are often nonlinear and nonstationary. The great revolutions of twentieth century physics were supported by nineteenth century mathematics, but the emergent concepts of psychophysiology, from the very outset, lead to new mathematics. We seem to be experiencing the type of scientific revolution wherein new intuitive physical concepts and new mathematics must both be simultaneously developed, each fertilizing the other. Such scientific developments bring with them special challenges, but also special intellectual rewards.

In this article, I shall discuss nonlinear and nonstationary concepts that arise from consideration of some basic questions about brain design, and I will indicate how the answers help to unify several of the contributions reported elsewhere in the book. These questions include: How can an organism's adaptive mechanisms be stable enough to resist environmental fluctuations which do not alter its behavioral success, but plastic enough to change rapidly in response to environmental demands that do alter its behavioral success? How is a balance between stability and adaptability achieved in a nonstationary environment? More simply expressed, how do internal representations of the environment

---

1980 *Mathematics Subject Classification*. Primary 92A25; Secondary 92A09.
[1]Supported in part by the National Science Foundation (NSF IST-80-00257).

develop through experience and thereafter maintain themselves? In particular, how are coding errors corrected, or adaptations to a changing environment effected, if individual nerve cells cannot detect that these errors or changes have occurred?

I will indicate how limitations in the information available to individual cells can be overcome when the cells act together in suitably designed feedback networks. In these networks, nonlinear competition between feedforward data patterns and learned feedback templates, or expectancies, helps to stabilize the developing code. This competition triggers events that buffer already committed populations against continual erosion by incompatible environmental fluctuations without sacrificing the adaptability of uncommitted populations. The buffering mechanism sets the stage for hypothesis testing and fast parallel search for uncommitted populations.

Environmental events lead to resonant activity within the system when their feedforward data and feedback templates reach concensus through a matching process. The resonant state embodies the perceptual event, or attentional focus, and its amplified and sustained activities are capable of driving slow adaptive changes in system structure. In other words, resonant activity within the short term memory (STM) system causes alterations in long term memory (LTM). These LTM changes, in turn, alter the resonant STM patterns, and the cycle continues until STM and LTM equilibrate. This dynamic exchange between STM and LTM is called an *adaptive resonance*. The "code" of a network is suggested to be the set of stable resonances which the network can support in response to a prescribed input environment (Grossberg **[1976a]**, **[1976b]**, **[1978a]**, **[1980a]**).

Many adult perceptual and cognitive properties emerge within this framework as manifestations of the developmental mechanisms that are needed to ensure a balance between code stability and adaptability. Adaptive resonances are also suggested to be functional coding units in nonneural cellular tissues and play a prominent role in efforts to discover a universal developmental code (Grossberg **[1978b]**).

This article is built around a thought experiment that shows us, in simple stages, how to build up these nonlinear feedback networks. The thought experiment does this by indicating that several principles of neural design are needed to construct the networks. Each principle solves an environmentally imposed problem to which a surviving species must adapt. The mechanisms that realize each principle imply a variety of psychophysiological and pharmacological properties.

An important aspect of understanding brain design is to realize which data properties are consequences of which principles. Wherever we can recognize a general principle at work in a given body of data, we can compare and contrast these data with other data wherein the same principle is operating. In this way, we can regroup the data in terms of underlying principles, rather than in terms of experimental techniques. Each experimental technique can probe only certain aspects of a principle, but by pooling the results from several of the techniques that are used in seemingly distinct, but mechanistically related, situations, we can understand the underlying mechanisms much better than we could have if we relied only on the techniques applicable in one situation.

From a mathematical viewpoint, such a classification is of independent interest, since the parametric changes which transform one data property into another are often due to bifurcations within a nonlinear dynamical system of high dimension.

2. Relations to other articles. I will illustrate the principles that arise in the thought experiment by discussing some of the data and other models that are included in this book. I believe that comparative analysis can greatly clarify the substantial experimental and theoretical progress that has been made by suggesting where each approach is strongest and where it might fail, where contributions probe an important principle and where they replicate prior work in a different setting, and where certain directions for future work promise to yield the greatest conceptual insights. This section sketches some comparative themes which will be clarified by reading the main theoretical arguments of the paper.

Freeman [**1981**] discovered that a resonance phenomenon occurs during olfactory perception. When a cat smells an expected scent, its neural potentials are amplified until a synchronized oscillation of activity is elicited across the olfactory bulb. The oscillation organizes the neural activity into a temporal sequence of spatial patterns. The spatial patterns of activity across cells carry the olfactory code. Freeman's model lumps together effects of presynaptic habituation and postsynaptic cell activity into a second-order system of differential equations. The thought experiment suggests how to unlump these effects, which are distinct in their mechanisms as well as in their spatial and temporal scales. When this is done, we can ask whether olfactory cells which are capable of habituating can also reset the olfactory code. They will have this capability if they are driven by a suitable arousal source and if their outputs thereupon compete. Or is code reset driven only by the reset of motivational mechanisms? The answer to this question has important implications for how many smells can be simultaneously encoded by olfactory feedback templates (Grossberg [**1981a**]).

When the olfactory bulb is active, its population activities oscillate rapidly. Travelling waves of activity across the neural tissue could not represent the spatial patterns of activity that encode a prescribed smell; only standing waves can do this. Are standing waves of neural activity achieved because the network's inhibitory interneurons rapidly average their excitatory signals? Or does cortical feedback, acting like a template, organize travelling waves into standing waves even when the inhibitory interneurons possess a slow averaging rate that creates a slowly varying adaptational baseline against which excitatory signals can be evaluated? This is a general alternative that must be decided whenever neural tissues encode a spatial pattern while they undergo sustained oscillations.

Travelling waves of neural activity are functionally disastrous when they traverse a network of feature detectors, as in a grand mal epileptic seizure. These waves are functionally desirable when they carry a spike potential down the nerve cell axon. Carpenter's [**1981**] work reveals that the circumstances which produce spike trains in the Hodgkin-Huxley equations are analogous to the circumstances which produce travelling waves across a network. Two general conclusions follow from these observations. The same formal principle can act on several levels of neural organization and properties of this principle which are selected for on one level can be selected against on a different level. This is one reason why a classification theory is indispensable to the theorist of mind. The study of how to select against travelling waves in a network made this comparison possible (Ellias and Grossberg [**1975**]). Also living tissue sometimes fails to maintain normal parameters, as when epileptic waves occur, and then a classification theory provides parametric insight into how the breakdown occurs.

Freeman's [**1980**] analysis of how an olfactory feedback template alters the processing of afferent input patterns can be recast in terms of the dynamics of mass action (or shunting) competitive networks. When such a network is designed to suppress noise, it automatically is capable of matching two or more input patterns (Grossberg [**1976b**]). If the patterns match, network activity is amplified. If the patterns mismatch, network activity is attenuated. The patterns to be matched alter the gains, or averaging rates, of the network cells. Such gain changes are also reported in Freeman's [**1975**] data, although they do not automatically occur in a model that omits shunting interactions.

Graham's [**1981**] analysis of multiple channels in the visual system can also be discussed in terms of mass action competitive networks that possess the noise suppression property. Both Freeman's work on olfactory coding in the cat and Graham's work on visual coding in humans seem to probe aspects of a common design principle, although this is not apparent from the formalisms that these authors have found it convenient to use.

In particular, a mass action competitive network with the noise suppression property is capable of filtering prescribed spatial frequencies. This property generalizes the ability of such networks to detect edges and curvature changes in an input pattern (Ellias and Grossberg [**1975**]), and is the basis for building up a model of medium bandwidth channels. To make this construction, it will be crucial to distinguish between feedforward and feedback competitive networks. Both types of networks share certain properties, such as matching properties and the ability to compute lightnesses, or the reflectances, of their input patterns (Grossberg [**1976b**]). However feedback networks possess disinhibitory, filling-in, choice, and hysteresis properties that are not enjoyed by their feedforward counterparts.

One reason for the existence of feedback networks in visual processing is the need to match the data in two monocular representations to generate globally self-consistent depth information. A good match yields a binocular resonance. I claim that a medium bandwidth channel is a feedback network, not a feedforward Fourier analysis of input data. When a feedback network of prescribed spatial scale (in a sense to be defined) achieves binocular resonance, it can then automatically compute lightnesses within this scale. Objects that are interpreted to lie at a prescribed distance thereby influence each other's perceived brightness (Beck [**1972**]). When a feedback network is activated, whether by monocular or binocular viewing, it also automatically fills in the spaces between the edge information that it extracts with disinhibitory patterns of activity. The Craik-O'Brien effect (Hamada [**1980**], O'Brien [**1958**]) thus naturally emerges in a feedback network. One might say the feedback network resolves the following dilemma. The noise suppression property tends to pick out edges, but then how can solid objects be perceived? A feedback network can fill in the edges with patterned, often periodic, activity that constitutes a medium bandwidth reaction to the input pattern, as it also computes lightness, depth, and other relevant aspects of this pattern.

To achieve binocular resonance, positive feedback between monocular competitive networks is needed. All competitive feedback networks must be nonlinear to avoid noise amplification (Grossberg [**1973**]). The simplest intercellular signal function that avoids noise amplification is a sigmoid, or S-shaped, signal function. Sigmoid signals occur in Freeman's data and modelling efforts (Freeman [**1979**]). One reason that Freeman builds gain control into his sigmoid as he does is that he uses additive, rather than shunting, interactions. Shunting interactions already possess gain control properties, whereas additive interactions do not. The sigmoid signal function also suggests a mechanistic interpretation of the sigmoid decision rule which Graham applies in her article. Since the sigmoid signal is nonlinear it seems doubtful that a linear Fourier analysis can be taking place in visual networks. Graham's model separates linear pattern processing from nonlinear decision rules, but in neural network dynamics, nonlinear feedback signalling helps to decide which spatial scales will resonate, and how patterns will be transformed within these scales before resonance occurs. The probability summation rules which Graham cautiously treats as effects due to independent channels might ultimately be traceable to such nonlinear signalling within and between spatial scales. These interactive effects should be easier to detect in experiments using moving objects, since the interactions between scales that help to create the impression of object permanence are then called into play.

Sperling [**1981**] also uses a special choice of sigmoid function to achieve choice behavior in his model. The Sperling model is a variant of a class of competitive networks whose dynamics have been completely analysed (Grossberg [**1973**]). Sperling's model omits cellular saturation terms, but the method of proof shows that these terms do not disturb the sigmoid's tendency to choose one population over another. Actually the sigmoid does not always make a choice: a *quenching threshold* (QT) exists such that the activities of populations which initially exceed the QT will be stored, whereas the activities of populations which are less than the QT will be quenched. A power law signal function $f(w) = Cw^n$, $n > 1$, can make choices. This kind of distinction cannot be understood without a classification theory.

Vision research provides an excellent forum wherein theories might profitably be grounded on a principled basis. So many properties covary in visual data that, in the absence of principled theories, one must often make a new hypothesis to account for each new visual fact. For example, nonlinear neural networks can perform computations like cross-correlations by using their filtering and matching properties; yet one would not wish to say that the brain is built to compute cross-correlations. If a model were constructed just to compute cross-correlations, then the model would later have to include a separate mechanism to compute lightness scales, a separate mechanism to compute sigmoids, and so on. There would be no internal coherence in the overall theory. The same thing is true if discrete Laplacians are used to model receptive field computations, as is now popular in Artificial Intelligence. Each tool admits certain short-run conveniences, but it soon leads to paradoxes and unwieldy extra hypotheses if it does not express a principle of neural design.

Mass action competitive networks also shed some light on Malsburg and Willshaw's [**1981**] model of retinotectal map development. Such networks are capable of conserving, or normalizing, their total activity as another manifestation of their ability to compute reflectances. Malsburg uses additive, rather than shunting, interactions in his model. Since additive interactions do not imply a normalization property, Malsburg assumes normalization as an extra hypothesis. It will be interesting to decide experimentally whether either one, or both, of these normalization properties are at work during retinotectal map formation. The mass action normalization effect is due to postsynaptic lateral inhibition. It can be disrupted by postsynaptic lateral cuts across the neural tissue, or by poisoning postsynaptic inhibitory transmitters. Malsburg's normalization hypothesis should survive these manipulations because it is due to an intracellular normalization mechanism. Intracellular normalization can occur when the receptor sites on the cell membrane form part of a mass action competitive network. The decision between the two normalization schemes therefore reduces to the question: Is the time scale of intracellular competition fast enough to affect map formation? In any case, Malsburg's normalization rule leads to examples of adaptation level systems; namely, his equations (6)–(8). Such systems naturally arise from mass action competitive rules.

Malsburg's similarity rules for matching two maps can be recast as mass action rules arising from a noise suppression property. Then the map matching process can also be viewed as a variation on the general theme of noise suppression. When this is done, both map matching and normalization emerge as two manifestations of a ubiquitous mass action principle, rather than as separate hypotheses in a particular example. One can then begin to compare matching processes during the development of spatial maps with other matching processes, such as the matching of olfactory templates or of visual spatial frequencies, and to organize these examples as part of a universal developmental code.

In a similar fashion, Malsburg uses rules for the growth of neural connections which have the same formal structure as laws for the differential sensitization of extant neural connections and for the regulation of transmitter production rates (Grossberg **[1964]**, **[1968]**, **[1971]**, **[1978b]**, **[1980b]**). Freeman's rules for the formation of olfactory templates are also a special case of these formal laws. The same formal developmental laws can thus occur at successive developmental stages, say a stage of prenatal map formation followed by a stage of map tuning by postnatal experience, despite the fact that each stage is realized by distinct chemical mechanisms. Successive developmental stages often refine the structural arrangements of previous stages. That these stages often obey similar formal laws prevents a later stage from contradicting the substrate established by a previous stage due to a merely formal, as opposed to a functional, incompatibility. Expressed in another way, it is wrong to believe that significant progress concerning the collective principles which organize developmental events cannot be made until we discover the chemicals which serve in each example. Quite the contrary is true. Without a collective analysis, knowing the chemicals teaches us very little. With a collective analysis, we can hope to classify how distinct local properties of the chemicals control distinct collective properties of the whole developing tissue.

Iverson and Pavel's **[1981]** analysis of human psychoacoustic data provides a wealth of parametric information about steady-state auditory processing as well as information about reaction rates. Their parametric data can be described by the steady-states of a suitably chosen feedback competitive network. The covariation of reaction rates on asymptotes in human psychoacoustic data is a basic property of these feedback networks. The covariation of gain changes and matching properties in Freeman's olfactory data is another example of the same property. Of special interest is the occurrence of distinct powers x^n and N^m $(n < m)$ of signal x and broad-band noise N in their matching data, as well as their shift property. Such powers approximate the suprathreshold behavior of a sigmoid law $f(w)$, evaluated at $f(x + \mu N)$ and $f(\mu N)$, and a shift property easily obtains in shunting networks. Their data might therefore be viewed as a suprathreshold manifestation of noise suppression by nonlinear signalling, but of course no power law is valid at the very high input intensities where saturation effects cannot be entirely avoided. General averaging rules, such as those described by Luce and Narens **[1981]** also arise as the steady-states of competitive networks. Properties such as reflectance processing and choice-making are often interpreted in terms of decision theoretic concepts within Luce's framework.

Many of our results can be interpreted as properties of neural parallel processors, although adaptive resonances are the result both of serial and parallel operations carried out in a suitable order. Vorberg **[1981]** discusses some reaction time distributions predicted by a serial self-terminating memory search model, whereas our results provide examples wherein some properties of a parallel process might seem serial, but the parallel process exhibits other properties which are not predicted by a serial theory.

Geman [**1981**] clarifies the circumstances under which a stochastic differential equation can be approximated by a deterministic equation that describes its average behavior through time. This problem is of importance in several sciences. In psychophysiology, its solution justifies the deterministic formalism of Freeman, Malsburg, and myself. My x_i's are mean activities, or potentials, of a population of sites whose input statistics mix rapidly enough to justify a deterministic approximation. My signals $f(x_i)$ are signal densities, and my products of terms such as $(B - x_i)f(x_i)$ say that the two processes with means $B - x_i$ and $f(x_i)$ are statistically independent ($E(AB) = E(A)E(B)$).

These comments briefly indicate some of the multifaceted and multilayered connections among the phenomena discussed within this book. The sections below provide the background needed to make these connections more explicit.

3. A thought experiment: the need for learned feedback templates. We now start to build a framework in which to discuss environmentally driven code development. Wherever possible, the *minimal* structure capable of achieving our ends will be defined. This procedure will clarify what mathematical problems have to be solved, what their relationship is to each other, and what types of thematic variations on the minimal structures can be anticipated in different species and different tissues in the same individual. This exposition is based on the one given in §§4 through 11, Grossberg [**1980a**].

Our central theoretical theme will be: How can a coding error be corrected if no individual cell knows that one has occurred? The importance of this issue becomes clear when we realize that erroneous cues can accidentally be incorporated into a code when our interactions with the environment are simple, and will only become evident when our environmental expectations become more demanding. Even if our code perfectly matched a given environment, we would certainly make errors as the environment itself fluctuates. Furthermore, we never have an absolute criterion of whether our understanding of a fixed environment is faulty, or the environment that we thought we understood is no longer the same. The problem of error correction is fundamental whenever either the environment fluctuates, or the individual keeps testing his ever deepening interpretation of the environment using ever sharper criteria of behavioral success.

Figure 1 introduces the functional elements on which our argument will build; namely, a collection of cells or cell populations $v_1, v_2, \ldots, v_n$ each of which has an activity, or potential, $x_1(t), x_2(t), \ldots, x_n(t)$ at every time t. The activity $x_i(t)$ of v_i is imagined to be due to inputs $I_i(t)$ to v_i from a prior stage of processing, or the external environment, or endogenous sources within v_i itself. At every time t, these activities form a *pattern* $x(t) = (x_1(t), x_2(t), \ldots, x_n(t))$ across the cells $v_1, v_2, \ldots, v_n$, to which we will refer collectively as a *field* of cells $\mathcal{F}$. Henceforth the time variable t will often be suppressed, since we will always take for granted that we are studying the system at a prescribed time.

Now consider two successive fields $\mathcal{F}^{(1)}$ and $\mathcal{F}^{(2)}$ of cells. Suppose that a pattern $x^{(1)}$ is active across $\mathcal{F}^{(1)}$ (Figure 2). Suppose that the signal-carrying pathways from $\mathcal{F}^{(1)}$ to $\mathcal{F}^{(2)}$ act to filter the pattern $x^{(1)}$, and that due to prior developmental experience, this filter "codes" pattern $x^{(1)}$ by eliciting pattern $x^{(2)}$ across $\mathcal{F}^{(2)}$. Knowing the detailed structure of this code is unnecessary to make our argument. However, we must be able to show how signal pathways can act as a filter that can be tuned by experience (Grossberg **[1976a]**, **[1976b]**).

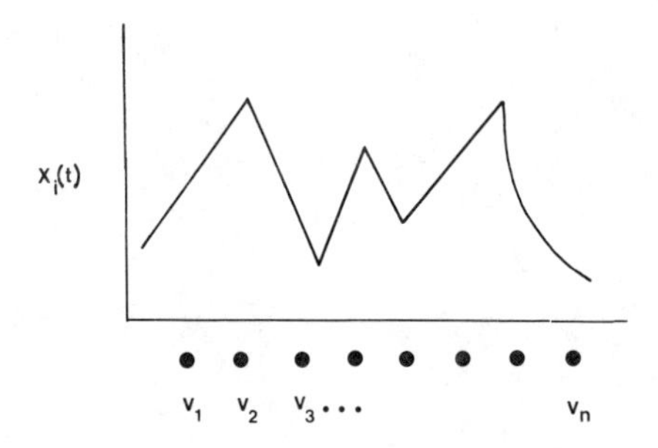

FIGURE 1. Each cell (or cell population) v_i possesses an activity or potential $x_i(t)$ at every time t, $i = 1, 2, \ldots, n$. The vector $(x_1(t), x_2(t), \ldots, x_n(t))$ is the spatial pattern of activity at time t.

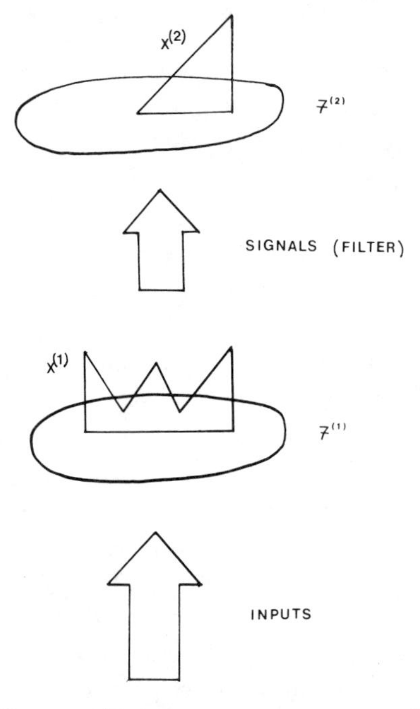

FIGURE 2. The activity pattern $x^{(1)}$ across $\mathcal{F}^{(1)}$ is filtered to elicit a pattern $x^{(2)}$ across $\mathcal{F}^{(2)}$.

Suppose after the system learns to code $x^{(1)}$ by $x^{(2)}$ that another pattern is presented to $\mathcal{F}^{(1)}$ and is erroneously coded at $\mathcal{F}^{(2)}$ by $x^{(2)}$. In order to describe this situation conveniently, we introduce some subscripts. Denote $x^{(1)}$ and $x^{(2)}$ by $x_1^{(1)}$ and $x_1^{(2)}$, respectively, and denote the erroneously coded pattern at $\mathcal{F}^{(1)}$ by $x_2^{(1)}$. Then we find Figure 3. In Figure 3, the pattern $x_2^{(2)}$ that codes $x_2^{(1)}$ is drawn to equal $x_1^{(2)}$. Equality is meant to imply functional equivalence rather than actual identity. We now ask the central question: How can this coding error be corrected if no individual cell knows that an error has occurred?

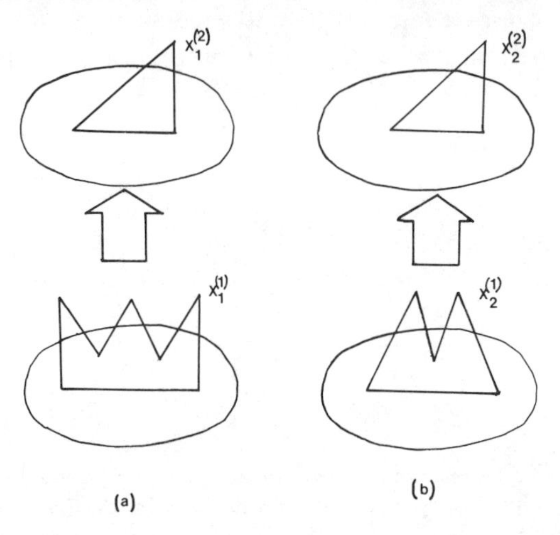

FIGURE 3. In (a), pattern $x_1^{(1)}$ at $\mathcal{F}^{(1)}$ elicits the correct pattern $x_1^{(2)}$ across $\mathcal{F}^{(2)}$. In (b), pattern $x_2^{(1)}$ elicits the incorrect pattern $x_2^{(2)}$, which is functionally equivalent to $x_1^{(2)}$, across $\mathcal{F}^{(2)}$.

Our first robust conclusion is now apparent: Whatever the mechanism is that corrects this error, it cannot exist within $\mathcal{F}^{(2)}$, since by definition $x_1^{(2)}$ and $x_2^{(2)}$ are functionally equivalent. In principle, $\mathcal{F}^{(2)}$ does not have the ability to distinguish the fact that $x_1^{(1)}$, and not $x_2^{(1)}$, should elicit $x_1^{(2)}$, since so far as $\mathcal{F}^{(2)}$ knows, $x_1^{(1)}$ *is* active at $\mathcal{F}^{(1)}$ rather than $x_2^{(1)}$.

It is important to realize that this argument is independent of coding details. It is based only on the type of information that $\mathcal{F}^{(2)}$ cannot, in principle, possess. Much of our argument will be based on similar limitations in the types of information which particular processing stages can, in principle, possess. The robustness of this argument suggests why the design that overcomes these limitations seems to occur ubiquitously, in one form or another, in so many neural structures.

Where in the network can this error be detected, in principle? At the time when $x_2^{(1)}$ elicits $x_1^{(2)}$, there exists no trace within the network that during prior learning trials, it was $x_1^{(1)}$ that elicited $x_1^{(2)}$, not $x_2^{(1)}$. Somehow this fact must be represented within the network dynamics. Otherwise $x_1^{(2)}$ could become associated with $x_2^{(1)}$, just as $x_1^{(1)}$ was on previous developmental trials. The only times when $x_1^{(1)}$ was active in the network were the developmental trials during which the filter from $\mathcal{F}^{(1)}$ to $\mathcal{F}^{(2)}$ was learning to code $x_1^{(1)}$ by $x_1^{(2)}$. In order to be, in principle, capable of testing whether the correct pattern $x_1^{(1)}$ elicits $x_1^{(2)}$ on later trials when $x_1^{(1)}$ is not presented, it must be true that during the developmental trials, $x_1^{(2)}$ activates a feedback pathway from $\mathcal{F}^{(2)}$ to $\mathcal{F}^{(1)}$ that is capable of learning the active pattern $x_1^{(1)}$ at $\mathcal{F}^{(1)}$. Then when $x_2^{(1)}$ erroneously activates $x_1^{(2)}$ on future trials, $x_1^{(2)}$ can read out the correct pattern $x_1^{(1)}$ across $\mathcal{F}^{(1)}$. When this happens, the two patterns $x_1^{(1)}$ and $x_2^{(1)}$ will be simultaneously active across $\mathcal{F}^{(1)}$, and they can be compared, or matched, to test whether or not the correct pattern has activated $x_1^{(2)}$ (Figure 4).

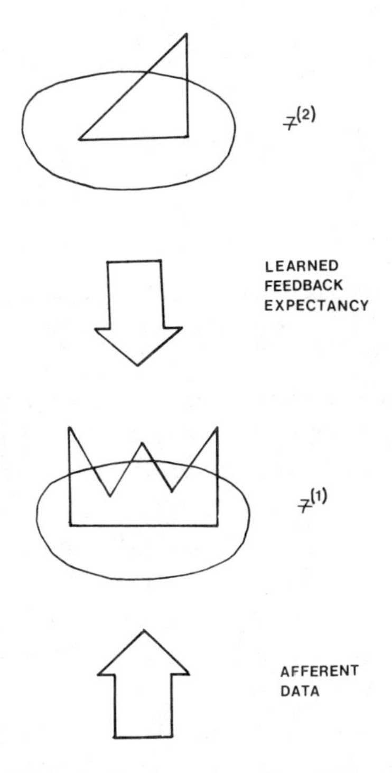

FIGURE 4. Pattern $x_1^{(2)}$ across $\mathcal{F}^{(2)}$ elicits a feedback pattern $x_1^{(1)}$ to $\mathcal{F}^{(1)}$, which is the pattern that is sampled across $\mathcal{F}^{(1)}$ during previous developmental trials. Field $\mathcal{F}^{(1)}$ becomes an interface when afferent data and learned feedback expectancies are compared.

In summary, if in principle it is possible to correct a coding error at $\mathcal{F}^{(2)}$, then there must exist learned feedback from $\mathcal{F}^{(2)}$ to $\mathcal{F}^{(1)}$. This learned feedback represents the pattern that $x_1^{(2)}$ *expects* to be at $\mathcal{F}^{(1)}$ due to prior developmental trials. The feedforward data to $\mathcal{F}^{(1)}$ and the learned feedback expectancy, or template, from $\mathcal{F}^{(2)}$ to $\mathcal{F}^{(1)}$ are thereupon compared at $\mathcal{F}^{(1)}$. Figure 5 illustrates this sequence of events as a series of snapshots that can occur at a very fast rate; e.g., on the order of hundreds of milliseconds.

At this point, we also recognize two more design problems for mathematics. The first problem is: How do feedback pathways from $\mathcal{F}^{(2)}$ learn a pattern of activity across $\mathcal{F}^{(1)}$? See Grossberg [**1972**], [**1974**] for a summary of this mechanism.

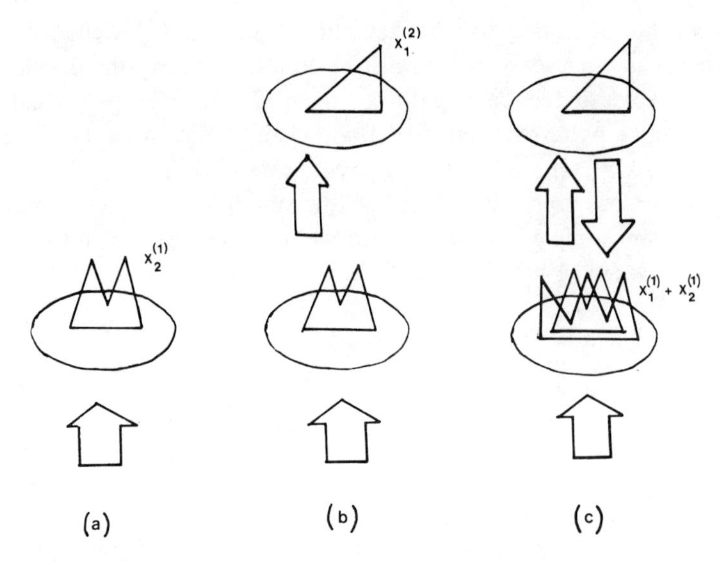

FIGURE 5. The stages (a), (b), and (c) schematize the rapid sequence of events whereby afferent data is filtered and activates a feedback expectancy that is matched against itself.

4. Noise suppression, pattern matching, and spatial frequency detection. The second design problem that we must face is this: Somehow the mismatch between the patterns $x_1^{(1)}$ and $x_2^{(1)}$ must rapidly shut off activity across $\mathcal{F}^{(1)}$. Otherwise, $x_1^{(2)}$ would learn to code $x_2^{(1)}$ much as $x_1^{(2)}$ learned to code $x_1^{(1)}$ on the preceding developmental trials. Pattern $x_1^{(2)}$ must also be rapidly shut off if only to prevent behavioral consequences of $x_1^{(2)}$ from being triggered by further network processing. Moreover, $x_1^{(2)}$ must be shut off in such a fashion that $x_2^{(1)}$ can thereupon be coded by a more suitable pattern across $\mathcal{F}^{(2)}$.

The only basis on which these changes can occur is the mismatch of $x_1^{(1)}$ and $x_2^{(1)}$ across $\mathcal{F}^{(1)}$. We must therefore ask: How does the mismatch of patterns across a field $\mathcal{F}^{(1)}$ of cells inhibit activity across $\mathcal{F}^{(1)}$? The mathematical details will be summarized in §13. Here, however, it is useful to make the important distinction between mechanisms that develop due to evolutionary pressures and properties that are merely consequences of these mechanisms. One might well worry that the design of a mismatch mechanism is a rather sophisticated evolutionary task. We now indicate that such a mechanism is a consequence of a more basic property, namely noise suppression, and that noise suppression is itself a variation of a basic evolutionary principle. Moreover, other useful properties follow from noise suppression, such as edge detection.

The environmental problem out of which the noise suppression property emerges is the *noise-saturation dilemma*. This dilemma has been discussed in detail elsewhere (e.g., Grossberg **[1978b]**, **[1980a]**). The dilemma confronts all noisy cellular systems that process input patterns as in Figure 1. If the inputs are too small, they can get lost in the noise. If the inputs are amplified to avoid the noisy range, they can saturate all the cells by activating all of their excitable sites, and thereby reduce to zero the cells' sensitivity to differences in the input intensities (Figure 6). §11 reviews how competitive interactions among the cells automatically retune their sensitivity to overcome the saturation problem. In a neural context, the competitive interactions are said to be shunting interactions and they are carried by an on-center off-surround anatomy. The retuning of sensitivity is due to automatic gain control by the inhibitory off-surround signals. §13 shows how the automatic gain control mechanism can inhibit a uniform pattern of inputs, no matter how intense the inputs are. This is the property of noise suppression that we seek.

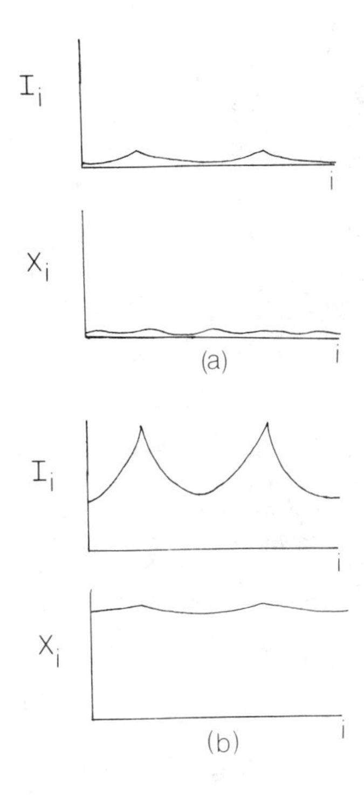

FIGURE 6. The noise-saturation dilemma: (a) At low background intensities, the input pattern $(I_1, I_2, \ldots, I_n)$ is poorly registered in the activity pattern $(x_1, x_2, \ldots, x_n)$ because of noise; (b) at high background intensities, the input pattern is poorly registered in the activity pattern because of saturation.

Figure 7a depicts this noise suppression property. A uniform pattern does not distinguish any cell from any other cell. For example, where the cells are feature detectors of one kind or another, a uniform input pattern contains no information that can distinguish one feature from any other feature. Noise suppression eliminates this irrelevant activity and allows the network to focus on informative discriminations.

Once noise suppression is guaranteed, several consequences automatically follow. For example, Figure 7b shows that such a network responds to the edges of a rectangular input, or to spatial gradients in more general input patterns.

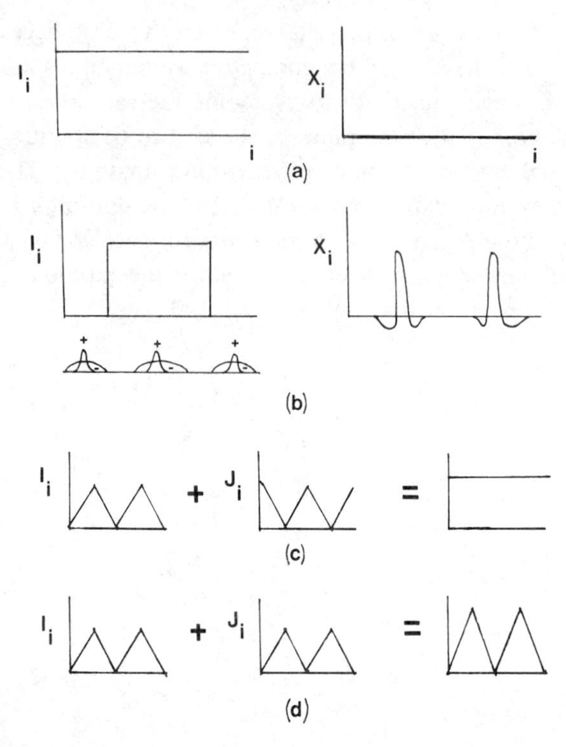

FIGURE 7. In (a), noise suppression converts a uniform (or zero spatial frequency) input pattern into a zero activity pattern. In (b), a rectangular input pattern elicits differential activity at its edges because the cells within its interior and beyond its boundary perceive uniform fields. In (c), two mismatched patterns add to generate an approximately uniform total input pattern, which will be suppressed by the mechanism of (a). In (d), two matched patterns add to yield a total input pattern that can elicit more vigorous activation than either input pattern taken separately.

This is because cells whose inhibitory surrounds fall outside the rectangle perceive a uniform field, and cells with inhibitory surrounds that are near the center of the rectangle also perceive a uniform field. Both types of cells suppress their inputs. Only cells near the edges of the rectangle do not perceive a uniform pattern. Consequently only the edges of the rectangle elicit large activation. This argument tacitly supposes that the lateral inhibitory interactions affecting each cell have a prescribed spatial extent, and that the width of the rectangle exceeds this spatial scale. More generally, spatial gradients in an input pattern are matched against the spatial scale of each cell's excitatory and inhibitory interactions. Only those spatial gradients in the input pattern that are nonuniform with respect to the cell's interaction scales generate large activities. By varying the inhibitory scales across cells, one can tune different cells to respond to different spatial frequencies. Thus spatial frequency detectors are a natural consequence of noise suppression properties within cells having a prescribed inhibitory scale. However, §24 will argue that spatial frequency detection *in vivo* is due to feedback in a competitive network, since a feedforward network suppresses the interior of a figure in its effort to suppress noise.

Finally, Figures 7c and 7d indicate how a noise suppression mechanism can accomplish pattern matching. Figure 7c supposes that two mismatched patterns feed into $\mathcal{F}^{(1)}$, where they add before coupling into the shunting dynamics. Because of the mismatch, the peaks of $x_1^{(1)}$ fill in the troughs of $x_2^{(1)}$. The total input pattern is approximately uniform, and is consequently quenched as noise. By contrast, in Figure 7d, the two patterns match. Their peaks and troughs mutually reinforce each other, so the resultant activities can be amplified beyond the effect of just one pattern. In summary, mismatched input patterns quench activity whereas matched patterns amplify activity across a field $\mathcal{F}^{(1)}$ that is capable of noise suppression. Moreover, spatial frequency detection and pattern matching go hand-in-hand when these properties are based on noise suppression.

A subsidiary mathematical question is now evident: How uniform must a pattern be in order for it to be suppressed? Part of the answer is determined by the choice of structural parameters, such as the strength and spatial distribution of lateral inhibitory coefficients (§16). However, the field $\mathcal{F}^{(1)}$ can also be dynamically tuned, or sensitized, by fluctuations in the level of nonspecific arousal that perturbs it through time. Such mechanisms will arise in a natural fashion as our argument continues.

5. Triggering of nonspecific arousal by unexpected events. Having suppressed $x_2^{(1)}$ at $\mathcal{F}^{(1)}$ due to mismatch with the feedback expectancy $x_1^{(1)}$, we must now use this suppression to inhibit $x_1^{(2)}$ at $\mathcal{F}^{(2)}$, since the mismatch at $\mathcal{F}^{(1)}$ is the only mechanism in the network that can, in principle, distinguish that an error has occurred at $\mathcal{F}^{(2)}$. Until $x_1^{(2)}$ is quenched, it will continue to read out the template $x_1^{(1)}$ to $\mathcal{F}^{(1)}$, which will prevent $x_2^{(1)}$ from eliciting a new signal to $\mathcal{F}^{(2)}$. Moreover, mismatch at $\mathcal{F}^{(1)}$ cannot quench $x_1^{(2)}$ simply by inhibiting the signals from $\mathcal{F}^{(1)}$ to $\mathcal{F}^{(2)}$ that originally activated $x_1^{(2)}$. Were this the only mechanism, then quenching $x_1^{(2)}$ would also quench its feedback template, $x_2^{(1)}$ would be reinstated, the same signals from $\mathcal{F}^{(1)}$ to $\mathcal{F}^{(2)}$ that caused the error would reoccur, and the error would perseverate. Some other mechanism, based on mismatch at $\mathcal{F}^{(1)}$, must inhibit $x_1^{(2)}$ and circumvent the error perseveration problem.

We were led to the mismatch mechanism at $\mathcal{F}^{(1)}$ by noting that $\mathcal{F}^{(2)}$ could not discriminate whether or not an error had occurred. Now we note that $\mathcal{F}^{(1)}$'s information is also limited. At $\mathcal{F}^{(1)}$, it cannot be discerned *which* pattern across $\mathcal{F}^{(2)}$ caused the mismatch at $\mathcal{F}^{(1)}$. It could have been any pattern whatsoever. All $\mathcal{F}^{(1)}$ knows is that a mismatch has occurred. Whatever pattern across $\mathcal{F}^{(2)}$ caused the mismatch must be inhibited. Consequently a mismatch at $\mathcal{F}^{(1)}$ must have a *nonspecific* effect on all of $\mathcal{F}^{(2)}$, since any of the cells in $\mathcal{F}^{(2)}$ might be one of the cells that must be inhibited.

We are therefore led to the following questions: How does mismatch and subsequent quenching of activity across $\mathcal{F}^{(1)}$ elicit a nonspecific signal (arousal!) to $\mathcal{F}^{(2)}$? Where does the activity that drives this nonspecific arousal pulse come from? Before answering these questions, we should realize that we have been led to a familiar conclusion: Unexpected, or novel, events are arousing. Now we will consider how such arousal is initiated and how it contributes to attentional processing.

Where does the activity that drives the arousal come from, and why is it released when quenching of activity at $\mathcal{F}^{(1)}$ occurs? There are two possible answers to the first part of the question, but only one of them survives closer inspection. The activity is either endogenous (internally and persistently generated) or the activity is elicited by the sensory input. If the activity were endogenous, then arousal would occur whenever $\mathcal{F}^{(1)}$ was inactive, whether this inactivity was due to active quenching by mismatched feedback from $\mathcal{F}^{(2)}$ or to the absence of sensory inputs. This leads to the unpleasant conclusion that $\mathcal{F}^{(2)}$ is tonically flooded with arousal whenever nothing interesting is happening at $\mathcal{F}^{(1)}$ or $\mathcal{F}^{(2)}$. We therefore conclude that sensory inputs to $\mathcal{F}^{(1)}$ bifurcate before they reach $\mathcal{F}^{(1)}$. One pathway is *specific*: it delivers information about the sensory event to $\mathcal{F}^{(1)}$. The other pathway is *nonspecific*: it activates the arousal mechanism that is capable of nonspecifically influencing $\mathcal{F}^{(2)}$.

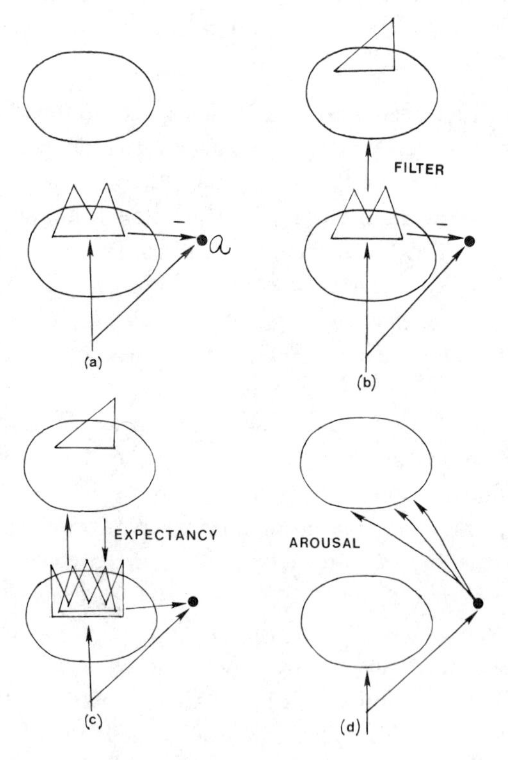

FIGURE 8. In (a), afferent data elicit activity across $\mathcal{F}^{(1)}$ and an input to the arousal source $\mathcal{A}$ that is inhibited by $\mathcal{F}^{(1)}$. In (b), the pattern at $\mathcal{F}^{(1)}$ maintains inhibition of $\mathcal{A}$ as it is filtered and activates $\mathcal{F}^{(2)}$. In (c), the feedback expectancy from $\mathcal{F}^{(2)}$ is matched against the pattern at $\mathcal{F}^{(1)}$. In (d), mismatch attenuates activity across $\mathcal{F}^{(1)}$ and thereby disinhibits $\mathcal{A}$, which releases a nonspecific arousal signal to $\mathcal{F}^{(2)}$.

Given that the sensory inputs to $\mathcal{F}^{(1)}$ also activate an arousal pathway, what prevents this pathway from being activated unless activity at $\mathcal{F}^{(1)}$ is quenched? The answer is now clear: Activity at $\mathcal{F}^{(1)}$ inhibits the arousal pathway, and quenching of this activity disinhibits the arousal pathway. Figure 8 schematizes the sequence of events to which we have been led. First a sensory event elicits a pattern $x_2^{(1)}$ across $\mathcal{F}^{(1)}$ as it begins to activate the arousal pathway $\mathcal{Q}$. This activation at $\mathcal{Q}$ is inhibited by activity from $\mathcal{F}^{(1)}$. Simultaneously, pattern $x_2^{(1)}$ activates pathways to $\mathcal{F}^{(2)}$ which act as a filter that erroneously activates $x_1^{(2)}$. Pattern $x_1^{(2)}$ reads out the learned feedback expectancy $x_1^{(1)}$ to $\mathcal{F}^{(1)}$. Mismatch of $x_1^{(1)}$ and $x_2^{(1)}$ at $\mathcal{F}^{(1)}$ quenches activity across $\mathcal{F}^{(1)}$. The inhibitory signal from $\mathcal{F}^{(1)}$ to $\mathcal{Q}$ is also quenched, and the arousal pathway is disinhibited. A nonspecific arousal pulse is hereby unleashed upon $\mathcal{F}^{(2)}$.

6. Parallel hypothesis testing in real time: the probabilistic logic of complementary categories. The next design problem is now clearly before us: How does the increment in nonspecific arousal differentially shut off the active cells in $\mathcal{F}^{(2)}$? The active cells are the cells that elicited the feedback expectancy to $\mathcal{F}^{(1)}$, and since mismatch occurred at $\mathcal{F}^{(1)}$, these cells must have been erroneously activated. Consequently, they should be shut off. Furthermore, inactive cells at $\mathcal{F}^{(2)}$ should not be inhibited, because these cells must be available for possible coding of $x_2^{(1)}$ during the next time interval. Thus a differential suppression of cells is required: The cells that were most active when arousal occurs should be most inhibited. This property realizes a kind of probabilistic logic in real time. If activating cell v_i in $\mathcal{F}^{(2)}$ to a given degree leads to a certain degree of error, or mismatch, at $\mathcal{F}^{(1)}$, then cell v_i should be inhibited to a degree that is commensurate both with its prior activation and with the size of the arousal increment, or the amount of error. Since cells that were only minimally active could have contributed only a small effect to the feedback expectancy, their inhibition will consequently be less, and they can contribute more to the correct coding of $x_2^{(1)}$ during the next time interval.

The arousal-initiated inhibition of cells across $\mathcal{F}^{(2)}$ must be enduring as well as selective. Otherwise, as soon as $x_1^{(2)}$ is inhibited, the feedback expectancy $x_1^{(1)}$ would be shut off, and $x_2^{(1)}$ would be free to reinstate $x_1^{(2)}$ across $\mathcal{F}^{(2)}$ once again. The error would perseverate, and the network would be locked into an uncorrectable error. The inhibited cells must therefore stay inhibited long enough for $x_2^{(1)}$ to activate a different pattern across $\mathcal{F}^{(2)}$ during the next time interval. The inhibition is therefore slowly varying compared to the time scale of filtering, feedback expectancy, and mismatch.

Once this selective and enduring inhibition is accomplished, the network has a capability for rapid hypothesis testing. By enduringly and selectively inhibiting $x_1^{(2)}$, we "renormalize" or "conditionalize" the field $\mathcal{F}^{(2)}$ to respond differently to pattern $x_2^{(1)}$ during the next time interval. If the next pattern elicited by $x_2^{(1)}$ across $\mathcal{F}^{(2)}$ also creates a mismatch at $\mathcal{F}^{(1)}$, then it will be suppressed, and $\mathcal{F}^{(2)}$

will be renormalized again. In this fashion, a sequence of rapid pattern reverberations between $\mathscr{F}^{(1)}$ and $\mathscr{F}^{(2)}$ can successively conditionalize $\mathscr{F}^{(2)}$ until either a match occurs, or a set of uncommitted cells is found with which $x_2^{(1)}$ can build a learned filter from $\mathscr{F}^{(1)}$ to $\mathscr{F}^{(2)}$, and a learned expectancy from $\mathscr{F}^{(2)}$ to $\mathscr{F}^{(1)}$.

7. The parallel dynamics of recurrent competitive networks: contrast enhancement, normalization, quenching threshold, tuning. At this point, one can justifiably wonder how $x_2^{(1)}$ elicits a supraliminal pattern across $\mathscr{F}^{(2)}$ after $x_1^{(2)}$ is inhibited. If $x_1^{(2)}$ is the pattern that $x_2^{(1)}$ originally excites, and $x_1^{(2)}$ is inhibited, then will not the next pattern elicited by $x_2^{(1)}$ across $\mathscr{F}^{(2)}$ have very small activity? Otherwise, why was not the second pattern also active when $x_1^{(2)}$ was active? See Figure 9.

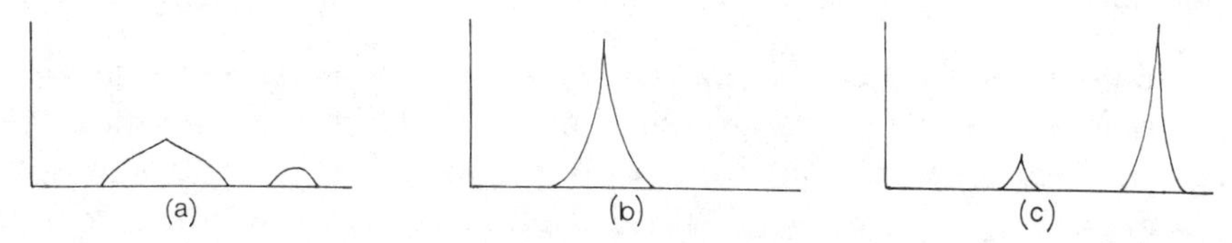

FIGURE 9. The input pattern in (a) elicits activity pattern $x_1^{(2)}$ in (b) as small input activities are suppressed. After $x_1^{(2)}$ is suppressed, the small input activities inherit normalized activity from $x_1^{(2)}$ to elicit the distinct activity pattern in (c).

This would be the case if the anatomy within $\mathscr{F}^{(2)}$ contained only feedforward, or nonrecurrent, pathways. Thus we are led to consider that the anatomy within $\mathscr{F}^{(2)}$ contains feedback, or recurrent, pathways. Since all cellular systems face the noise-saturation dilemma, these pathways are distributed in a competitive geometry, or an on-center off-surround anatomy (Figure 10). When these competitive networks are designed to overcome noise amplification and saturation, they also enjoy several other properties that we need (§17).

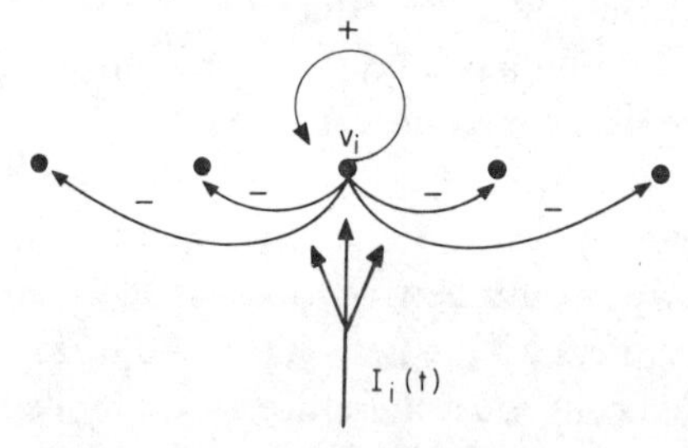

FIGURE 10. A recurrent shunting on-center off-surround network is capable of contrast-enhancing its input pattern, normalizing its total activity, and storing the contrast enhanced pattern in short term memory (STM). If the feedback signals are sigmoid functions of activity, then a quenching threshold (QT) exists that defines the activity level below which activity is treated as noise and quenched, and above which activity is contrast enhanced and stored in STM.

A basic property of such a network is its tendency to conserve, or adapt, the *total* activity that it stores in STM. This is the *normalization* property. If certain cells in the network are prevented from sharing the STM activity, say due to arousal-initiated inhibition, then the total activity is renormalized by being distributed to the other cells. Thus after $x_1^{(2)}$ is inhibited across $\mathcal{F}^{(2)}$, the network will respond to the signals due to $x_2^{(1)}$ by differentially amplifying them in a way that tends to preserve the total STM activity across $\mathcal{F}^{(2)}$. This new STM pattern will inherit much of the STM activity that $x_1^{(2)}$ had before it was suppressed, but the new STM pattern across $\mathcal{F}^{(2)}$ will be a quite different pattern from $x_1^{(2)}$, since it is built from $\mathcal{F}^{(1)}$ signals that previously fared poorly in the competition for STM activity.

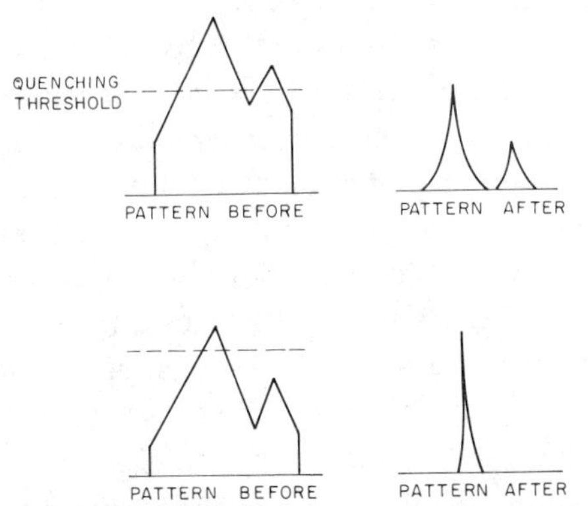

FIGURE 11. If the QT is variable, say due to shunting signals that nonspecifically control the size of the network's inhibitory feedback signals (§22), then the network's sensitivity can be tuned to alter the ease with which inputs are stored in STM.

Competitive feedback networks are also capable of contrast-enhancing small differences in initial pattern activities into large and easily discriminable differences that are thereupon stored in short term memory (STM). See Figure 11. This property is necessary to build up the codes for $\mathcal{F}^{(1)}$ patterns at $\mathcal{F}^{(2)}$. Before an $\mathcal{F}^{(1)}$ pattern is coded by $\mathcal{F}^{(2)}$, it might elicit an almost uniform activity pattern across $\mathcal{F}^{(2)}$. The recurrent dynamics within $\mathcal{F}^{(2)}$ quickly contrast-enhances and stores the contrast-enhanced pattern in STM, where it can be sampled and stored in LTM by the pathways from $\mathcal{F}^{(1)}$ to $\mathcal{F}^{(2)}$. When the next occurrence of the same pattern at $\mathcal{F}^{(1)}$ occurs, these pathways therefore elicit a more differentiated pattern across $\mathcal{F}^{(2)}$, which is again contrast-enhanced and stored in STM. The feedback enhancement between STM and LTM continues until the two processes equilibrate, other things being equal.

These recurrent networks also possess a *quenching threshold* (QT), which is a parameter whose size determines what activities will be suppressed, or quenched, and what activities will be stored in STM. Activities in populations that start below the QT will be suppressed; activities that exceed the QT will be contrast-enhanced and stored in STM. Thus the QT defines the cut-off point that defines "noise" in a recurrent network. All networks which possess a QT can be "tuned", that is, by varying the QT, the criterion of which data shall be stored in STM and which data shall be quenched can be altered through time. Several parameters work together to determine QT size, notably the strength of recurrent lateral inhibitory pathways within the network. For example, if a nonspecific arousal pulse multiplicatively inhibits, or shunts, the inhibitory interneurons of a recurrent network, then its QT will momentarily decrease–the network's inhibitory "gates" will open–to facilitate STM storage.

8. Antagonistic rebound within on-cell off-cell dipoles. We are now faced with a subtle design problem: How can a nonspecific event, such as arousal, have specific consequences of any kind, let alone generate an exquisitely graded, enduring, and selective suppression of active cells?

First let us consider some familiar behavioral facts that help to motivate the mechanism. Suppose that I wish to press a lever in response to the offset of a light. If light offset simply turned off the cells that code for light being on, then there would exist no cells whose activity could selectively elicit the lever press response after the light was turned off. Clearly, offset of the light not only turns off the cells that are turned on by the light, but also selectively turns on cells which will transiently be active after the light is shut off. The activity of these "off"-cells–namely the cells that are turned on by light offset–can then activate the motor commands leading to the lever press. Let us call the transient activation of the off-cell by cue offset *antagonistic rebound.*

When such on-cell off-cell interactions are modelled, one finds examples akin to Figure 12. In Figure 12a, a nonspecific arousal level input I is delivered equally to both channels, whereas a test input J is delivered to the on-cell channel. These inputs create signals S_1 and S_2 in both channels, and the signals are multiplicatively gated by slowly varying chemical transmitters z_1 and z_2, respectively. The gated signals S_1z_1 and S_2z_2 thereupon compete, and yield the on-cell off-cell responses that are depicted in Figure 12a. Grossberg [**1981b**] describes the details that are needed for a better understanding, but the main idea behind antagonistic rebound is easy to describe. Consider Figure 12a. Here each transmitter z_1 and z_2 is depleted by being released at a rate proportional to S_1z_1 and S_2z_2, respectively. More depletion of z_1 than z_2 occurs if the signal S_1 exceeds S_2. While the test input J is on, the on-channel receives a larger input than the off-channel, since its total input is J plus the nonspecific input I, whereas the off-cell channel only receives the input I. Consequently $S_1 > S_2$, so that depletion of transmitter leads to the inequality $z_1 < z_2$. Despite this fact, one proves that the gated signals satisfy the inequality $S_1z_1 > S_2z_2$. Consequently, the on-channel receives a larger gated signal than the off-channel, so that after competition takes place, there is a net on-reaction.

What happens when the test input is shut off? Then both channels receive only the equal nonspecific input I. The signals S_1 and S_2 rapidly equalize ($S_1 = S_2$). However, the transmitters are more slowly varying in time, so that the inequality $z_1 < z_2$ continues to hold. The gated signals therefore satisfy $S_1z_1 < S_2z_2$. Now the off-channel receives a larger signal, so that after competition takes place, there is an antagonistic rebound in response to offset of the test input.

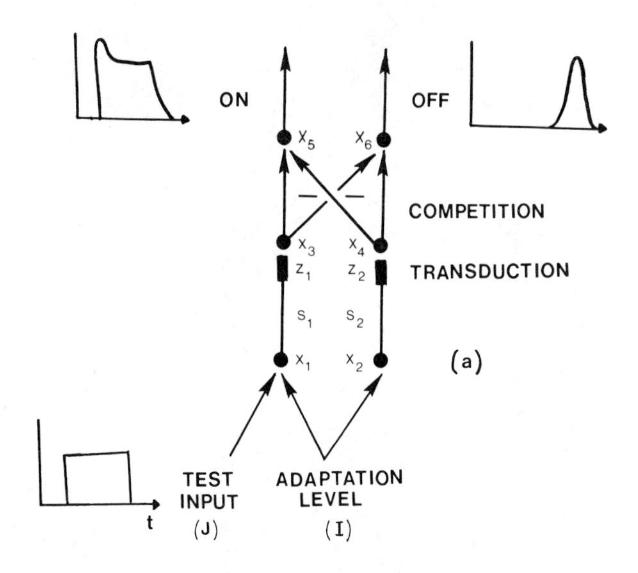

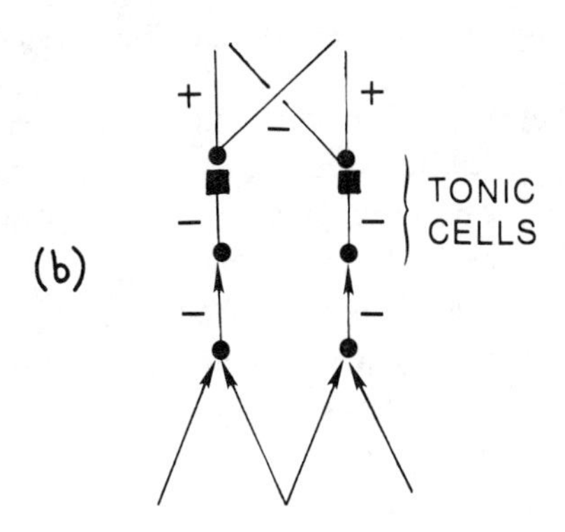

FIGURE 12. Two examples of on-cell off-cell dipoles. In (a), the test input J and arousal level input I add in the on-channel. The arousal level input also perturbs the off-channel. Each input is gated by a slowly varying excitatory transmitter (square synapses). Then the channels compete before eliciting a net on-response or off-response. In (b), the slowly varying transmitters are inhibitory, and the net effect of two successive inhibitory transmitters (e.g., dopamine and/or GABA) is a net disinhibitory effect.

Why is the rebound transient in time? The equal signals S_1 and S_2 continue to drive the depletion of the transmitters z_1 and z_2. Gradually the amounts of z_1 and z_2 also equalize, so that $S_1 z_1$ and $S_2 z_2$ gradually equalize. As the gated signals equalize, the competition shuts off both the on-channel and the off-channel. These facts are summarized in Figure 13.

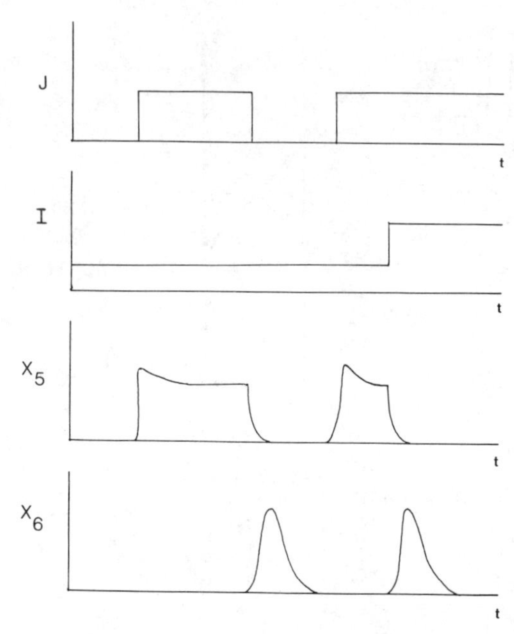

FIGURE 13. Antagonistic rebound at offset of a specific input or onset of nonspecific arousal.

9. Arousal elicits antagonistic rebound: surprise and counterconditioning. A surprising feature of the on-cell off-cell dipole is its reaction to rapid temporal fluctuations in the arousal level. This reaction allows us to answer the question posed in §8: How can a nonspecific event, such as arousal, selectively suppress active on-cells? Grossberg **[1981b]** shows that arousal fluctuations can reset the dipole, despite the fact that they generate equal inputs to the on-cell and off-cell channels. In particular, a sudden increment in arousal can, by itself, cause an antagonistic rebound in the relative activities of the dipole. Moreover, the size of the arousal increment that is needed to cause rebound can be independent of the size of the test input that is driving the on-channel. When this occurs, an arousal increment that is sufficiently large to rebound any dipole will be large enough to rebound all dipoles in a field. In other words, if the mismatch is "wrong" enough to trigger a large arousal increment, then all the errors will be simultaneously corrected. This cannot, in principle, happen in a serial processor. Moreover, the size of the rebound is an increasing function of the size of the on-cell test input. Thus the amount of antagonistic rebound is precisely matched to the amount of on-cell activation that is to be inhibited. Finally, in previously inactive dipoles, no rebound occurs but the arousal increment can sensitize the dipole to future signals by changing by equal amounts the gain, or temporal averaging rate, of the on-cell and off-cell. In summary, the on-cell off-cell dipole is superbly designed to selectively reset $\mathcal{F}^{(2)}$, and to do so in an enduring fashion because of the slow fluctuation rate of the transmitter gates.

The above mechanisms indicate how critical periods might be terminated and dynamically maintained by learned feedback expectancies. These expectancies modulate an arousal mechanism that buffers already coded populations by shutting them off so rapidly in response to erroneous STM coding that LTM recoding is impossible. In other words, the mechanism helps to stabilize the LTM code against continual erosion by environmental fluctuations as it drives the search for new codes. The theory does not deny the possibility that structural changes, like long-term receptor modification, also help to maintain network properties after they develop under dynamical guidance. The theory does suggest, however, that where network properties develop in response to external environmental events, the structural changes will be triggered by resonant activity, since resonant activity signifies that a behaviorally meaningful representation has been successfully coded.

The thought experiments from which these conclusions follow are purely abstract: one experiment describes how limitations in the types of information available to individual cells can be overcome when the cells act together in suitably designed feedback schemes; another experiment describes a solution to the noise-saturation dilemma; another experiment describes how to design a chemical transducer and how dipoles formed when such transducers compete in parallel channels can achieve antagonistic rebound. The remainder of this article will discuss aspects of the noise-saturation dilemma including classification of some pattern transformation and storage capabilities within feedforward and feedback on-center off-surround networks. This classification has become a subject in itself during the last ten years, and many perceptual phenomena can be grouped around formal properties of these networks. Grossberg [**1981b**] discusses how dipoles built up from chemical gates behave, and will apply those results towards understanding data about schedule interactions and neuropharmacology that are usually studied as separate subjects.

10. Brightness constancy and contrast. To motivate the most elementary construction that overcomes the noise-saturation dilemma, I will review some data about visual perception, notably data about brightness constancy and contrast (Cornsweet [**1970**]). I will then show that constancy and contrast phenomena are implied when the noise-saturation dilemma is solved. Since the noise-saturation dilemma confronts all cellular systems, these visual properties should have analogs in nonvisual cellular tissues. I will indicate that variations on this construction imply other visual properties, such as edge detection, spatial frequency detection, and pattern matching properties, that on the surface seem to have little to do with brightness constancy and contrast, and which also can be anticipated to occur in nonvisual cellular systems. Some of these nonneural cellular analogs are described in Grossberg [**1978b**]. Since that article was written, the article of Mimura and Murray [**1978**] has clarified the formal connection between patchiness phenomena in ecosystems and the disinhibitory bumps in nonlinear feedback networks (Ellias and Grossberg [**1975**]) which are suggested in this article to play the role of medium bandwidth visual channels with a filling-in capability. This survey will progress from simpler feedforward network properties towards deeper feedback network properties. Even brightness constancy percepts are based on both types of properties acting on several levels. To deeply understand the aggregate effect of these several processes, one must first understand each process with quantitative precision on its own terms.

An example of brightness constancy occurs in the following situation. A white light of prescribed intensity is shined on a gray disk and its surrounding darker gray annulus. Each point in this picture reflects a fraction of the incident light back to the observer's eyes. The points in the circle reflect a larger fraction of the incident light than the points in the annulus. The observer's task is to match the apparent brightness of the gray disk with the apparent brightness of one of a series of comparison disks, which are separately illuminated.

Now double the intensity of the white light shining on the disk and annulus. Each point in the picture now reflects twice as much light back to the observer's eyes. The points in the disk again reflect more light than the points in the annulus, because each point reflects a fixed fraction, or ratio, of the light that reaches it. These ratios are called *reflectances*, and they are a property of the paper from which the disk and the annulus are made. The observer is again asked to match the apparent brightness of the disk with the apparent brightness of a comparison disk. Although each point in the disk now reflects twice as much light, the observer chooses the same comparison disk as before. In fact, the same comparison disk is chosen even if the light illuminating the picture is varied over a surprisingly wide intensity range.

Thus the observer does not perceive the absolute amount of light that is reflected from each point. Instead, the observer perceives a quantity that is independent of the illuminating light intensity over a wide range. Such a quantity is the reflectance of the gray disk, or the relative amount of light that is reflected by each point in the disk. The observer's ability to compute this reflectance depends on the fact that the same light illuminates the disk and the annulus. If only the disk were illuminated, it would look brighter as the light intensity increased. The observer's ability to estimate reflectances must therefore be based on a comparison of the relative amounts of light reflected from the disk and the annulus, since this relative quantity does not change when the picture is illuminated at successively higher levels of illumination. Brightness constancy refers to the observer's perception that the brightness does not change as the illumination of the picture increases.

Brightness contrast illustrates the same principle as brightness constancy, but in a slightly different experimental paradigm. Brightness constancy teaches us that an observer often overlooks the absolute level of illumination and computes instead a collection of ratios, or reflectances, at each point in a picture. If we interpret this idea literally, and in the simplest way, we might say that the observer computes a collection of ratios θ_i from each point v_i in the picture, and that the sum of these ratios, namely $\Sigma_{k=1}^{n} \theta_k = 1$, is the same no matter what the light intensity is. Another way to say this is as follows: the *total* brightness of a scene tends to be independent of the light intensity. The total brightness tends to be "conserved", or is "invariant" under changes in light intensity. The conservation cannot, however, be absolute or we could not tell the difference between very dark or very bright pictures.

This observation suggests why brightness constancy and brightness contrast are related. To observe brightness contrast, prepare two pictures in which identical gray disks are surrounded by annuli of equal size but different brightness. If the total brightness tends to be conserved, then the disk that is surrounded by a lighter annulus should look darker whereas the disk that is surrounded by a darker annulus should look lighter. This is brightness contrast.

11. Competition in feedforward mass action systems: reflectances, Weber law, shift property. The noise-saturation dilemma points out that cellular systems can easily become insensitive to patterned inputs both at low and high background input intensities. The simplest mass action competitive network that solves this problem is the following one (Grossberg [**1970**]). Let a spatial pattern $I_i = \theta_i I$ of inputs be processed by the cells v_i, $i = 1, 2, \ldots, n$. Each θ_i is the constant relative size, or reflectance, of its input I_i and I is the variable total input size. In other words, $I = \sum_{k=1}^n I_k$ because $\sum_{k=1}^n \theta_k = 1$. How can each cell v_i maintain its sensitivity to θ_i when I is parametrically increased? How is saturation avoided?

To compute $\theta_i = I_i(\sum_{k=1}^n I_k)^{-1}$, each cell v_i must have information about all the inputs I_k, $k = 1, 2, \ldots, n$. Moreover, since $\theta_i = I_i(I_i + \sum_{k \neq i} I_k)^{-1}$, increasing I_i increases θ_i whereas increasing any I_k, $k \neq i$, decreases θ_i. When this observation is translated into an anatomy for delivering feedforward inputs to the cells v_i, it suggests that I_i excites v_i and that all I_k, $k \neq i$, inhibit v_i. This rule represents the simplest feedforward on-center off-surround anatomy (Figure 14).

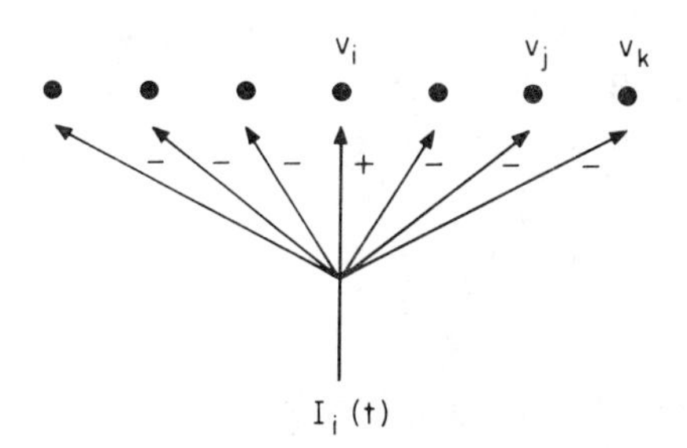

FIGURE 14. A feedforward on-center off-surround anatomy in which each input I_i excites its population v_i and inhibits other populations v_k, $k \neq i$.

How does the on-center off-surround anatomy activate and inhibit the cells v_i via mass action? Let each v_i possess B excitable sites of which $x_i(t)$ are excited and $B - x_i(t)$ are unexcited at each time t. Then at v_i, I_i excites $B - x_i$ unexcited sites by mass action, and the total inhibitory input $\sum_{k \neq i} I_k$ inhibits x_i excited sites by mass action. Moreover excitation x_i can spontaneously decay at a fixed rate A, so that the cell can return to an equilibrium point (arbitrarily set equal to 0) after all inputs cease. These rules say that

$$\frac{d}{dt}x_i = -Ax_i + (B - x_i)I_i - x_i \sum_{k \neq i} I_k. \tag{1}$$

If a fixed spatial pattern $I_i = \theta_i I$ is presented and the background input I is held constant for awhile, each x_i approaches an equilibrium value. This value is easily found by setting $dx_i/dt = 0$ in (1). It is

$$x_i = \theta_i BI/(A + I). \tag{2}$$

Note that the relative activity $X_i = x_i(\Sigma_{k=1}^n x_k)^{-1}$ equals θ_i no matter how large I is chosen; there is no saturation. This is due to automatic gain control by the inhibitory inputs. In other words, $\Sigma_{k \neq i} I_k$ multiplies x_i in (1). The total gain in (1) is found by writing

$$dx_i/dt = -(A + I)x_i + BI_i. \tag{3}$$

The gain is the coefficient of x_i, namely $-(A + I)$, since if $x_i(0) = 0$,

$$x_i(t) = \theta_i \frac{BI}{A + I}(1 - e^{-(A+I)t}). \tag{4}$$

Both the steady state and the gain of x_i depend on the input strengths. This is characteristic of mass action, or shunting, networks but not of additive networks, such as the Hartline-Ratliff equations (Ratliff [**1965**]), or of additive networks in a different coordinate system (Wilson and Cowan [**1972**]), or of discrete Laplacian approximations to network dynamics. These alternative models cannot retune themselves in response to parametric shifts of background intensity.

The simple law (2) combines two types of information: information about pattern θ_i, or "reflectances", and information about background activity, or "luminance". In a visual setting, the tendency towards reflectance processing helps to explain brightness constancy, and the rule $I(A + I)^{-1}$ helps to explain the Weber-Fechner law. In Cornsweet's [**1970**] interesting book, rules of the form $J(A + I)^{-1}$ are discussed, but they do not include pattern processing. This is perhaps why Cornsweet uses logarithms to explain reflectance processing, despite the fact that logarithmic singularities at small and large arguments have no physical significance.

Another property of (2) is that the total activity

$$x = \sum_{k=1}^{n} x_k = \frac{BI}{A + I} \tag{5}$$

is independent of the number of active cells. This *normalization* rule is a conservation law which says that, in a network that receives a fixed total luminance, making one part of the field brighter tends to make another part of the field darker. This property helps to explain brightness contrast. Brightness constancy and contrast are two sides of a coin: on one side is Weber-law modulated reflectance processing and on the other side is a normalization rule.

Equation (2) can be written in another form that expresses a different physical intuition. If we plot the intensity of an on-center input in logarithmic coordinates K_i, then $K_i = \ln I_i$ and $I_i = \exp(K_i)$. Also write the total off-surround input as $J_i = \sum_{k \neq i} I_k$. Then (2) can be written in logarithmic coordinates as

$$x_i(K_i, J_i) = Be^{K_i}/\left(A + e^{K_i} + J_i\right). \tag{6}$$

How does the response x_i at v_i change if we parametrically change the off-surround input J_i? The answer is that x_i's entire response curve to K_i is shifted, and thus its dynamic range is not compressed. A shift occurs in bipolar cells of the Necturus retina (Werblin [**1971**]) and in a modified form in Iverson and Pavel's [**1981**] psychoacoustic data. The shift property says that

$$x_i\left(K_i + S, J_i^{(1)}\right) = x_i\left(K_i, J_i^{(2)}\right) \tag{7}$$

for all $K_i \geqslant 0$, where the amount of shift S caused by changing the total off-surround input from $J_i^{(1)}$ to $J_i^{(2)}$ is predicted to be

$$S = \ln\left(\frac{A + J_i^{(1)}}{A + J_i^{(2)}}\right). \tag{8}$$

In particular, linear increments in the off-surround, such as $J_i, 2J_i, 3J_i, \ldots,$ cause progressively smaller shifts in the response curve of the on-center cell v_i. Although logarithmic coordinates are used to plot (6), the shift property is not caused by logarithmic processing and certainly not by additive processing. The shift property is due to shunting interactions. Werbin [**1974**] himself does not seem to have realized this, because he writes: "that surround antagonism in the bipolar is a subtractive phenomenon" (p. 78). He chose this possibility instead of the possibility that "the surround acts to *attenuate* the signal reaching the bipolar cell by a constant multiplicative factor." This latter notion is a form of gain control, but it is not the form that occurs in mass action networks. Thus Werblin chose the lesser of two evils between two nonexhaustive alternatives, despite the fact that his data are incompatible with a subtractive mechanism.

The ease with which we found equations (1) and (2) is due to a deceptively simple fact. We considered the processing of patterns by cellular systems in a continuously fluctuating environment. By framing our questions on a conceptual level which has evolutionary and behavioral significance, the equations flowed out easily. If we considered single cells rather than patterns, or discrete rather than continuous inputs, or systems with an infinite number of sites rather than cells, we would have failed.

12. Membrane equations and symmetry-breaking. Equation (1) is a special case of a law that occurs *in vivo*; namely, the membrane equation on which all cellular neurophysiology is based. In particular, the membrane equation is the voltage equation that appears in the Hodgkin-Huxley equations, which Carpenter analyses in her chapter.

The membrane equation describes the voltage $V(t)$ of a cell by the law

$$C\partial V/\partial t = (V^+ - V)g^+ + (V^- - V)g^- + (V^p - V)g^p. \tag{9}$$

In (9), C is a capacitance; V^+, V^-, and V^p are constant excitatory, inhibitory, and passive saturation points, respectively; and g^+, g^-, and g^p are excitatory, inhibitory, and passive conductances, respectively. We will scale V^+ and V^- so that $V^+ > V^-$. Then *in vivo*, $V^+ \geqslant V(t) \geqslant V^-$ and $V^+ > V^p \geqslant V^-$. Often V^+ represents the saturation point of a Na^+ channel and V^- represents the saturation point of a K^+ channel. There is also symmetry-breaking in (9) because $V^+ - V^p$ is usually much larger than $V^p - V^-$. We will see how this symmetry-breaking operation, which is usually mentioned in the experimental literature without comment, achieves the noise suppression property. Symmetry-breaking is a single cell property which becomes more meaningful when it is considered from the viewpoint of pattern processing.

First we must see why (1) is a special case of (9). Suppose that (9) holds at each cell v_i. Then at v_i, $V = x_i$. Set $C = 1$ (rescale time), $V^+ = B$, $V^- = V^p = 0$, $g^+ = I_i$, $g^- = \sum_{k \neq i} I_k$, and $g^p = A$. These identifications show that (1) is special in several ways. First, the hypothesis that $V^- = V^p$ does not always hold. It means that an inhibitory input can shut off excitatory activity, but cannot hyperpolarize the cell. Often $V^p > V^-$, and then the inequalities $V^+ - V^p \gg V^p - V^- > 0$ achieve noise suppression. The anatomy in Figure 14 is also special. Let us first see how the symmetry-breaking inequalities $V^+ - V^p \gg V^p - V^- > 0$ achieve noise suppression.

13. Adaptation level and noise suppression. Replace equation (1) by

$$\frac{d}{dt} x_i = -Ax_i + (B - x_i)I_i - (x_i + C) \sum_{k \neq i} I_k, \tag{10}$$

where $-C \leqslant 0$. The constant $-C$ plays the role of V^- in (9). In (10), x_i can fluctuate between B and $-C$ since $B - x_i = 0$ when $x_i = B$ and $x_i + C = 0$ when $x_i = -C$. The passive saturation point V^p is chosen equal to 0 without loss of generality. In response to a sustained spatial pattern $I_i = \theta_i I$, the equilibrium response of (10) is

$$x_i = \frac{(B + C)I}{A + I}\left[\theta_i - \frac{C}{B + C}\right]. \tag{11}$$

Again a Weber law $(B + C)I(A + I)^{-1}$ modulates the processing of reflectances θ_i. Again the relative activity $X_i = x_i(\sum_{k=1}^n x_k)^{-1}$ is independent of total activity I. The new term $C(B + C)^{-1}$ is called an *adaptation level* because $x_i > 0$ if $\theta_i > C(B + C)^{-1}$ whereas $x_i \leqslant 0$ if $\theta_i \leqslant C(B + C)^{-1}$. In other words, only reflectances which exceed the adaptation level can elicit excitatory cellular responses.

How does symmetry-breaking influence the size of the adaptation level? Equality $V^+ - V^p \gg V^p - V^-$ becomes $B \gg C$. Thus the adaptation level $C(B + C)^{-1}$ is a number much smaller than 1. The most perfect choice of B and C is one in which there exists a match between intracellular and intercellular symmetry-breaking operations. By this I mean the following. Just as there is symmetry-breaking between the excitatory and inhibitory saturation points B

and C, there is also symmetry-breaking between the distribution of excitatory and inhibitory intercellular connections; viz., the narrow on-center and broad off-surround. Suppose that the ratio of C to B equals the ratio of on-center to off-surround connections, namely $1/(n-1)$. Then $C(B+C)^{-1} = 1/n$.

In this situation, suppose that the input pattern $I_i = \theta_i I$ is uniform. Then all $\theta_i = 1/n$. By (11), all $x_i = 0$ no matter how intense the total input I is. This is the simplest version of noise suppression: the network computes how well reflectances deviate from the noise level, or adaptation level, and modulates this comparison with a Weber law. Increasing CB^{-1} increases the adaptation level and thereby makes the criterion for excitatory cellular response more stringent. This is a type of contrast control that is due entirely to feedforward competitive interactions. Analogous effects occur in feedback networks, and have been predicted to control the size of such visual illusions as angle expansion (Levine and Grossberg [**1976**]).

14. Pattern matching. The noise suppression property implies a pattern matching property that is one substrate of adaptive resonance. Suppose that two input patterns $\hat{J} = (J_1, J_2, \ldots, J_n)$ and $\hat{K} = (K_1, K_2, \ldots, K_n)$ add before they are coupled into the shunting dynamics. The total input pattern $\hat{I} = (I_1, I_2, \ldots, I_n)$ then has components $I_i = J_i + K_i$, $i = 1, 2, \ldots, n$. If the two patterns are maximally mismatched, then the peaks of $\hat{J}$ fill in the troughs of $\hat{K}$, and conversely. The total pattern $\hat{I}$ is approximately uniform, and is therefore quenched by (11). If the two patterns match, then they have the same reflectances $\hat{\theta} = (\theta_1, \theta_2, \ldots, \theta_n)$. Consequently, $J_i = \theta_i J$ and $K_i = \theta_i K$, so the equilibrium response in (11) is

$$x_i = \frac{(B+C)(I+J)}{A+I+J}\left[\theta_i - \frac{C}{B+C}\right]. \tag{12}$$

Note that x_i in (12) exceeds x_i in (11) due to the added activity J in the second pattern, although both equilibrium responses compute the same reflectances and the same adaptation level. Thus a mismatch attenuates activity whereas a match amplifies activity due to an interaction between reflectance processing and Weber law modulation. Freeman's [**1980**] data on olfactory template matching also suggests such an interaction.

Any theory in which pattern matching is not related to activity scaling will run into severe difficulties when it tries to explain how multiple spatial scales interact with a data pattern to select the scales that provide a globally self-consistent interpretation of the data. Many Artificial Intelligence models of visual processing have this deficiency.

15. Power law invariance. As our classification continues, it will become increasingly important to ask: what types of parametric changes or input transformations leave important properties invariant? Which ones change these properties, and how? In (10), for example, suppose that the inputs I_i are the outputs $I_i = f(J_i)$ of a prior processing stage whose inputs equal J_i. What transformations $f(w)$ allow (10) to process the same reflectances at all total input levels? Power laws $f(w) = Dw^m$ have this property, since if $J_i = \phi_i J$, then (11) holds with $\theta_i = \phi_i^m(\Sigma_{k=1}^n \phi_k^m)^{-1}$ and $I = DJ^m\Sigma_{k=1}^n \phi_k^m$. Input laws which differ from a power law, such as sigmoid laws, can cause a shift in reflectance processing as J increases. Correspondingly, power law approximations to sigmoid processing require different powers at different activity levels, as occurs in Iverson and Pavel's [**1981**] model of psychoacoustic matching data. Are there other causes of such reflectance shifts? The answer is "yes": a different anatomy, notably distance-dependent interactions, can cause such shifts even if the input law is a power law. The next section indicates why this is so.

16. Edge detection and chromatic shifts. Equation (10) is a special case of equation (9) because each input excites only one population and inhibits all other populations with equal strength. In general, both excitatory and inhibitory influences can depend on the distances between the cells v_i. Then the above properties hold in a modified form and new properties also emerge. These modified properties can be ascribed to the anatomy once we clearly understand how they differ from the properties of equation (10).

Let (10) be generalized to

$$\frac{dx_i}{dt} = -Ax_i + (B - x_i) \sum_{k=1}^{n} I_k D_{ki} - (x_i + C) \sum_{k=1}^{n} I_k E_{ki} \tag{13}$$

where the excitatory interaction coefficients D_{ki} and the inhibitory interaction coefficients E_{ki} can both depend on the distance between v_k and v_i. The excitatory coefficients D_{ki} typically decrease as a function of the distance between v_k and v_i faster than do the inhibitory coefficients E_{ki}. The equilibrium activity of (13) in response to spatial pattern $I_i = \theta_i I$ is

$$x_i = IF_i/(A + IG_i) \tag{14}$$

where

$$F_i = \sum_{k=1}^{n} \theta_k(BD_{ki} - CE_{ki}) \tag{15}$$

and

$$G_i = \sum_{k=1}^{n} \theta_k(D_{ki} + E_{ki}). \tag{16}$$

Equation (14) is an averaging rule similar to that mentioned by Luce and Narens [**1981**] when $I_i = f(J_i)$.

Noise suppression occurs in system (13) if

$$B \sum_{k=1}^{n} D_{ki} \leqslant C \sum_{k=1}^{n} E_{ki}, \qquad i = 1, 2, \ldots, n, \tag{17}$$

since then all $x_i \leqslant 0$ if all $\theta_i = 1/n$. Inequalities (17) generalize the concept of matched symmetry-breaking to the distance-dependent case. How a network might be clever enough to grow inequalities like (17) during its development is discussed in Grossberg [**1978a,** §35]. How a network might dynamically tune its inequalities by using a pattern-contingent modulation of its shunting arousal level is discussed in Grossberg [**1978a**, §59]. The concept of the *quenching threshold* of a feedback competitive network, which is reviewed in §17 below, is relevant to this analysis.

When interaction coefficients D_{ki} and E_{ki} decrease with distance, a given cell v_i need not be influenced by certain inputs I_k, $k \neq i$. Such a cell's activity x_i will be suppressed whenever the pattern weights θ_k which it can detect are uniformly distributed. In particular, let a rectangular input perturb such a network and let the length of the rectangle exceed the bandwidth of the coefficients D_{ki} and E_{ki} (Figure 7b). Then cells near the middle of the rectangle will detect a uniform field; hence their activities will be suppressed no matter how intense the input is. Cells well outside the rectangle's influence will also detect a uniform field; their activities will also be suppressed. Only cells near the edges of the rectangle can respond. These cells appear to be edge detectors, but they possess no special anatomy to filter edges per se. They react vigorously to any pattern whose spatial variation appears nonuniform to their spatial interaction coefficients. Thus, the noise suppression property in a network whose interaction coefficients have a restricted spatial bandwidth implies an edge enhancement property. This is a type of "medium bandwidth channel," to use Graham's terminology. When this simple property acts in feedback networks, it provides a basis for spatial frequency detection, active filling-in processes, and binocular resonance, rivalry, and hysteresis, as the following sections will suggest.

The equilibrium response in (14) does not in general have the property that the ratios $X_i = x_i(\Sigma_{k=1}^n x_k)^{-1}$ are independent of the total input I. Unless G_i is independent of i, as when $D_{ki} + E_{ki}$ is independent of i in (10), increments in total activity I can alter the responses X_i to the reflectances θ_i. This deformation is due to the finite interaction range of the anatomy, since it does not occur in (11). A complex interaction between the reflectance pattern $(\theta_1, \theta_2, \ldots, \theta_n)$, the interaction coefficients D_{ki} and E_{ki}, and the total input I controls the change in X_i, much as in chromatic shifts at high luminances (Cornsweet [**1970**]). In particular,

$$\frac{\partial X_i}{\partial I} = H_i \sum_k \frac{F_k(G_k - G_i)}{(A + IG_k)^2} \tag{18}$$

where

$$H_i = \frac{AF_i}{[\Sigma_k F_k(A + IG_i)/(A + IG_k)]^2}. \tag{19}$$

The interpretation of (18) is simplified if we restrict attention to ratios $X_i = x_i(\Sigma_{k \in L} x_k)^{-1}$, $i \in L$, where L ranges over all the cells whose activities x_i are positive. Correspondingly, interpret all sums in (18) and (19) to range over

$k \in L$. Then all $F_i > 0$ and all $H_i > 0$, the quantities

$$P_k = \frac{F_k(A + IG_k)^{-2}}{\Sigma_m F_m(A + IG_m)^{-2}}, \tag{20}$$

are probabilities, and by (18), inequality

$$G_i > \sum_k G_k P_k \tag{21}$$

implies that X_i decreases as I increases, whereas the reverse inequality implies that X_i increases as I increases. Inequality (21) says that the shift in X_i can be expressed as a linear function of the G_k only by hiding a great deal of nonlinearity in the probabilities P_k. By (21), at the cell v_i whose G_i is maximal, X_i decreases as I increases, whereas at the cell v_i whose G_i is minimal, X_i increases as I increases. This implies a tendency for cells near the regions of highest reflectance to have decreasing relative responses as I increases. The cells can begin to saturate at high I values if the off-surround is too weak. By contrast, the relative size of edge effects at the boundary of a region can increase as I increases. Ellias and Grossberg [**1975**] describe parametric studies of this type, and also show how extra edges at the boundary of a rectangle, peak splitting in response to a localized spot, and curvature detection can occur when the interaction coefficients D_{ki} and E_{ki} are Gaussian functions of distance. The spurious edge bumps and peak splits vanish when the inhibitory coefficients E_{ki} decrease with distance slower than the excitatory coefficients D_{ki}.

17. Competitive feedback networks: short term memory and bifurcations. Activity patterns can be stored after inputs cease if positive feedback signals exist. To solve the noise-saturation dilemma, these positive feedback signals must be supplemented by competitive negative feedback signals. The simplest mass action competitive feedback network is

$$\frac{d}{dt}x_i = -Ax_i + (B - x_i)[I_i + f(x_i)] - x_i\left[J_i + \sum_{k \neq i} f(x_k)\right]. \tag{22}$$

Equation (22) is the feedback analog of equation (1). Term $(B - x_i)f(x_i)$ describes how a positive feedback signal $f(x_i)$, derived from activity x_i, excites v_i's unexcited sites $B - x_i$ by mass action. Term $-x_i\Sigma_{k \neq i} f(x_k)$ describes the switching-off of excited sites x_i by competitive feedback signals $f(x_k)$, $k \neq i$. Term J_i is the inhibitory input to v_i. In (1), $J_i = \Sigma_{k \neq i} I_k$.

System (22) confronts us with a general question and with a specific question. The general question concerns the network's ability to transform a fixed pattern of inputs and initial data $(x_1(0), x_2(0), \ldots, x_n(0))$ into a definite asymptotic pattern $(x_1(\infty), x_2(\infty), \ldots, x_n(\infty))$. In other words, can the network store the transformed pattern in STM? This is a question about the network's stability. Otherwise expressed, is the network capable of global pattern formation?

The specific question concerns the nature of this transformation and is, mathematically speaking, a question about bifurcation theory. How does the transformation depend on the choice of the signal function $f(w)$? In particular, how does the network use its positive feedback signals to store behaviorally useful patterns in STM, yet prevent positive feedback from amplifying noise and flooding the network with self-generated signals that will hamper the registration of behaviorally important signals?

The specific problem was solved in Grossberg [**1973**]. I will summarize only the main features here, first in physical terminology and then to emphasize their bifurcation properties. Suppose that inputs I_i and J_i acting before $t = 0$ establish an arbitrary initial activity pattern $(x_1(0), x_2(0), \ldots, x_n(0))$ before being shut off at $t = 0$. How does the choice of $f(w)$ control the transformation of this pattern as $t \to \infty$? This problem was originally studied by transforming (22) into a system describing the variables $X_i = x_i(\sum_{k=1}^n x_k)^{-1}$ and $x = \sum_{k=1}^n x_k$. Defining $g(w) = w^{-1}f(w)$, this system takes the form

$$\frac{d}{dt}X_i = BX_i \sum_{k=1}^{n} X_k[g(X_i x) - g(X_k x)] \tag{23}$$

and

$$\frac{d}{dt}x = x\left[-A + (B - x)\sum_{k=1}^{n} X_k g(X_k x)\right]. \tag{24}$$

By (23), a linear signal ($g(w) = C =$ constant) can remember an arbitrary pattern of reflectances ($X_i(t) =$ constant), but by (24), such a system either cannot store any pattern, or it amplifies noise (either $x(\infty) = 0$ if $B - AC^{-1} < 0$, or $x(\infty) = B - AC^{-1} > 0$ no matter how small $x(0) > 0$ is chosen). A slower-than-linear signal ($g(w)$ decreasing) amplifies noise and creates a uniform asymptotic pattern ($X_i(\infty) = 1/n$). A faster-than-linear signal ($g(w)$ increasing) suppresses noise (if $A > Bg(0)$, then given sufficiently small $x(0)$, it follows that $x(\infty) = 0$), but chooses only the populations with maximal initial activity for STM storage and suppresses all other populations. A sigmoid, or S-shaped, signal ($g(w)$ an inverted U) suppresses noise because it is faster-than-linear at small activity values, but can store a partially contrast-enhanced pattern in STM because it is approximately linear at intermediate activity values. The network possesses a *quenching threshold* (QT) if a sigmoid signal function is used: if an initial activity $x_i(0)$ is less than the quenching threshold, then $x_i(\infty) = 0$. The spatial pattern of suprathreshold initial activities is contrast-enhanced and stored. The QT defines the noise level in such a network. Freeman [**1981**] chooses his parameters so that the QT equals 2; if his Q_m parameter exceeds 2, a burst occurs, otherwise not.

A network that possesses a QT can be tuned: Any operation that increases the QT helps to choose a population for storage, or prevents storage if the QT is too high. Any operation that lowers the QT abets storage, and can flood the network with noise if the QT falls below the level of arousal or endogenous noise fluctuations. A sigmoid signal function appears in Freeman's data and modelling efforts and in Graham's probability summation rules. A sigmoid signal also suggests why more than one power function $f(w) = Cw^n$ is needed to fit psychophysical data like that of Iverson and Pavel with different powers n at different intermediate suprathreshold ranges of population activities. A sigmoid signal can be interpreted as the total output of a large number of cells, or cell sites, in a population whose output thresholds are distributed around a mean threshold value with a Gaussian or other bell-shaped distribution. Then the power law approximates the law whereby more cell sites are recruited to emit outputs as a population's mean activity increases.

In more mathematical terms, when $g(w)$ is strictly increasing, the system trajectories approach equilibrium points that lie on the coordinate axes. The equilibrium points equal 0 or solutions of the equations $g(x_i) = A(B - x_i)^{-1}$, $x_j = 0$, $i \neq j$. Typically these equations possess only finitely many solutions. When $g(w)$ is deformed to equal a constant C, then all trajectories approach either 0 or the hyperplane $\Sigma_{k=0}^n x_k = B - AC^{-1}$ of equilibrium points. Every point in this infinite set can be reached by some trajectory, since all the ratios $x_i(t)x_j^{-1}(t)$ are constant as $t \to \infty$. If $g(w)$ is deformed even further until it is strictly decreasing, then all trajectories approach either 0 or the unique positive solution of $g(x_i/n) = An(Bn - x_i)^{-1}$, $i = 1, 2, \ldots, n$. If $g(w)$ is deformed (to combine all these tendencies) into a concave function, then the trajectories can approach a continuum of equilibrium points that include, but are not restricted to, the coordinate axes.

18. Global decision-making and storage in STM. The general stability question forces a deeper mathematical investigation than was necessary to analyse the effects of simple signal functions $f(w)$ on pattern transformations. It has been shown that every competitive system induces a decision scheme that can be used to globally analyse its oscillatory or limiting behavior (Grossberg **[1978c]**, **[1978d]**, **[1980c]**). Persistent oscillations in a competitive system can often be traced to a global contradiction in its decision scheme (Grossberg **[1978c]**), as in the voting paradox. What is the largest class of systems, that includes (22), which can reach a global decision and store it in STM?

All competitive systems of the form

$$\frac{d}{dt}x_i = -A_i x_i + (B_i - x_i)[I_i + f_i(x_i)] - (x_i + C_i)\left[J_i + \sum_{k \neq i} f_k(x_k)\right] \quad (25)$$

can be rewritten in the form

$$dx_i/dt = a_i(x)[b_i(x_i) - c(x)], \quad (26)$$

where $x = (x_1, x_2, \ldots, x_n)$, by setting

$$a_i(x) = x_i, \quad (27)$$

$$b_i(x_i) = -A_i - I_i - J_i + x_i^{-1}[A_i C_i + I_i + (B_i + C_i)f_i(x_i - C_i)] \quad (28)$$

and

$$c(x) = \sum_{k=1}^{n} f_k(x_k - C_k). \quad (29)$$

Function $c(x)$ in (26) is called a state-dependent *adaptation level* by analogy with the feedforward equation (11). I have shown under mild conditions on the functions $a_i(x)$, $b_i(x_i)$ and $c(x)$ that every competitive adaptation level system (26) can reach a global decision and store it in STM (Grossberg [**1978d**]). Because the parameters in (25), or more generally (26), can be freely altered without destroying the system's ability to reach a global decision, I call the class of competitive systems which admit an adaptation level *absolutely stable*. The absolute stability property does not imply that these systems contain no instabilities. Indeed, networks (22) with $f(w) = Cw^n$, $n > 1$, can make choices because they possess unstable equilibrium points. The absolute stability property refers to a global property of the phase portraits of these systems: they always approach equilibrium points.

19. Classification of absolutely stable systems using Liapunov functionals. Once we reach this level of generality, we begin to probe a general design principle that takes on a life of its own. For example, once the concept of adaptation level systems is defined, such systems are readily identified in other sciences. They occur in Eigen's theory of macromolecular evolution (Eigen and Schuster [**1978**]). They occur in Lacker's theory of control of ovulation number in mammals (Lacker [**1980**]). They occur in Volterra-Lotka systems of population biology whose interaction coefficients are products of statistically independent factors (Grossberg [**1978d**]). These examples illustrate *descriptive* appearances of adaptation level systems. *Prescriptive* appearances are also useful, since they suggest rules whereby absolutely stable interactions can be guaranteed, by the consent of competing individuals, even if these individuals know very little about each other's behavior. For example, due to Moe Hirsch's interest in economic applications of these models, I defined a class of absolutely stable production strategies for an economic market. If all competitors produce the same product, and each competitor chooses a strategy from this class, then even without knowledge of the other competitors' choices, each competitor will realize his expected profit without disturbing the absolute stability of the market. I leave to the reader the exercise of designing such a market!

The classification theory takes on a life of its own in a mathematical sense as well. The mathematical method generalizes beyond competitive systems to all dynamical systems which admit a Liapunov functional $L^+(x_t)$ of a certain type. The Liapunov functional must be the time integral of a maximum function, as in $L^+(x_t) = \int_0^t M^+(x(v))\,dv$. The maximum function, in turn, is defined in terms of a dynamical system $dx_i/dt = a_i(x)M_i(x)$, $a_i(x) \geq 0$, $i = 1, 2, \ldots, n$, by $M^+(x) = \max_i M_i(x)$. The Liapunov property that $dL^+(x_t)/dt \geq 0$ for $t \geq T$ is derived from the property that the region $R^+ = \{x: M^+(x) \geq 0\}$ is positively invariant: if $x(T) \in R^+$ then $x(t) \in R^+$ for all $t \geq T$. Within R^+, decisions can occur on the hypersurfaces $J_{ij} = \{x \in R^+: M^+(x) = M_i(x) = M_j(x)\}$, $i, j = 1, 2, \ldots, n$. Such a Liapunov functional exists, for example, in all population models whose total population $x(t) = \sum_{k=1}^n x_k(t)$ is conserved, or increases monotonically through time, whether or not these models are competitive. The following basic problem can be stated for these systems.

Global consensus problem. What is the largest class of absolutely stable dynamical systems which includes the competitive adaptation level systems? The larger the class of absolutely stable systems that is at our disposal, the richer will be the class of adaptive resonances (cognitive codes) that we can design.

20. Masking, developmental or attentional biases, and gain control. The classification of competitive adaptation level systems includes many physically interesting special cases, which can also be mathematically interpreted as bifurcation phenomena. One comparison shows how different anatomical distributions of shunting signals can control a masking mechanism or a gain control mechanism, but not both. Another comparison shows why a linear feedback signal function produces physically implausible masking properties, whereas a sigmoid feedback signal produces physically useful properties. Yet another comparison shows how changing the anatomical distribution of inhibitory feedback interactions from long-range interactions to distance-dependent interactions converts a fast masking reaction into a slow normative drift that helps to explain certain visual illusions and pattern completion effects.

Masking occurs in systems

$$\frac{d}{dt}x_i = -Ax_i + (B_i - x_i)f(x_i) - x_i \sum_{k \neq i} f(x_k) \tag{30}$$

in which not all population v_i have the same number B_i of excitable sites, or equivalently in systems

$$\frac{d}{dt}x_i = -Ax_i + (B - x_i)f(C_i x_i) - x_i \sum_{k \neq i} f(C_k x_k) \tag{31}$$

whose population activities or signals are differentially amplified by shunting factors C_i. System (31) can be formally transformed into system (30) by a change of variables. Nonetheless, the physical interpretations of these systems can differ significantly. In (30), the differential numbers of sites can be interpreted as a development bias; for example, more feature detectors might code vertical and horizontal lines than oblique lines. In (31), the differential amplification of population signals can be interpreted as an attentional shunt that gates all the feedback interneurons, both excitatory and inhibitory, of each population v_i by its own shunting parameter C_i. Where both developmental and attentional biases occur, as in

$$\frac{d}{dt}x_i = -Ax_i + (B_i - x_i)f(C_i x_i) - x_i \sum_{k \neq i} f(C_k x_k), \tag{32}$$

masking is controlled by the relative sizes of the products $B_1C_1, B_2C_2, \ldots, B_nC_n$. For definiteness, order these products such that $B_1C_1 > B_2C_2 \geq \cdots \geq B_nC_n$. Let the signal function be linear, as in $f(w) = Ew$. In this situation, a masking phenomenon occurs such that $x_i(\infty) = 0$ for all i such that $B_iC_i < B_1C_1$, whereas $x_i(\infty)x_j^{-1}(\infty) = x_i(0)x_j^{-1}(0)$ for all i and j such that $B_iC_i = B_jC_j = B_1C_1$. In other words, the activity pattern across the subfield of populations v_i with maximal parameters $B_iC_i = B_1C_1$ is faithfully stored, but all other population activities are masked. This type of masking is unphysical, because the salience of a feature in an external display, as reflected by a large $x_i(0)$ value, cannot overcome internal biases $B_iC_i < B_1C_1$ even if $x_i(0) \gg x_1(0)$. A more plausible form of masking occurs if $f(w)$ is a sigmoid signal function. Then a subfield of cells with a common B_iC_i value has its activity pattern contrast enhanced before being stored in STM. Which subfield is stored depends on the initial values $x_i(0)$. Relatively large initial values $x_i(0)$ can successfully compete with larger parameters B_1C_1 if $x_1(0)$ is sufficiently small. A tug-of-war occurs between cue salience, developmental biases, and attentional shunts to determine which subfield will be stored in STM (Grossberg and Levine [**1975**]). Once again, a linear signal proves unequal to a task which a sigmoid signal can accomplish.

Another adaptation level system in which shunts occur is the system

$$\frac{d}{dt}x_i = -Ax_i + (B - x_i)Cf(x_i) - Dx_i \sum_{k \neq i} f(x_k). \tag{33}$$

By contrast with (31), the shunts in (33) influence either *all* the excitatory interneurons of all the populations, via C, or *all* the inhibitory interneurons, via D, rather tha the excitatory and inhibitory interneurons of each population taken separately. In such a system, gain control can truly be said to occur, because varying the ratio CD^{-1} above or below the value 1 can alter the contrast with which the initial pattern is transformed before it is stored. For example, if $f(w)$ is a linear signal $f(w) = Ew$, then the network chooses the population with maximal initial activity for storage if $C < D$. If $C > D$ the network transforms any initial pattern into a uniform pattern before storing it. If $C = D$, the stored pattern has the same reflectances as the initial pattern. A similar tendency towards higher contrast as CD^{-1} increases holds if $f(w)$ is a nonlinear function (Ellias and Grossberg [**1975**]). This example illustrates how periodic nonspecific shunting of the network's interneurons can define time intervals during which the network enhances patterns before storing them, interspersed with time intervals during which the network resets itself by smoothing out and attenuating previous activity patterns.

The anatomies subserving shunting laws for masking by developmental or attentional biases vs. shunting laws for opening and closing STM storage gates by gain control are wholly distinct. This is the type of distinction that requires a mathematical analysis to be clearly understood.

The above examples all suppose that the feedback inhibitory interactions are long-range. Basic problems remain open concerning systems which do not possess an adaptation level, notably feedback networks with narrow on-centers and distance-dependent off-surrounds, such as the systems

$$\frac{d}{dt}x_i = -Ax_i + (B_i - x_i)[I_i + f(x_i)] - (x_i + C_i)\left[J_i + \sum_{k=1}^{n} f(x_k)D_{ki}\right] \quad (34)$$

where $D_{ki} = D(|k - i|)$. Are all mass action distance-dependent competitive systems absolutely stable? This possibility presently stands as a conjecture.

21. Normative drifts, pattern completion, behavioral contrast in discrimination learning, temporal order information in free recall. An interesting aspect of the classification theory is to understand how network properties change when long-range inhibitory interactions, as in (10), are replaced by distance-dependent inhibitory interactions, as in (13). Edge detection and pattern shifts at high input levels are two consequences of this change in feedforward networks. Much more subtle effects can occur in feedback networks, and lead to a mechanistic comparison of data that seem on the surface to be unrelated.

In the long-range networks

$$\frac{dx_i}{dt} = -Ax_i + (B_i - x_i)[I_i + f(C_i x_i)] - x_i\left[J_i + \sum_{k \neq i} f(C_k x_k)\right], \quad (35)$$

populations with larger numbers of excitable sites B_i (due to development), or larger signal amplifications C_i (due to attentional tuning) can mask the activities of other populatons. The masking process can be intuitively thought of as very rapid sucking of the normalized total activity into the populations with maximal parameters.

In the distance-dependent networks

$$\frac{dx_i}{dt} = -Ax_i + (B_i - x_i)\left[I_i + \sum_{k=1}^{n} f(x_k)C_{ki}\right] - x_i\left[J_i + \sum_{k=1}^{n} f(x_k)D_{ki}\right], \quad (36)$$

the masking process is weakened, and the process of sucking activity into populations with maximal parameters can be very slow, leading to a slow drift of the locus of maximal activity towards the populations with maximal parameters. The slow drift is thus a masking phenomenon due to distance-dependent inhibitory feedback. This mechanism has been suggested to explain the visual illusion of line neutralization (Levine and Grossberg [**1976**]) as well as other normative drifts. For example, ambiguous data can activate local codes whose activity spontaneously drifts to the most global code in their vicinity, whereupon the global code reads out its feedback template thereby leading to pattern completion of the data base (Grossberg [**1978a**, §§40–42]).

The normalization property in a long-range network can help to explain behavioral contrast in discrimination learning experiments (Bloomfield [**1969**], Grossberg [**1975**]). In the distance-dependent case, the normalization property is weakened since not all cells inhibit each other. When this weakened normalization property is joined to a shunting law, one can derive a class of STM codes for temporal order information in free recall experiments (Grossberg [**1978e**]). These codes explain free recall data using parallel mechanisms rather than a serial STM buffer (cf. Vorberg [**1981**]), and predict a primacy effect in STM that controls the correct free recall of short lists that have not yet been stored in LTM. The primacy effect in STM is a behavioral contrast effect, but one that develops in time, as in the case of normative drifts, rather than across space, as in the case of masking and behavioral contrast during discrimination learning. It is suggested that a primacy effect in STM is not easily measured using interference experiments because it is masked by normalization due to competition from the STM recency effect on the noninterference trials. One way to test for the existence of a primacy effect in STM is to do interference experiments in which list length is parametrically varied.

22. Nonspecific gain changes, travelling waves and the Hodgkin-Huxley analogy. Theorems concerning the existence of limiting STM patterns as $t \to \infty$ tell us that a network does not contain internal contradictions which prevent it from ever reaching a global decision. Of course no decision is stored *in vivo* for an infinite amount of time. The limit theorems are based on two hypotheses that do not always hold *in vivo*: all the feedback signals are produced instantaneously, and the sensitivity of the system to inputs and its own feedback signals remains constant through time. The latter property can fail if the QT oscillates periodically due to nonspecific arousal bursts from an endogenous oscillator, or if the QT is momentarily reset by arousal bursts that are triggered by inputs. These arousal bursts can feed into a gain control mechanism, as in (33). Such a QT modulation seems to occur in the olfactory system, since the olfactory bulb reacts in phase with the breath cycle that delivers new scents to the olfactory mucosa. When network sensitivity is periodically reset, a limit theorem's validity is restricted to the time intervals when the QT is low enough to permit STM storage. The existence of a limiting pattern during a brief time interval is just as important as the existence of a limiting pattern during a long time interval.

Feedback signalling in a system such as

$$\frac{dx_i}{dt} = -Ax_i + (B - x_i)\left[I_i + \sum_{k=1}^{n} f(x_k)D_{ki}\right] - (x_i + C)\left[J_i + \sum_{k=1}^{n} g(x_k)E_{ki}\right] \tag{37}$$

occurs instantaneously. *In vivo* signals are never propagated instantaneously. If the signals act very rapidly, then we can anticipate that a limit theorem will again obtain, although the limiting patterns might be modified. For example, a system such as

$$\frac{d}{dt}x_i = -Ax_i + (B - x_i)\left[I_i + \sum_{k=1}^{n} F(w_k)D_{ki}\right] - (x_i + C)\left[J_i + \sum_{k=1}^{n} G(y_k)E_{ki}\right], \tag{38}$$

$$\frac{d}{dt}w_i = \mu H(x_i)[-w_i + J(x_i)], \tag{39}$$

and

$$\frac{d}{dt}y_i = \nu K(x_i)[-y_i + L(x_i)] \tag{40}$$

will behave increasingly like (37) with $f(x_i) = F(J(x_i))$ and $g(x_i) = G(L(x_i))$ as $\mu \to \infty$ and $\nu \to \infty$. The functions w_i and y_i are the activities of excitatory and inhibitory feedback interneurons, respectively. Often excitatory interneurons react more rapidly than inhibitory interneurons. Then system (38)–(40) can be approximated by

$$\frac{d}{dt}x_i = -Ax_i + (B - x_i)\left[I_i + \sum_{k=1}^{n} f(x_k)D_{ki}\right] - (x_i + C)\left[J_i + \sum_{k=1}^{n} G(y_k)E_{ki}\right] \tag{41}$$

and

$$dy_i/dt = \nu K(x_i)[-y_i + L(x_i)]. \tag{42}$$

Replacing (38) and (39) by (41) is a singular perturbation approximation, since it assumes that w_i reacts infinitely quickly ($\mu \to \infty$) to fluctuations in x_i.

A remarkable analogy exists between system (38)–(40) and Carpenter's version of the Hodgkin-Huxley equations. The n potentials $(x_1(t), x_2(t), \ldots, x_n(t))$ in the network are analogous to a spatially discrete version of the partial differential equation for the Hodgkin-Huxley potential $V(u, t)$. Equation (39) for the fast $w_i(t)$ variable is analogous to the Hodgkin-Huxley equation for the fast $m(t)$ variable in the Na^+ conductance. Replacing $w_k(t)$ by $J(x_k)$ in (41) is formally identical to replacing $m(t)$ by $m_\infty(V)$ in the membrane equation for V. In fact, both $J(x_k)$ and $m_\infty(V)$ are typically sigmoid functions of their variables.

This striking analogy leads to a natural question. When the y_i reaction is slow (ν is small), do travelling waves occur in the network, as they do in the Hodgkin-Huxley equation when $n(t)$ is slow? Ellias and Grossberg [**1975**] demonstrated the existence of such travelling waves in a one dimensional ring of cells. These waves were found before Carpenter's papers appeared, and makes the formal analogy between cell membranes and network dynamics seem more remarkable to me.

Ellias and Grossberg [**1975**] also studied how these travelling waves can be prevented. In an axon, travelling waves are desirable because they carry spikes reliably from one end of the axon to the other. In a network that is trying to store an internal representation in STM, travelling waves are disastrous, for much the same reason that a grand mal epileptic seizure is disastrous. Travelling waves represent a functional instability which cannot store a pattern in STM. Two rules arose from this work that should be clarified by more mathematical analysis. Either decreasing the inhibitory-to-excitatory gain or decreasing the inhibitory-to-excitatory spatial bandwidth can destabilize network dynamics by first causing a series of disinhibitory bumps in the activity pattern before the pattern breaks up into travelling waves.

23. The standing wave alternative: perturb the fast network or template match a slow adaptation level. The search for a general scheme to achieve limiting patterns or standing waves within the active time windows of a neural tissue has led me to conjecture the following alternative, which also needs more experimental and mathematical study. One way to achieve standing waves is first to prove a limit theorem in networks with instantaneous feedback signals. Then perturb off the fast network, but keep rates of feedback signalling fast enough to ensure that any oscillations that appear approximate the limiting pattern for the duration of the network's active phase (low QT). This option is called: perturbing off an absolutely stable fast network.

A complementary option can hold even when the inhibitory averaging rate is slow (Grossberg [**1978f**]). One might desire a slow inhibitory gain to achieve a slowly varying adaptation level against which excitatory fluctuations can be evaluated. How can one marry the two requirements of slow adaptation and STM storage? In particular, how are travelling waves prevented? Template matching suggests a solution.

Suppose that an input template is kept on. Let the pattern of feedback signals caused by the inputs begin to mismatch the input template as a travelling wave starts to form. Then the total pattern of input template plus feedback signals becomes more uniform. The network's activity is attenuated by the noise suppression property, and the feedback signals are eliminated. The input template can then reinstate itself in the activity pattern, and the cycle starts again. A standing wave is hereby achieved by periodically suppressing any travelling wave that might start to develop using the input pattern as a template.

These two options suggest a fundamental alternative for the control of standing waves in neural tissues, or for that matter in all cellular tissues. Is the standing wave held together by the absolute stability of its fast network, or is it held together by template matching? One way to test this alternative empirically is to cut the template feedback pathways and test if the wave begins to travel, or to slow down the inhibitory interneurons and test if the wave begins to travel. A deeper and still more fascinating question is this: Where two networks act as templates for each other, do their feedback signal bursts get entrained well enough to attenuate the other network's travelling waves without suppressing all resonant activity?

The standing wave alternative emphasizes the importance of stability questions in understanding perceptual dynamics. The fast manifold option is based on finding an absolute stability theorem. The template matching option arises because templates are needed to stabilize a developing code against erosion by later coding possibilities. Once the template is there, it can also help to preserve the form of the coded pattern during each time frame.

24. Functional vs. structural length scales in feedback networks. Template matching processes in shunting networks shed light on some issues concerning binocular vision, notably how global correspondences between object sizes, disparities, and velocities can be computed, how active filling in processes can occur, and how object permanence is achieved, rather than series of discrete snapshots through time of a moving object. Dev's [**1975**] interesting network model for computing disparities does not contain these properties, nor do later versions of the Dev model (Marr, Palm and Poggio [**1978**]).

An instructive starting point is to contrast the reaction of a feedforward competitive network to a rectangular input with the reaction of a feedback competitive network to a rectangular input. Both types of networks are assumed to possess the noise suppression proerty. As discussed in §16, the feedforward network enhances the edges of the rectangle and suppresses its interior. This is not generally true in feedback networks. If a feedback network's spatial bandwidth is smaller than the rectangle width, then the feedback network enhances edges, but it can also generate a series of regular disinhibitory bumps throughout the extent of its excited cells (Figure 15).

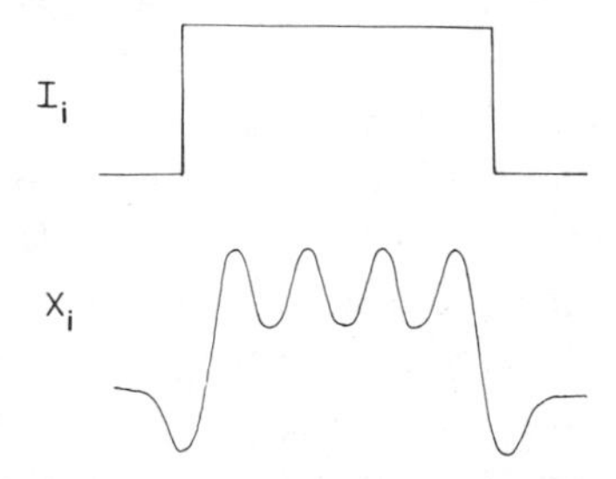

FIGURE 15. A feedback network can respond to a rectangular input with a regular series of disinhibitory bumps that fills the input-perturbed region with spatially patterned activity.

At least two distinct spatial scales must be distinguished in a feedback network: a *structural* scale which describes how rapidly its feedback interaction coefficients decrease as a function of distance, and a *functional* scale which describes the bandwidth of disinhibitory bumps in response to prescribed input patterns. These bandwidths are not generally the same. For example, increasing the intensity of a rectangle can cause an increase followed by a decrease in the number of disinhibitory bumps. However, the scales are related, since an increase in the structural inhibitory bandwidth causes a reduction in the number of disinhibitory bumps, and thus an increase in the functional bandwidth (Ellias and Grossberg [**1975**]).

The functional bandwidth that is generated by an input pattern is a global property of the network. It is the spatial scale that has behavioral significance, and it actively fills in the network with patterned activity, notably periodic activity, that is not to be found in the input pattern itself. I suggest that the functional bandwidth of a feedback competitive shunting network is the physiological analog of spatial channels in the Fourier approach to visual perception.

In particular, if several feedback networks, all with distinct structural bandwidths, receive the same input pattern, then they will each generate distinct patterns of disinhibitory bumps, and more bumps will occur in the activity patterns of networks with shorter structural bandwidths.

The existence of functional bandwidths in feedback networks partially supports the intuition that visual processing includes a spatial frequency analysis of visual data (Robson [**1976**]). However the traditional view of this concept uses a feedforward mechanism to accomplish the analysis, whereas networks require feedback. Also the structural bandwidth, which is the basis of the traditional analysis, does not equal the functional bandwidth, which alone has behavioral significance. In particular, increasing the intensity of an input pattern can change a network's functional bandwidth, although its structural bandwidth remains unaltered. In the traditional view, such a change might be interpreted as the activation of a new visual channel, which would be false.

25. Quantization and hysteresis. A more profound and exciting difference between structural and functional scales is this. The length L of a rectangular input might equal a nonintegral multiple of the structural bandwidth, but there can exist only an integral number of disinhibitory bumps in the activity pattern induced by the rectangle. The feedback network hereby quantizes its activity in a way that depends on the global structure of the input pattern. For example, rectangular inputs of length $L, L + \Delta L, \ldots, L + m\Delta L$ might all elicit M_L disinhibitory bumps. Not until a rectangle of length $L + (m + 1)\Delta L$ is presented might the network respond with $M_L + 1$ bumps. How this quantization and hysteresis property influences estimates of length and depth when only a single spatial channel is active does not seem to have been studied experimentally.

Another way to describe this quantization property is as follows. Contrary to theories like the Fourier theory of pattern vision or the typical Artificial Intelligence approach, the patterned activity that fills a feedback network is a property not only of the network, but also of *global* features of the input data, such as input length or symmetry, as they interact with the network. Another issue which should be studied experimentally is how to see these disinhibitory bumps. During normal foveal vision, several spatial scales are often simultaneously active. The bumps in scales of high spatial frequency can overlay the spaces between low frequency bumps, and retinal tremor can spatially average over the high frequency bumps. One possible experimental approach is to adapt out the high spatial frequencies before steadily fixating a display which strongly activates a low spatial frequency scale and which possesses a uniform interior luminance that can elicit periodic network activity.

26. Cross-correlations, feedback enhancement, and numerosity. How do feedforward and feedback networks react to patterns more complex than rectangles, such as a periodic pattern of high spatial frequency bars superimposed on a periodic pattern of low spatial frequency bars (Figure 16a)? A network's structural scales group input data before it is transformed into functional scales; the structural scales perform a "statistical analysis" on the input data. Indeed, input terms of the form $\Sigma_{k=1}^{n} I_k D_{ki}$ cross-correlate the input pattern $(I_1, I_2, \ldots, I_n)$ with the kernel $D(j)$ when the interaction coefficients $D_{ki} = D(k - i)$ are distance-dependent. These statistics of the input pattern, rather than the pattern itself, are the local data which are synthesized into a global functional scale. Networks with different structural scales compute different statistics of the same input pattern *and* compute different functional scales from input patterns that have identical statistics. Structural scales are a computational interface between input data and functional scales.

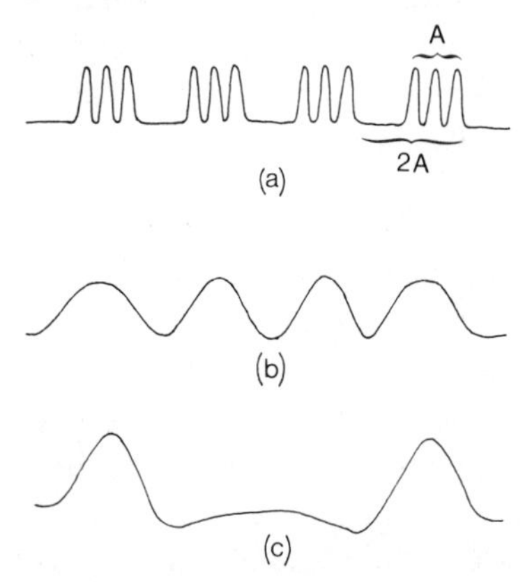

FIGURE 16. In (a), an input pattern of high spatial frequency bumps is superimposed on low spatial frequency bumps. In (b), the activity pattern of a feedforward network with excitatory bandwidth equal to A senses the input pattern's low spatial frequency. In (c), the activity pattern of a feedforward network with excitatory bandwidth at least $2A$ senses only the edges of the input pattern.

First let us consider the reaction of *feedforward* networks to the input pattern in Figure 16a. Suppose that the excitatory on-center $(\Sigma_{k=1}^{n} I_k D_{ki})$ is much narrower than the inhibitory off-surround $(\Sigma_{k=1}^{n} I_k E_{ki})$ to avoid spurious peak splits and multiple edge effects. Then the excitatory structural bandwidth determines a unit length over which input data is pooled, whereas the inhibitory structural bandwidth determines a unit length over which the pooled data of nearby populations are evaluated for their uniformity. A network whose excitatory bandwidth equals width A (as depicted in Figure 16a) can react to the input pattern with a periodic series of bumps which are not due to disinhibition (Figure 16b), but a network whose excitatory bandwidth at least equals the period $2A$ (see Figure 16a) can react only to the pooled "edges" of the input pattern (Figure 16c). The interior of the input pattern is *statistically uniform* with respect to the low spatial frequency network, and therefore the interior of the pattern is inhibited by noise suppression.

How does a *feedback* network react to the input pattern in Figure 16a? Two main comments are relevant to this summary. The first concerns the interaction between a sigmoid signal function $f(w)$ with feedback excitatory coefficients D_{ki}, as in the term $\Sigma_{k=1}^{n} f(x_k)D_{ki}$. The structural bandwidth of feedback excitatory coefficients pools excitatory feedback activity, much as the structural bandwidth of feedforward excitatory coefficients pools excitatory input activity. The power law behavior of the sigmoid signal at low activities attenuates the effect of low activities on the feedback pooling process. Populations v_i which receive small inputs due to the feedforward pooling process will have a negligible effect on the feedback pooling process.

This latter property manifests itself in a variety of data wherein a percept depends on the occurrence of a sufficiently large density of similar features in a picture. Notable examples are found in the demonstrations by Glass and Switkes [**1976**] of perceived correlations between two juxtaposed transparencies of identical random dots. If the two transparencies are slightly shifted or rotated with respect to each other, an impression of horizontal lines or circles is achieved only if sufficiently many dots occur within a unit area of the transparency. A feedforward pooling process corresponding to each retinal region can excite all orientation detectors that receive pooled activity, and a feedback pooling process, augmented by competitive feedback interactions, can suppress all but the most favored orientations in a given region, as it simultaneously defines a functional scale for this region.

This latter comment leads us to the second point that needs to be made about feedback networks. Even though the feedforward low spatial frequency network responds to Figure 16a with edge detection, the feedback low frequency network responds with a functional scale that fills the network with patterned activity. This property suggests a way out of a basic dilemma: If noise suppression tends to quench the interior of visual objects, and does so most vigorously within the low spatial frequency scales that might have been able to span the entire object, then how can we ever achieve an impression of object solidity? This seems to me to be the essence of the filling-in dilemma, although it cannot be stated clearly unless one confronts the present Fourier approach to pattern vision with the noise suppression property. The answer seems to be: use a *feedback* network of competitive mass action type.

27. Binocular resonance and the fixation point. These concepts can be used to design binocular resonances which correlate disparity and length information, execute a filling-in process, and compute a lightness scale. My starting point is a simple but basic fact whose significance is often overlooked. Suppose that the two eyes fixate on a long edge (Figure 17). By definition, the point of fixation is a zero disparity point. Both eyes receive essentially identical visual data at such a point. Moreover, an edge changes so rapidly across space that all structural spatial scales can detect it. (In this discussion, I will ignore factors which modify the feedforward pooling process, such as orientation selectivity and cortical expansion factor, but which do not alter our general conclusions.) Since the fixation point is a zero disparity point, all possible scales in the monocular representations of each eye can match their data with the corresponding data from the other eye at the fixation point. At the fixation point, all possible monocular spatial scales could resonate to form a binocular representation, other things being equal. Other things are not equal, however, and this is why certain monocular scales are chosen above others for binocular resonance.

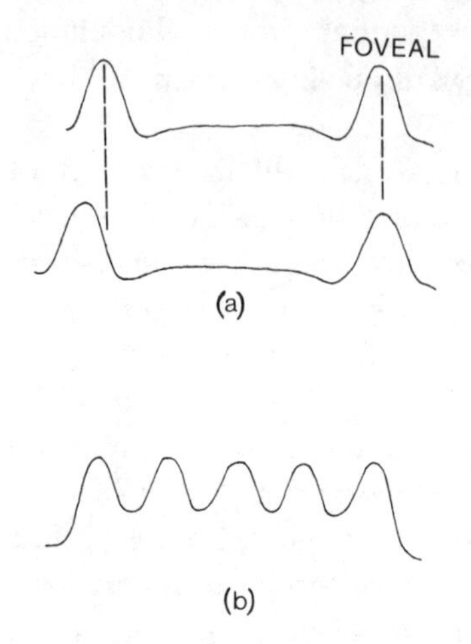

FIGURE 17. In (a), feedforward monocular representations enhance the edges of a rectangular input. Only the foveally fixated edges correspond. In (b), both pairs of edges can resonate due to positive feedback exchange, creating a new pooled nonfoveal edge, and filling in the network with patterned activity between these edges.

I will illustrate this idea in the simplest possible network, since the goal herein is merely to introduce the concept. In Figure 17a, two feedforward monocular representations react to a rectangular input with edge enhancement. Since one edge of the rectangle is fixated by each eye, the two monocular representations correspond at one edge. Due to the asymmetric position of the other edge with respect to the two eyes, this edge activates a different retinal position in each eye. The monocular representations therefore have different widths. (The monocular reactions in Figure 17a can alternatively be interpreted as structural low spatial frequency reactions to a regular pattern, as in Figure 16a.)

Let the two monocular representations send each other topographically organized positive feedback signals. Assume that the structural scales of these feedback signals monotonically increase with the structural scales within their monocular representations. In other words, a low (high) spatial frequency monocular structural scale elicits low (high) spatial frequency feedback signals. Variants of this self-similarity constraint are often needed to synthesize globally self-consistent representations out of locally ambiguous data (Grossberg [**1978a**]). If the nonfixated monocular edges fall within the excitatory structural scale of this feedback signalling process, then a new pooled edge is created (Figure 17b). The total network is now a feedback network with two matched edges. A functional bandwidth is hereby defined and the network is filled with patterned activity. This feedback process correlates the disparity and spatial position of the unmatched edge, the network's structural scale and its ability to define a matched edge via feedback signalling, and the functional scale which fills the network with patterned activity up to the zero disparity edge.

An interesting difference in this network's dynamics occurs in response to monocular viewing of the same input pattern. One monocular representation is excited and generates an edge-enhanced activity pattern. This pattern is transferred to the other monocular representation by binocular signalling. The second monocular representation then returns edge-enhanced feedback signals to the first monocular representation. The total network is again a feedback network with matched edges. A functional scale is therefore defined which fills the network with patterned activity up to the zero disparity fixation edge. Several points are worth noticing in this example. Both the monocularly viewed and the binocularly viewed input can elicit a feedback signalling process. However, the width of the internal representation and the functional scale need not be the same in the two situations. Also of importance is the fact that the disparity matching process selects out a whole structural scale–a whole "depth plane" if you will–and fills it with patterned activity–creates solidity–as it automatically computes a lightness scale throughout this depth plane.

That the above discussion is overly simple can be seen from the following remark. A feedback network can fill-in between the bounding edges of a statistically uniform region only if a source of input activity is available in this region on which the disinhibitory process can feed. This property is achieved if the monocular representations are themselves competitive feedback networks, as in a dipole field (Grossberg, [**1980a**]). Then binocularly discordant filling-in within the monocular representations can be quickly attenuated by binocular matching (§13) before binocularly coherent filling-in is induced by the pooled edge.

A mature theory of binocular vision must show how multiple structural scales influence each other through feedback signalling to determine which scales can achieve binocular resonance, and how the resonant data can elicit accurate commands to motor maps. In particular, the feedback signals from low spatial frequency structural scales can be averaged away by high spatial frequncy scales using their noise suppression property. The converse statement is not true, however. High spatial frequency feedback signalling can modulate the resonant patterns of low spatial frequency detectors. Such interscale interactions are suggested to be the dynamical "glue" which ties together the several resonating structural scales and controls the visual impression of object permanence and continuity as an object moves before us.

28. Concluding remarks. The thought experiment with which this article began demonstrates that the problem of error correction by a self-organizing (learning, developing) system in a nonstationary environment leads us to consider a small number of design principles and their mechanistic realizations. The article then reviewed some of the varied mathematical results which arise when we classify manifestations of one such principle: automatic self-tuning by competitive networks. My other chapter in the book illustrates some applications of a second principle: gating properties of chemical transmitters. A third principle is reviewed elsewhere (Grossberg **[1971]**, **[1974]**, **[1978b]**): pattern learning and growth properties by networks containing fast (STM) and slow (LTM) feedback interactions. These several principles are joined together to generate adaptive resonances and reset operations in the service of global code development and error-correction.

Various other articles in this book have probed aspects of these concepts from several experimental and theoretical perspectives. Thus the principles are starting to enjoy the multiple emergence that often characterizes major conceptual progress in a science. We are, happily, approaching an era when it will be appropriate to identify a small number of laws on which the brain sciences can be based. It is imperative that we not confuse the invariant structure of these laws with the endless list of minor experimental or numerical variations that can draw both experimental and theoretical neuroscience to the brink of conceptual solipsism. A law must be identified before it can be classified, just as a single Schrödinger or Laplace equation can be identified despite its appearance in a vast number of physically distinct examples. The possibility of achieving such coherence is vested in mathematics for the brain sciences no less than for science in general.

References

Beck, J., 1972. *Surface color perception*, Cornell Univ. Press, Ithaca, N.Y.

Bloomfield, T. M., 1969. *Behavioral contrast and the peak shift*, Animal Discrimination Learning (R. M. Gilbert and N. S. Sutherland, eds.), Academic Press, New York.

Boring, E. G., 1950. *A history of experimental psychology*, 2nd ed., Appleton-Century-Crofts, New York.

Campbell, L. and W. Garnett, 1882. *The life of James Clerk Maxwell*, Macmillan, London.

Cornsweet, T. N., 1970. *Visual perception*, Academic Press, New York.

Carpenter, G. A., 1981. *Normal and abnormal signal patterns in nerve cells*, these Proceedings.

Dev, P. 1975. *Computer simulation of a dynamic visual perception model*, Internat. J. Man-Machine Studies 7, 511–528.

Eigen, M. and P. Schuster, 1978. *The hypercycle: A principle of natural self-organization. B. The abstract hypercycle*, Naturwissenschaften **65**, 7–41.

Ellias, S. A. and S. Grossberg, 1975. *Pattern formation, contrast control, and oscillations in the short term memory of shunting on-center off-surround networks*, Biol. Cybernet. **20**, 69–98.

Freeman, W. J., 1975. *Mass action in the nervous system*, Academic Press, New York.

______, 1979. *Nonlinear gain mediating cortical stimulus-response relations*, Biol. Cybernet. **33**, 237–247.

______, 1980. *EEG analysis gives model of neuronal template-matching mechanism for sensory search with olfactory bulb*, Biol. Cybernet. **35**, 221–234.

______, 1981. *A neural mechanism for generalization over equivalent stimuli in the olfactory system*, these Proceedings.

Geman, S., 1981. *The law of large numbers in neural modelling*, these Proceedings.

Glass, L. and E. Switkes, 1976. *Pattern recognition in humans: Correlations which cannot be perceived*, Perception **5**, 67–72.

Graham, N., 1981. *The visual system does a crude Fourier analysis of patterns*, these Proceedings.

Grossberg, S., 1964. *The theory of embedding fields with applications to psychology and neurophysiology*, Rockefeller Institute for Medical Research, New York.

______, 1968. *Some physiological and biochemical consequences of psychological postulates*, Proc. Nat. Acad. Sci. U.S.A. **60**, 758–765.

______, 1970. *Neural pattern discrimination*, J. Theoret. Biol. **27**, 291–337.

______, 1971. *Pavlovian pattern learning by nonlinear neural networks*, Proc. Nat. Acad. Sci. U.S.A. **68**, 828–831.

______, 1972. *Pattern learning by functional-differential neural networks with arbitrary path weights*, Delay and Functional-Differential Equations and their Applications (K. Schmitt, ed.), Academic Press, New York.

______, 1973. *Contour enhancement, short term memory, and constancies in reverberating neural networks*, Stud. Appl. Math. **52**, 217–257.

______, 1974. *Classical and instrumental learning by neural networks*, Progress in Theoretical Biology (R. Rosen and F. Snell, eds.), vol. 3, Academic Press, New York.

______, 1975. *A neural model of attention, reinforcement, and discrimination learning*, Internat. Rev. Neurobiol. **18**, 263–327.

______, 1976a. *Adaptive pattern classification and universal recoding*. I. *Parallel development and coding of neural feature detectors*, Biol. Cybernet. **23**, 121–134.

______, 1976b. *Adaptive pattern classification and universal recoding*. II. *Feedback, expectation, olfaction, and illusions*, Biol. Cybernet. **23**, 187–202.

______, 1978a. *A theory of human memory: Self-organization and performance of sensory-motor codes, maps, and plans*, Progress in Theoretical Biology (R. Rosen and F. Snell, eds.), vol. 5, Academic Press, New York.

______, 1978b. *Communication, memory, and development*, Progress in Theoretical Biology (R. Rosen and F. Snell, eds.), vol. 5, Academic Press, New York.

______, 1978c. *Decisions, patterns, and oscillations in the dynamics of competitive systems with applications to Volterra-Lotka systems*, J. Theoret. Biol. **73**, 101–130.

______, 1978d. *Competition, decision, and consensus*, J. Math. Anal. Appl. **66**, 470–493.

______, 1978e. *Behavioral contrast in short term memory: Serial binary memory models or parallel continuous memory models*? J. Math. Psych. **17**, 199–219.

______, 1978f. *A theory of visual coding, memory, and development*, Formal Theories of Visual Perception (E. L. J. Leeuwenberg and H. F. J. M. Buffart, eds.), Wiley, New York.

______, 1980a. *How does a brain build a cognitive code*? Psych. Rev. **87**, 1–51.

______, 1980b. *Intracellular mechanisms of adaptation and self-regulation in self-organizing networks: the role of chemical transducers*, Bull. Math. Biol. **3**, 365–396.

______, 1980c. *Biological competition: Decision rules, pattern formation, and oscillations*, Proc. Nat. Acad. Sci. U.S.A. **77**, 2338–2342.

______, 1981a. *A psychophysiological theory of normal and abnormal motivated behavior* (submitted).

______, 1981b. *Psychophysiological substrates of schedule interactions and behavioral contrast*, these PROCEEDINGS.

Grossberg, S. and D. S. Levine, 1975. *Some developmental and attentional biases in the contrast enhancement and short term memory of recurrent neural networks*, J. Theoret. Biol. **53**, 341–380.

Hamada, J., 1980. *Antagonistic and non-antagonistic processes in the lightness perception*, 22nd Internat. Congr. Psych. (Leipzig, GDR, July 1980).

Iverson, G. and M. Pavel, 1981. *Invariant properties of masking phenomena in psychoacoustics and their theoretical consequences*, these PROCEEDINGS.

Koenigsberger, L., 1906. *Hermann von Helmholtz* (F. A. Welby, trans.), Clarendon, Oxford, England.

Lacker, H. M., 1980. *Regulation of ovulation number in mammals*, lecture, Gordon Conf. Theoret. Math. Biol.

Levine, D. S. and S. Grossberg, 1976. *Visual illusions in neural networks: Line neutralization, tilt aftereffect, and angle expansion*, J. Theoret. Biol. **61**, 477–504.

Luce, R. D. and L. Narens, 1981. *Axiomatic measurement theory*, these PROCEEDINGS.

Marr, D., G. Palm and T. Poggio, 1978. *Analysis of a cooperative stereo algorithm*, Biol. Cybernet. **4**, 223–239.

Mimura, M. and J. D. Murray, 1978. *On a diffusive pre-predator model which exhibits patchiness*, J. Theoret. Biol. **75**, 249–262.

von der Malsburg, Ch. and D. J. Willshaw, 1981. *Differential equations for the development of topological nerve fibre projections*, these PROCEEDINGS.

O'Brien, V., 1958. *Contour perception, illusion and reality*, J. Opt. Soc. Amer. **48**, 112–119.

Ratliff, F., 1965. *Mach bands: Quantitative studies of neural networks in the retina*, Holden-Day, San Francisco, Calif.

Robson, J. G., 1976. *Receptive fields: Neural representation of the spatial and intensive attributes of the visual image*, Handbook of Perception (E. C. Carterette and M. P. Friedman, eds.), vol. 5, Academic Press, New York.

Sperling, G., 1981. *Mathematical models of binocular vision*, these PROCEEDINGS.

Vorberg, D., 1981. *Reaction time distributions predicted by serial self-terminating models of memory search*, these PROCEEDINGS.

Werblin, F. S., 1971. *Adaptation in a vertebrate retina. Intracellular recordings in Necturus*, J. Neurophysiology **34**, 228–241.

______, 1974. *Control of retinal sensitivity*. II. *Lateral interactions at the outer plexiform layer*, J. General Physiology **63**, 62–87.

Wilson, H. R. and J. D. Cowan, 1972. *Excitatory and inhibitory interactions in localized populations of model neurons*, Biophys. J. **12**, 1–24.

DEPARTMENT OF MATHEMATICS, BOSTON UNIVERSITY, BOSTON, MASSACHUSETTS 02215

SIAM-AMS Proceedings
Volume **13**
1981

Psychophysiological Substrates of Schedule Interactions and Behavioral Contrast

STEPHEN GROSSBERG[1]

1. Introduction: An interdisciplinary approach to schedule interactions. This chapter analyses some properties of schedule interactions, notably behavioral contrast effects, in terms of concepts that are suggested by the thought experiment in Grossberg [**1981a**]. None of these concepts was derived with schedule interactions in mind, and in that sense these behavioral properties, albeit interesting, are not fundamental constraints on neural design. On the other hand, without a knowledge of a few basic mechanistic concepts, questionable conclusions about schedule interactions can be generated. I will discuss Hinson and Staddon's [**1978**] recent article on schedule interactions to illustrate this point. Schedule interactions are nonetheless challenging both as complex data and in terms of their historical treatment by Skinnerian psychologists. I will argue that schedule interactions are strongly influenced by processes which are currently studied as if they were parts of different disciplines, whereas really these processes work together to help determine the cues to which we pay attention. Chemical transmitters are usually discussed in neuropharmacology or neurophysiology. Expectancies are often discussed in cognitive or social psychology. Extinction, reinforcement, and behavioral contrast effects are often analysed in operant conditioning experiments. I will suggest below how these seemingly disparate concepts work together during reinforced behavior.

A conceptual roadblock in psychopharmacological no less than learning studies has been due to insufficient understanding of how chemical transmitters act as gates. I will review below how chemical gates acting in competitive geometries can cause intracellular adaptation and habituation, antagonisitic rebound in response to specific cue offset or to nonspecific arousal onset that is driven by an unexpected event, and inverted U effects. I will then indicate how

---

1980 *Mathematics Subject Classification*. Primary 92A09, 92A25.
[1] Supported in part by the National Science Foundation (NSF IST-80-00257).

these formal properties control aspects of extinction, secondary reinforcement, emotional depression, and analgesia. To accomplish this interpretation, one also needs to know how the gating properties interact with properties of long-term memory to endow external cues with reinforcing and incentive motivational properties. When the STM normalization property acts on a cue's net incentive motivational feedback, behavioral contrast effects can occur. Otherwise expressed, behaviorial contrast occurs in a particular type of adaptive resonance. Historically, such reinforcement and discrimination learning properties were derived from learning postulates (Grossberg **[1971]**, **[1972a]**, **[1972b]**, **[1975]**) before the resonance idea was derived from code development postulates (Grossberg **[1976a]**, **[1976b]**). The reader might find it instructive to follow both roads, seemingly so disjoint, to their common mechanistic endpoint.

Although the behaviorial contrast effects are not fundamental in themselves, the behaviorial paradigms in which they occur might become a useful framework wherein to probe pharmacological substrates of behavior. This is because the gating properties are identified with a catecholaminergic transmitter system, whereas the LTM properties are identified with a cholinergic transmitter system. The theory thus suggests how cholinergic-catecholaminergic interactions subserve schedule interaction effects (Butcher **[1978]**, Friedhoff **[1975a]**, **[1975b]**, Grossberg **[1972a]**, **[1972b]**). In particular, the theory suggests what might happen to behavior if the positive vs. negative incentive motivational effects, or the gating vs. LTM effects, or the LTM vs. STM effects are dissected using pharmacological manipulations.

These remarks illustrate an alternative strategy for evaluating models than is often practiced in the experimental literature. When a model is constructed to explain a particular phenomenon, one test of its adequacy is the number of parameters needed to fit the data. When the data can be explained by design principles that are not derived to explain these data, and which are implicated in a wide variety of other data, then counting parameters is a naive approach to verifying model adequacy. For example, to count all the interaction strengths as parameters in a distance-dependent network is absurd, especially if the behavioral properties controlled by these interactions are qualitatively invariant under significant changes in the parameters, and if the existence of distance-dependent interactions can be directly tested or inferred from other considerations. Rather, one might classify the several ways by which the design principle manifests itself across a large body of data, and attempt to use in each experimental paradigm the experimental techniques that have been successfully used to probe the principle in other paradigms. Parametric studies, both within and between paradigms, are needed once a principle and its mechanistic realization have been identified.

2. Transmitters as gates. What is the simplest law whereby one nerve cell could conceivably send a signal to another nerve cell? The answer is obvious. If a signal S passes through a given nerve cell v_1, the simplest law says that the signal S has a proportional effect

$$T = SB, \tag{1}$$

where $B > 0$, on the next nerve cell v_2. Such a law would permit unbiased transmission of signals from one cell to another.

We are faced with a dilemma, however, if the signal from v_1 to v_2 is due to the release of a chemical $z(t)$ from v_1 that activates v_2. Such a chemical is called a *transmitter* in neurophysiology. If the transmitter is persistently released when S is large, what keeps the net signal T from getting smaller and smaller as v_1 runs out of the transmitter? Some means of replenishing, or accumulating, the transmitter must exist to counterbalance its depletion due to release from v_1.

Based on this discussion, we can rewrite (1) in the form

$$T = Sz \tag{2}$$

and ask how the system can keep z replenished so that

$$z(t) \cong B \tag{3}$$

at all times t. Equation (2) has the following interpretation. The signal S causes the transmitter z to be released at a rate $T = Sz$. Whenever two processes, such as S and z, are multiplied, we say that they interact by *mass action*, or that z *gates* S. Thus (2) says that z gates S to release a net signal T, and (3) says that the cell tries to replenish z to maintain the system's sensitivity to S.

What is the simplest law that joins together both (2) and (3)? It is the following differential equation

$$dz/dt = A(B - z) - Sz \tag{4}$$

for the net rate dz/dt of change of z. Term AB on the right-hand side of (4) says that z is produced at a constant rate AB. Term $-Az$ says that once z is produced, it inhibits the production rate by an amount proportional to z's concentration. In biochemistry, such an inhibitory effect is called *feedback inhibition* by the endproduct of a reaction. Without feedback inhibition, the constant rate AB of production would eventually cause the cell to burst. With feedback inhibition, the net production rate is $A(B - z)$, which causes $z(t)$ to approach the finite amount B, as we desire by (3). The term $A(B - z)$ thus enables the cell to accumulate a target level B of transmitter. Term $- Sz$ in (4) says that z is released at a rate Sz, as we desire by (2). In all, (4) is the simplest law that "corresponds" to the constraints (2) and (3). It describes four effects working together: production, feedback inhibition, gating, and release. Without a precise law such as (4), we could talk ourselves to death about the effects of these four processes without getting anywhere. With the law, we can derive some surprising and important facts, which is how a relationship between transmitters, expectancies, and reinforcement was first discovered (Grossberg **[1968]**, **[1969]**, **[1972b]**).

3. Intracellular adaptation and habituation. First let us determine how the net signal $T = Sz$ reacts to a sudden change in S. We will suppose that $z(t)$ reacts slowly compared to the rate with which S can change. For definiteness, suppose that $S(t) = S_0$ for all $t \leqslant t_1$ and that at time $t = t_1$, $S(t)$ suddenly increases to S_1. By (4), $z(t)$ reacts to the constant value $S(t) = S_0$ by approaching an equilibrium value $z(t_1)$. This equilibrium value is found by setting $dz/dt = 0$ in (4) and solving for

$$z(t_1) = AB/(A + S_0). \tag{5}$$

By (2), the net signal to v_2 at time t_1 is

$$S_0 z(t_1) = ABS_0/(A + S_0). \tag{6}$$

Now let $S(t)$ switch to the value $S_1 > S_0$. Because $z(t)$ is slowly varying, $z(t)$ approximately equals $z(t_1)$ for awhile after $t = t_1$. Thus the net signal to v_2 during these times is approximately equal to

$$S_1 z(t_1) = ABS_1/(A + S_0). \tag{7}$$

Equation (7) has the same form as a Weber law $J(A + I)^{-1}$. The signal S_1 is evaluated relative to the baseline S_0 just as J is evaluated relative to I. This Weber law is due to slow intracellular adaptation of the transmitter gate to the input level through time. It is not due to fast intercellular lateral inhibition across space as in equation (2) of Grossberg [**1981a**].

An important difference between intracellular and intercellular adaptation is this: the intracellular gate is sensitive only to the input passing through its own channel, whereas the intercellular lateral inhibition is sensitive to the entire pattern lying within its spatial bandwidth. Actually this distinction is weakened in cases wherein each synaptic knob can reabsorb a fraction of the transmitter that is released from neighboring knobs, as in the system of transmitter pathways

$$\frac{dz_i}{dt} = A\left(B - z_i - \sum_{k=1}^{n} S_k z_k C_{ki}\right) - S_i z_i. \tag{8}$$

The total amount of transmitter $\sum_{k=1}^{n} z_k$ obeys a normalization rule analogous to the rule obeyed by the total potential $\sum_{k=1}^{n} x_k$ of a distance-dependent competitive feedback network. The normalization rule takes on an important new meaning in this context. Destroying some of the transmitter pathways allows the surviving pathways to produce more transmitter by reducing the amount of feedback inhibition to which they are subjected. An analogous effect is due to lesions that disrupt dopaminergic neurons of the nigrostriatal bundle (Stricker and Zigmond [**1976**]).

As $z(t)$ in (4) begins to respond to the new transmitter level $S = S_1$, $z(t)$ gradually approaches the new equilibrium point that is determined by $S = S_1$, namely

$$z(\infty) = AB/(A + S_1). \tag{9}$$

The net signal consequently decays to the asymptote

$$S_1 z(\infty) = ABS_1/(A + S_1). \tag{10}$$

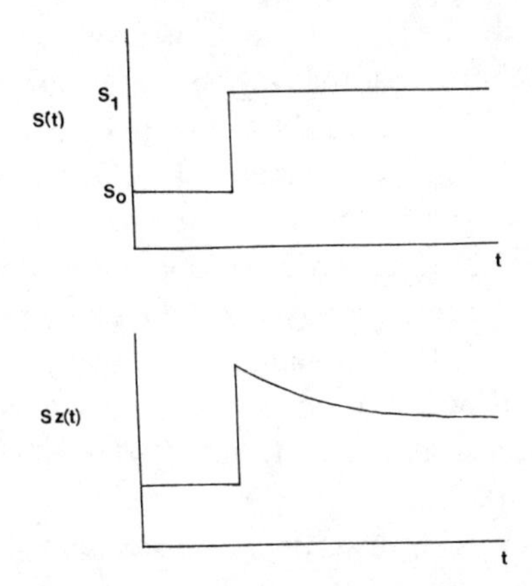

FIGURE 1. In response to a sudden increment in $S(t)$ from S_0 to S_1, the gated signal $S(t)z(t)$ overshoots and then slowly habituates to the new level S_1.

Thus after $S(t)$ switches from S_0 to S_1, the net signal Sz jumps from (6) to (7), and then gradually decays to (10) (Figure 1). The exact course of this decay is described by the equation

$$S_1z(t) = \frac{ABS_1}{A + S_0}e^{-(A+S_1)(t-t_1)} + \frac{ABS_1}{A + S_1}(1 - e^{-(A+S_1)(t-t_1)}) \tag{11}$$

for $t \geqslant t_1$, which shows that the gain of the response $A + S_1$ depends on the signal S_1 just as in the case of shunting lateral inhibition. The sudden increment followed by slow decay can be intuitively described as an overshoot followed by habituation to the new sustained signal level S_1. Both intracellular adaptation and habituation hereby occur whenever a transmitter fluctuates more slowly than the signals that it gates. The size of the overshoot can be found by subtracting (10) from (7). For definiteness, let $S_0 = f(I)$ and $S_1 = f(I + J)$, where $f(w)$ is a function that transmutes the inputs I and $I + J$ before and after the increment J into a net signal S_0 or S_1. Then the rebound size is approximately

$$S_1z(t_1) - S_1z(\infty) = \frac{ABf(I + J)[f(I + J) - f(I)]}{[A + f(I)][A + f(I + J)]}. \tag{12}$$

In §§5 and 6 below, I will show that the rebound size in response to specific cue offset or to nonspecific arousal onset is related to (12) in a way that allows us to estimate both $f(w)$ and the arousal level I.

4. Competition and antagonistic rebound. In many behavioral situations, two or more neural pathways compete before a signal is generated that has observable behavioral effects. For example, suppose that I wish to press a lever in response to the offset of a light. If light offset simply turned off the cells that code for light being on, then there would exist no cells whose activity could selectively elicit the lever press response after the light was turned off. Clearly, offset of the light not only turns off the cells that are turned on by the light, but it also selectively turns on cells that will transiently be active after the light is shut off. The activity of these "off"-cells–namely the cells that are turned on by light offset–can then activate the motor commands leading to the lever press. Let us call the transient activation of the off-cell by cue offset *antagonistic rebound*.

Antagonistic rebound also occurs in a variety of other behavioral situations. For example, shock can unconditionally elicit the emotion of fear and various autonomic consequences of fear (Dunham **[1971]**, Estes **[1969]**, Estes and Skinner **[1941]**). Offset of shock is (other things equal) capable of eliciting relief or a complementary emotional reaction (Denny **[1971]**, Masterson **[1970]**, McAllister and McAllister **[1970]**).

I will indicate below how, if transmitters gate signals before the gated signals compete, then antagonistic rebound can be elicited by offset of a specific cue, as in light-on vs. light-off, or fear vs. relief. I will also show how unexpected events can cause an antagonistic rebound. They do this by triggering an increase in the level of nonspecific arousal that is gated by all the transmitter pathways.

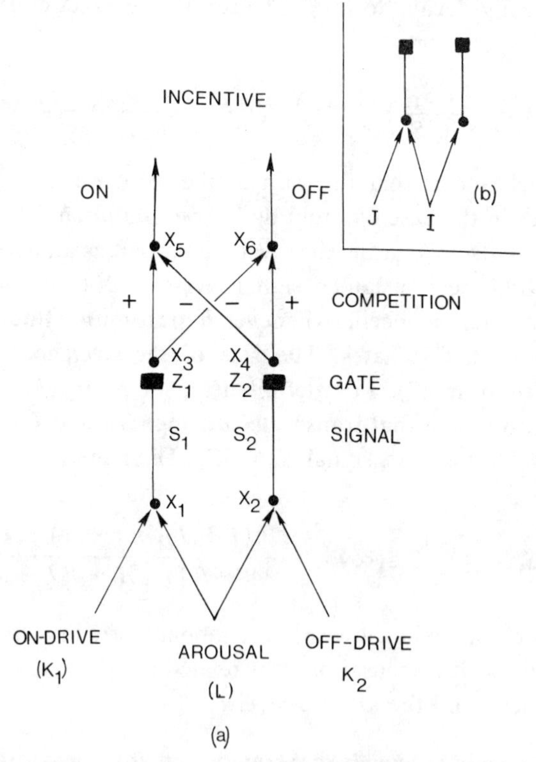

FIGURE 2. (a) The minimal dipole network in which specific inputs (K_1 and K_2) and a nonspecific input (L) to two parallel channels are gated by slowly varying transmitters before they compete to elicit net outputs. The inputs excite potentials x_1 and x_2 that elicit signals S_1 and S_2, respectively. These signals are gated by transmitters z_1 and z_2 before they activate potentials x_3 and x_4. The signals elicited by x_3 and x_4 compete before activating x_5 and x_6. The signals elicited by x_5 and x_6 are incentive motivational outputs, the on-reaction driven by x_5 and the off-reaction driven by x_6. (b) If $K_1 > K_2$, then the network can be redrawn with $I = K_2 + L$ and $J = K_1 - K_2$.

Figure 2 depicts the simplest network in which two channels receive inputs that are gated by slowly varying transmitters before the channels compete to elicit a net output response. Two types of inputs will be introduced: specific inputs that are turned on and off by internal or external cues, and nonspecific arousal inputs that are on all the time, even though their size can vary through time. Each channel can have its own specific input, K_1 or K_2, and both channels receive the same arousal input L. The total signals to the two channels are therefore $S_1 = f(K_1 + L)$ and $S_2 = f(K_2 + L)$, where the signal function $f(w)$ is monotone increasing. We will see that the relative sizes of S_1 and S_2 and their rates of change through time relative to the transmitter fluctuation rate determine whether an antagonistic rebound will occur. To emphasize this fact, I define

$$I = \min(K_1 + L, K_2 + L) \tag{13}$$

and

$$J = |K_1 - K_2|. \tag{14}$$

The quantity I determines the network's net arousal level, and J determines how asymmetric the inputs are to the two channels. Suppose for definiteness that $K_1 > K_2$. Then $S_1 = f(I + J)$ and $S_2 = f(I)$.

5. Rebound due to phasic cue offset. A rebound can be caused if, after the network equilibrates to the input J, the cue is suddenly shut off. This effect is analogous to the reaction that occurs when a light is shut off, or a fearful cue is shut off. To see how this rebound is generated, suppose that the arousal level is I and that the cue input is J. Let the total signal in the on-channel be $S_1 = f(I + J)$ and in the off-channel be $S_2 = f(I)$. Let the transmitter z_1 in the on-channel satisfy the equation

$$dz_1/dt = A(B - z_1) - S_1 z_1, \tag{15}$$

and the transmitter z_2 in the off-channel satisfy the equation

$$dz_2/dt = A(B - z_2) - S_2 z_2. \tag{16}$$

After z_1 and z_2 equilibrate to S_1 and S_2, $(d/dt)z_1 = (d/dt)z_2 = 0$. Thus by (15) and (16)

$$z_1 = AB/(A + S_1) \tag{17}$$

and

$$z_2 = AB/(A + S_2). \tag{18}$$

Since $S_1 > S_2$, it follows that $z_1 < z_2$; that is, z_1 is depleted more than z_2. However, the gated signal in the on-channel is $S_1 z_1$, and the gated signal in the off-channel is $S_2 z_2$. Since

$$S_1 z_1 = ABS_1/(A + S_1) \tag{19}$$

and

$$S_2 z_2 = ABS_2/(A + S_2) \tag{20}$$

it follows from the inequality $S_1 > S_2$ that $S_1z_1 > S_2z_2$ despite the fact that $z_1 < z_2$. Thus the on-channel gets a bigger signal than the off-channel. After the two channels compete, the input J produces a sustained on-output whose size is proportional to

$$S_1z_1 - S_2z_2 = \frac{A^2B[f(I+J) - f(I)]}{[A + f(I)][A + f(I+J)]}. \tag{21}$$

Division of (12) by (21) yields an interesting relationship between the size of the overshoot in the on-channel to the size of the steady-state on-output; namely,

$$\frac{\text{on-overshoot}}{\text{steady on-output}} = \frac{f(I+J)}{A}, \tag{22}$$

which provides an estimate of $f(w)$ if J is parametrically varied. In particular, if $f(w)$ is a linear signal $f(w) = w$, then (21) becomes

$$S_1z_1 - S_2z_2 = A^2BJ/(A+I)(A+I+J), \tag{23}$$

which is an increasing function of J (more fear given more shock) but a decreasing function of I (linear analgesic effect).

Now shut J off to see how an antagonistic rebound (relief) is generated. The cell potentials rapidly adjust until new signal values $S_1^* = f(I)$ and $S_2^* = f(I)$ obtain. However the transmitters z_1 and z_2 change much more slowly, so that (17) and (18) are approximately valid in a time interval that follows J offset. Thus the gated signals in this time interval approximately equal

$$S_1^*z_1 \cong \frac{ABf(I)}{A + f(I+J)} \tag{24}$$

and

$$S_2^*z_2 \cong \frac{ABf(I)}{A + f(I)}. \tag{25}$$

Thus $S_1^*z_1 < S_2^*z_2$. The off-channel now gets the bigger signal, so an antagonistic rebound occurs whose size is approximately

$$S_2^*z_2 - S_1^*z_1 = \frac{ABf(I)[f(I+J) - f(I)]}{[A + f(I)][A + f(I+J)]}. \tag{26}$$

Division of (26) by (21) yields an interesting relationship between the maximal off-output and the steady on-output; namely,

$$\frac{\text{off-output}}{\text{on-output}} = \frac{f(I)}{A} \tag{27}$$

which provides an estimate of $f(w)$ as I is parametrically varied. A comparison of (22) with (27) shows that, as I is parametrically varied, (22) should have the same graph as (27), shifted by J. This comparison provides an estimate of J (that is, of how the behavioral input is transformed into neural units) and also a strong test of the model. Once $f(w)$ is estimated, the equations (21) and (26) can be verified.

If $f(w) = w$ in (26), then

$$S_2^* z_2 - S_1^* z_1 = \frac{ABIJ}{(A + I)(A + I + J)}. \tag{28}$$

The rebound is then an increasing function of J (offset of larger shock elicits more relief) and an inverted U function of I (an optimal arousal level exists).

An important issue in conditioning experiments concerns the relative off-output to on-output size in (27) when the arousal level is optimally chosen. This quantity controls whether enough positive incentive can be generated by a positive conditioned reinforcer in the presence of a fearful cue to elicit escape behavior (Grossberg [**1972b**]). By (27), this relative size increases as a function of the arousal level, but by (26), the absolute amount of relief can decrease to zero if $f(w)$ is bounded and I is chosen too high (overaroused depression). When, for example, $f(w) = w$, the "relief" rebound (28) is maximized by choosing $I = [A(A + J)]^{1/2}$. At this arousal level, the "relief-to-fear" ratio (27) equals $(1 + J/A)^{1/2}$, which exceeds 1. Thus the arousal level that maximizes relief also produces more relief than fear, and consequently conditioned escape during a fearful cue is possible.

The rebound is transient because the equal signals $S_1 = S_2 = f(I)$ gradually equalize the z_1 and z_2 levels until they both approach $AB(A + f(I))^{-1}$. Then $S_1 z_1 - S_2 z_2$ approaches zero, so the competition between channels shuts off both of their outputs (Figure 3).

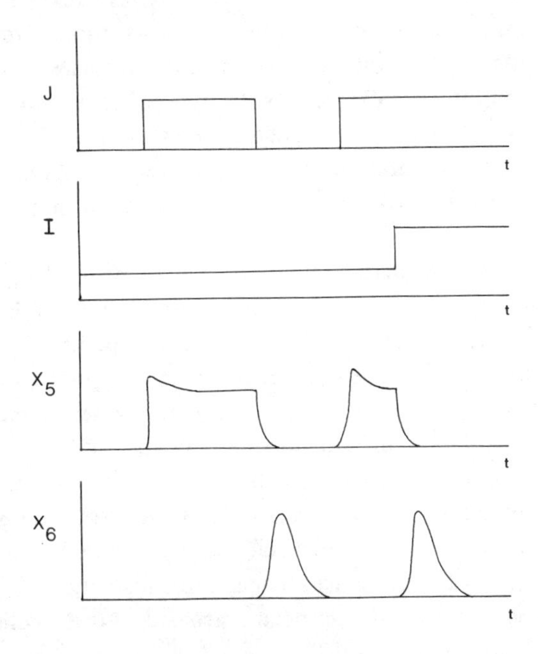

FIGURE 3. On-reactions x_5 and off-reaction rebounds x_6 in response to rapid fluctuations in the inputs I and J. Rapid decrements in J and increments in I tend to cause rebounds.

6. Rebound due to arousal onset. A surprising property of these dipoles of on-cell and off-cell pairs is their reaction to sudden increments in the arousal level I. Such increments are, for example, hypothesized to occur in response to unexpected events.

Suppose that the on-channel and the off-channel have equilibrated to the input levels I and J. Now suddenly increase I to I^*, thereby changing the signals to $S_1^* = f(I^* + J)$ and $S_2^* = f(I^*)$. The transmitters z_1 and z_2 continue to obey (17) and (18) for awhile with $S_1 = f(I + J)$ and $S_2 = f(I)$. A rebound occurs if $S_2^* z_2 > S_1^* z_1$. In general,

$$S_2^* z_2 - S_2^* z_1 = \frac{AB[f(I^*) - f(I^* + J)] + B[f(I^*)f(I + J) - f(I)f(I^* + J)]}{[A + f(I)][A + f(I + J)]}. \tag{29}$$

In particular, if $f(w) = w$ then a rebound occurs whenever

$$I^* > I + A \tag{30}$$

since then

$$S_2^* z_2 - S_1^* z_1 = \frac{ABJ(I^* - I - A)}{(A + I)(A + I + J)}. \tag{31}$$

Thus given a linear signal function, a rebound will occur if I^* exceeds $I + A$ no matter how J is chosen. If the event is so unexpected that it increments the arousal level by more than amount A, then all on-cell off-cell dipoles in the network will simultaneously rebound. This is a parallel processing effect that has no natural analog using serial mechanisms, such as those employed in a computer. Moreover, the size of the off-cell rebound increases as a function of the size of the on-cell input J, as (31) shows. In particular, no rebound occurs if the on-cell was inactive before the unexpected event occurs. Thus the rebound mechanism is *selective*. It rebounds most vigorously those cells which are most active ($J \gg 0$) and spares inactive cells ($J \cong 0$), as required in §9 of Grossberg [**1981a**].

Rebound in response to an increment in nonspecific arousal rapidly resets a cellular field by inhibiting active cells and preparing inactive cells for possible later excitation by increasing their gains. Dipole competition per se is not needed to reset a cellular field. You only need specific inputs and nonspecific arousal to be gated by slow transmitters before the gated signals compete. Dipole competition organizes the field into on-cells and off-cells. Competitive interactions with a broader spatial bandwidth can also reset a field, but the field need not then contain well-defined off-cells. How can you recognize such a field? A good start is to test whether some cells can habituate. From this one can often infer the existence of slow gates. Then one must check if the outputs compete and if there is a source of nonspecific arousal to these cells.

7. Inverted *U* in incentive motivation. Antagonistic rebound due to arousal increments is a basic property of competing transmitter-gated channels. Without the aid of mathematics, this property would remain obscure despite its importance in models of cognitive development and attention. Other important properties of competing transmitter-gated channels can also be derived by mathematical analysis. For example, often the signals S_1 and S_2 are not linear functions of $I + J$ and I, respectively. In particular, often $S_1 = f(I + J)$ and $S_2 = f(I)$, where $f(w)$ is a sigmoid, or S-shaped, function of w to prevent noise amplification in a feedback network (Grossberg **[1973]**). A linear signal function approximates a sigmoid function $f(w)$ at intermediate values of the independent variable w. Thus our previous computations would approximately hold in the sigmoid case at intermediate values of I and J. At extreme values of I or J, however, new properties occur. These properties show that the network is insensitive to fluctuations in either very small inputs or very large inputs. Another way of saying this is: there exists an intermediate range of input values to which the network is most sensitive. This property is called an *inverted U* effect, because network sensitivity is maximal at intermediate input values. Inverted U effects are familiar both in psychophysiology (Hebb **[1955]**) and in discrimination learning (Berlyne **[1969]**). An inverted U occurs in both the on-cell output and the off-cell output when a fixed J value is turned on and shut off at a succession of arousal levels I. In other words, the network becomes depressed either at underaroused or at overaroused levels. Underaroused depression enjoys very different formal properties than overaroused depression. The interested reader might wish to consult Grossberg **[1972b]**, **[1981b]** to see why this is so, and to correlate these distinct network syndromes with familiar mental pathologies; e.g., underaroused (Parkinson's disease, hyperphagia, juvenile hyperactivity) vs. overaroused (simple schizophrenia).

The inverted U effect holds if $f(0) = (df/dw)(0) = 0$, $(df/dw)(w) > 0$ if $w > 0$, $\lim_{t\to\infty} f(w) < \infty$, and $(d^2f/dw^2)(w)$ changes sign once from positive to negative as w increases. An inverted U occurs in the sustained on-output (21), as I is parametrically increased, if $f(w)$ is sigmoid, despite the fact that an inverted U does not obtain in (23) when $f(w)$ is linear. Here again, the sigmoid signal succeeds where the linear signal fails. To simplify the results, I use the signum function

$$\operatorname{sgn}\{w\} = \begin{cases} +1 & \text{if } w > 0, \\ 0 & \text{if } w = 0, \\ -1 & \text{if } w < 0. \end{cases} \tag{32}$$

First consider the on-reaction in (21), which is denoted by x_5 (Figure 2). Write the derivative of a function $g(I)$ with respect to I by $g'(I)$. Then by (21), for each fixed J,

$$\begin{aligned} \operatorname{sgn}\{x_5'(I)\} = \operatorname{sgn}\{&A^2[f'(I+J) - f'(I)] \\ &+2A[f(I)f'(I+J) - f(I+J)f'(I)] \\ &+[f^2(I)f'(I+J) - f^2(I+J)f'(I)]\}. \end{aligned} \tag{33}$$

Since $f(w)$ is sigmoid,

$$f(0) = f'(0) = 0. \tag{34}$$

Thus by (33),

$$\operatorname{sgn}\{x_5'(0)\} = \operatorname{sgn}\{A^2 f'(J)\} > 0. \tag{35}$$

At large values of I,

$$f(I + J) > f(I) \tag{36}$$

whereas

$$f'(I + J) < f'(I). \tag{37}$$

Consequently each term in brackets on the right-hand side of (33) is negative. Thus at large I values,

$$\operatorname{sgn}\{x_5'(I)\} < 0. \tag{38}$$

Inequalities (35) and (38) show that, for fixed J, $x_5(I)$ increases and then decreases as a function of I. This is the inverted U for the on-reaction. In fact, it follows directly from (21) that $\lim_{I\to\infty} x_5(I) = 0$ since $\lim_{w\to\infty} f(w) < \infty$.

The off-reaction x_6 in (26) can be similarly treated with one additional observation. By (27),

$$x_6 = x_5 f(I)/A. \tag{39}$$

Thus

$$\operatorname{sgn}\{x_6'(I)\} = \operatorname{sgn}\{x_5'(I) f(I) + x_5(I) f'(I)\}. \tag{40}$$

At $I = 0$, it follows from (34) that

$$\operatorname{sgn}\{x_6'(0)\} = 0 \tag{41}$$

although by (34) and (35), at slightly larger I values,

$$\operatorname{sgn}\{x_6'(I)\} > 0. \tag{42}$$

Inspection of (26) shows that $\lim_{I\to\infty} x_6(I) = 0$. Again an inverted U obtains.

8. Drive, incentive motivation, and conditioned reinforcer. This section indicates how external cues learn to control incentive motivational signals; in other words, how external cues become conditioned reinforcers. To discuss these learned changes, we need to introduce a mechanism of LTM (Grossberg [**1964**], [**1968**]) and to distinguish the gating properties of the LTM mechanism from the gating properties of the rebound mechanism. The LTM mechanism will be appended to a dipole in which the intuitive notions of drive and incentive motivation are distinguished (Figure 4a). The failure to clearly make this distinction has often led to confusion in the literature (Bolles [**1967**], Mackintosh [**1974**]). For example, incentive motivation need not be released even when drive is high if compatible cues are unavailable. Incentive motivation need not be released even when compatible cues are available if drive is low. This distinction is basic in explaining effects such as self-stimulation, wherein the stimulating electrode can act like an artificial drive input whose effects, via an incentive motivational output, are modulated by natural drive and conditioned reinforcer inputs (Grossberg [**1972a**], Olds [**1977**]).

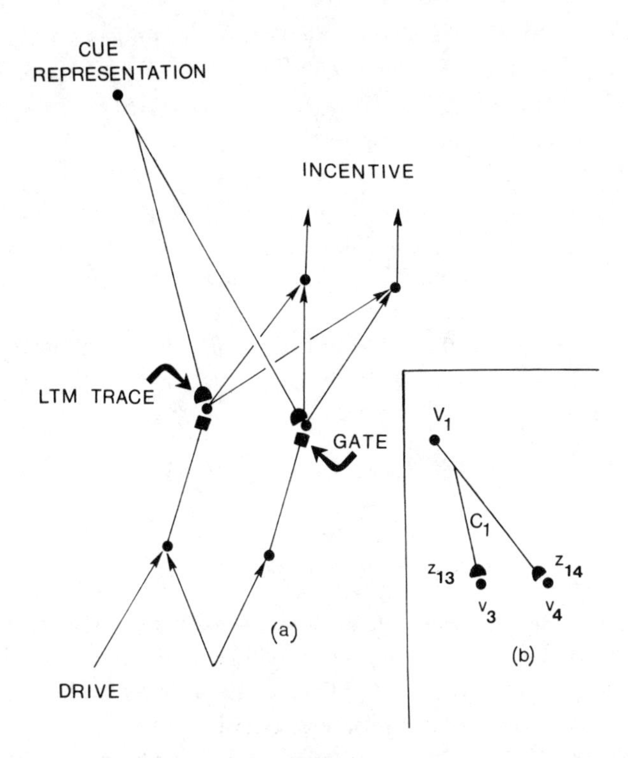

FIGURE 4. (a) A minimal network that distinguishes the concepts of drive, conditioned reinforcer, and incentive motivation; (b) the cue representation V_1 can sample v_3 and v_4 when its signal C_1 is positive. The LTM traces Z_{1i} are then attracted to the x_i values, $i = 3, 4$.

In Figure 4a, drives are the specific inputs to the dipole. For example, a positive drive input might correspond to a hunger signal and its corresponding negative drive input might correspond to a satiety signal (Novin, Wyrwicka and Bray [1976], Reichsman [1972]). Incentive motivations are the outputs of the dipole after the inputs are gated and compete. A positive incentive motivational output might support feeding behavior whereas a negative incentive motivational output might correlate with frustration. In another example of a dipole, the negative drive input is due to shock, the negative incentive output correlates with fear, and the positive incentive output signifies relief.

Where do conditionable pathways activated by external cues send their signals to a dipole? These pathways end after the transmitter gating stage, so they can sample both the on-reactions and the off-reactions of the dipole. These pathways also end before the competition stage so that only one of the competing incentive channels can be active at any time.

We now introduce some notation (Figure 4b) with which to describe the conditioning process. Suppose that presentation of a prescribed external cue activates a population V_1 which thereupon sends signals to both the on-channel and the off-channel of the dipole. Denote the dipole populations that occur after the transmitter gates but before the incentive competition by v_3 and v_4. Denote the signal from V_1 by $C_1(t)$ and the potentials of v_3 and v_4 by $x_3(t)$ and $x_4(t)$, respectively. Suppose that the signal $C_1(t)$ from V_1 to v_3 is gated by an LTM trace $Z_{13}(t)$ on its way to v_3. This LTM trace is computed at the synaptic knobs of V_1 which abut v_3. Similarly, suppose that the signal $C_1(t)$ from V_1 to v_4 is gated by an LTM trace $Z_{14}(t)$ on its way to v_4. In other words, the gated signal from V_1 that reaches v_3 is C_1Z_{13} and the gated signal from V_1 that reaches v_4 is C_1Z_{14}. This gating rule is the same as the gating rule that subserves antagonistic rebound, since both the rebound and LTM mechanisms are hypothesized to be due to chemical transmitters.

The simplest version of the LTM learning rule is the following (Grossberg [**1969**], [**1976a**]). Let each LTM trace Z_{1i} grow only if $C_1 > 0$ and $x_i > 0$, $i = 3, 4$. In particular, let

$$dZ_{13}/dt = C_1(-Z_{13} + x_3) \tag{43}$$

and

$$dZ_{14}/dt = C_1(-Z_{14} + x_4). \tag{44}$$

If (say) Z_{13} grows but Z_{14} does not, then when the cue is later turned on, v_3 will receive a large gated signal from V_1 but v_4 will receive a small gated signal from V_1, even if C_1 is large in both signal pathways. Let us now see how a conditioned reinforcer can be established using these associative laws.

Suppose that during early learning trials, the cue occurs just before a large phasic input K_1 to the on-channel. Then signal C_1 is large when x_3 is large, so by (43), the LTM trace Z_{13} grows, but Z_{14} grows much less, if at all. For example, suppose that a large K_1 represents a large shock input, and that a large potential x_5 at v_5 corresponds to a fear reaction. Then the large K_1 input automatically elicits fear even before learning occurs. By pairing the cue with shock, however, ultimately Z_{13} becomes much larger than Z_{14}. If the cue is then presented by itself, the gated signal to v_3 is much larger than the gated signal to v_4. The outputs from v_3 and v_4 compete before eliciting net reactions at v_5 and v_6. The on-channel wins the competition, so the cue elicits fear. The cue has hereby become a *conditioned reinforcer* that can elicit a *conditioned emotional response*, or CER (Estes and Skinner, [**1941**]).

9. Extinguishing irrelevant events and establishing conditioned avoidance responses. Now suppose that after the cue has become a CER, environmental contingencies change. In particular, suppose that the cue no longer reliably predicts future events, and that an unexpected event occurs while the cue is on. Suppose that the unexpected event triggers a sudden increase in arousal level, which thereupon causes an antagonistic rebound. Consequently $x_4(t)$ will become much larger than $x_3(t)$. Since $C_1(t)$ is still large, $Z_{14}(t)$ will begin to grow rapidly, by (44). If this occurs sufficiently often, $Z_{14}(t)$ can grow as large as $Z_{13}(t)$. When this happens, the gated signals from V_1 to v_3 and v_4 will become approximately equal. After these signals compete, the net inputs to v_5 and v_6 will both become very small. The cue no longer is a CER. It has been rapidly extinguished by unexpected events.

This cue is extinguished because it remains on both before and after the unexpected event. It is an irrelevant cue with respect to the contingency that triggered the unexpected event. By contrast, a cue that turns on right after the unexpected event occurs will sample only the off-reaction of the dipole. Only the LTM trace in the off-channel will grow. Later presentation of this cue will elicit a large off-reaction. Suppose that activating v_6 corresponds to a relief reaction, just as activating v_5 corresponds to a fear reaction. We hereby see how a cue can become a source of conditioned relief by being paired with offset of a source of fear. The subtle fact is that this cue becomes a source of relief, or a positive reinforcer, even though it is never paired with a positive reinforcer. This mechanism helps us to understand how avoidance behavior can be persistently maintained long after an animal no longer experiences the fear that originally motivated the avoidance learning (Grossberg [**1972b**], [**1975**], Maier, Seligman and Solomon [**1969**], Seligman and Johnston [**1973**], Solomon, Kamin and Wynne [**1953**]). Relief would appear to maintain asymptotic avoidance behavior, not fear.

10. Secondary conditioning, steady motivational baseline, medial forebrain bundle. We will now indicate how conditioned external cues can act as reinforcers in a secondary conditioning paradigm. A cue that is classically conditioned to become a CER can be used as a reinforcer in an instrumental conditioning paradigm. This possibility led to the development of two-factor theories (Dunham [**1971**]) which faced the difficulty that a clear definition of classical and instrumental mechanisms was not available. In the present framework, these distinctions melt away into a unified description of how inputs to a dipole alter its incentive outputs through time.

In particular, Figure 4 is inadequate because an external cue can elicit, as well as sample, a rebound. For example, offset of a source of conditioned fear can be positively reinforcing (Dunham [**1971**], Estes [**1969**], Maier et al. [**1969**], McAllister and McAllister [**1970**], McAllister et al. [**1971**]). To sample a rebound, the cue's pathways must come after the rebound. To elicit a rebound, the cue's pathways must come before the rebound, just like the drive input pathway. How can a cue's pathways come both before and after the rebound? To accomplish this, the rebound mechanism must be part of a recurrent pathway, or feedback loop (Figure 5). This feedback loop leads to meaningful comparisons with psychophysiological data when it is interpreted as a formal analog of the medial forebrain bundle (Grossberg [**1972b**], [**1975**]). Once we recognize the existence of a feedback loop, we also recognize the need to use a sigmoid signal function to avoid noise amplification within this loop (Grossberg [**1981a**]). The inverted U properties of §7 now emerge as consequences of a basic noise suppression mechanism. A feedback network also has a capability for choosing one feedback pathway above others for STM reverberation, and of maintaining a stable operating level against small input perturbations using its normalization and hysteresis properties. In a motivational context, these feedback properties describe sharp motivational switching between incompatible motivational alternatives and maintenance of a stable baseline of incentive motivation despite small momentary fluctuations in internal and external cues.

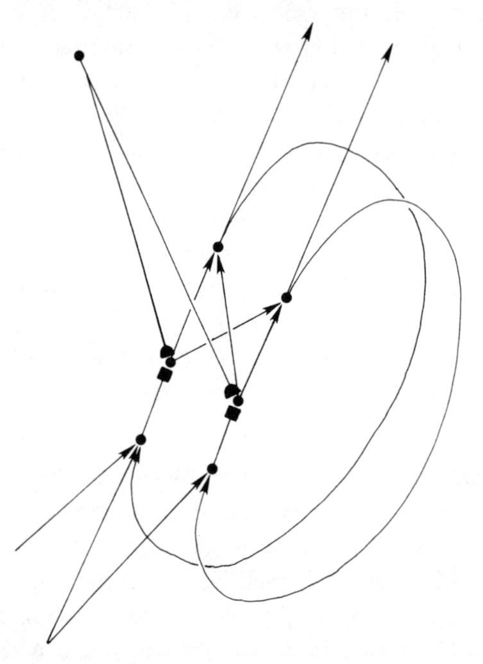

FIGURE 5. The incentive motivational outputs of the dipole feed back as excitatory inputs to the dipole, so that a cue can both sample and elicit a rebound.

The feedback loop is also important because it illustrates how an external cue which is correlated with offset of an aversive event can become a positive reinforcer that motivates stable nonchalant asymptotic avoidance. If this cue's internal representation turns on when the aversive cue representation turns off, then the LTM traces of the cue will sample the antagonistic rebound in the positive incentive channel that is caused by aversive event offset. On later trials, the indifferent cue is a source of positive incentive despite the fact that it was never correlated with a positive incentive cue.

This explanation of secondary conditioning leads to an intriguing prediction. If the feedback loop is transected at the correct place, then classical conditioning of emotional responses should still be possible via the direct conditionable pathways from internal representations to dipoles, but instrumental conditioning using offset of the classically conditioned cue as a reinforcer should be impossible since the feedback loop can no longer drive the rebound.

This explanation of secondary conditioning also uncovers a new design principle whose functional capabilities must be classified by mathematical analysis; namely, a feedback competitive network whose pathways are gated by slow transmitters and stimulated by specific inputs and expectancy-modulated nonspecific arousal.

11. Conditionable incentives and adaptive resonance. The incentive motivational outputs of the dipoles feed back to the internal representations of external cues, where they influence which cues will be stored in STM. The incentive motivational pathways to the cue representations are nonspecific so that any representation can, in principle, be modulated by any incentive. These pathways are conditionable to also prevent inappropriate cues from receiving the wrong incentives after conditioning occurs. I have elsewhere suggested that these conditionable pathways are formal analogs of the neural pathways *in vivo* which subserve the contingent negative variation (Grossberg [**1975**]).

The pathways from external cue representations to internal drive representations and the reverse pathways from drives to cues are both conditionable. The former pathways subserve conditioned reinforcer properties of the cues and the latter pathways subserve incentive motivational properties of the drives. Taken together, these feedback pathways constrain the adaptive resonances which can occur within these networks when the conditioned reinforcer properties of presently available cues are compatible with the incentive motivational properties of active drives.

12. Generalization gradient and STM normalization. The drive representations are joined together by competitive feedback pathways that choose the incentive pathway which has the most favorable combination of conditioned reinforcers, cue and internal drive inputs at each time for STM storage. The competitive feedback also maintains a steady level of incentive motivation (after initial reset transients such as overshoots die down) using the STM normalization property.

Competitive feedback interactions also join together the internal representations of external cues. These interactions choose for STM storage the populations which receive the most favorable combination of conditioned reinforcer and external cue inputs at each time. This STM competition is not always so sharply tuned that only one population at a time can reverberate in STM. Instead a quenching threshold determines which populations are active enough to be stored in STM, and which populations will be quenched (Grossberg [**1973**], [**1981a**]). Correspondingly, the STM normalization rule will not maintain a strictly constant total STM activity, but will nonetheless approximate a constant total STM activity.

In all, the external cue and internal drive representations are both designed according to similar rules that are the basis of their STM properties, the interfield pathways in both directions are both designed according to similar rules that are the basis of their LTM properties, and the whole structure is capable of resonant activity whereby STM and LTM mutually modify each other, albeit on different spatial and temporal scales.

In particular, net incentive signals can feed back to the cue representations as they compete to have their activities stored in STM. Cells that receive large positive incentive signals are favored in the STM competition because of the normalization property. The existence of this feedback pathway shows how a cue's energetic and conditioned reinforcer properties can interact with the network's drive properties at a given time to determine whether the cue's incentive feedback will abet or interfere with its STM storage and subsequent control of attentional processing. This interaction helps to explain how cues can be overshadowed (Grossberg [**1975**]) and how a subject's "information processing capacity" can vary with experimental conditions (Thomas et al. [**1970**]).

13. Behavioral contrast. Now we are ready to consider behavioral contrast. The behavioral contrast considered herein is related to a peak shift. Another type of contrast will be described when we review the data of Hinson and Staddon below.

Let a cue, such as a wavelength λ^+ of light, illuminate a key when the animal presses the key for food reinforcement. The internal representations of the cue λ^+, as well as of other key-related and situational cues, begin to acquire net positive incentive properties by being correlated with food. This happens because food presentation controls a large input to the positive incentive channel, which the LTM traces of the active cue representations sample. These attended cues also begin to learn a sensory expectancy of food via the LTM correlations that form along the pathways from their representations to the food-related cue representations. Then λ^+ is shut off and a new wavelength of light λ^- is turned on. Key pressing continues to be maintained by the active key-related and situational cues, other than λ^+, that have acquired net positive incentive properties.

Key pressing is not rewarded when λ^- is on. We can say that key pressing is unexpectedly nonrewarded in the following sense. The active cue representations activate the sensory expectancy that food will occur, in addition to their positive incentive pathway. The sensory expectancy is mismatched by the nonoccurrence of food. This "unexpected change for the worse" (Bloomfield [**1969**]) causes a negative incentive rebound. The negative rebound is caused by three types of competition acting together: the competition that controls expectancy mismatch (Grossberg [**1981a**, §14]), the competition between specific cue and nonspecific arousal channels (Grossberg [**1981a**, §5]), and the competition between on-cells and off-cells in the dipole. Just saying that "lateral inhibition is operating" is totally uninformative. The rebound is negative only because the active cue representations were eliciting net positive incentive when the expectancy mismatch occurred. In other words, "novelty" per se does not determine the sign of the rebound; the incentive value of active cues does this. A positive rebound would have been elicited by expectancy mismatch had the net incentive value been negative at the time of mismatch.

The negative rebound is sampled by all cue representations that are active when it occurs. The internal representations which are differentially sensitive to λ^- hereby acquire negative incentive properties. By contrast, representations which are active both when λ^+ and λ^- are presented sample both the positive and negative incentive pathways. Such cues include key-related and situational cues. The net incentive controlled by these cues tends to zero due to dipole competition. This is one mechanism whereby these cues become "irrelevant", and are overshadowed, even as λ^+ and λ^- learn to activate strong net positive and negative incentive reactions, respectively.

To see how STM normalization interacts with these incentive properties, suppose that each wavelength λ activates a generalization gradient $G_\lambda(\mu)$ of nearby wavelength-sensitive representations. In other words, a presentation of λ excites λ's representation with intensity $G_\lambda(\lambda)$ and excites μ's representation with intensity $G_\lambda(\mu)$, for every wavelength μ. For definiteness, let each gradient $G_\lambda(\mu)$ have a Gaussian shape centered at the representation of wavelength λ (Figure 6a). Suppose that on those learning trials when λ^+ occurs, each wavelength representation μ samples the incentive channels with an intensity proportional to its activity, namely $G_{\lambda^+}(\mu)$. Also suppose, for simplicity, that on these trials the positive incentive channel has unit intensity and the negative incentive channel has zero intensity. Then each wavelength representation μ eventually controls a positive incentive proportional to $G_{\lambda^+}(\mu)$. Similarly, suppose that on those learning trials when λ^- occurs, each wavelength representation μ samples the incentive channels with an intensity proportional to its activity, namely $G_{\lambda^-}(\mu)$. Again we suppose for simplicity that the negative incentive channel has unit intensity and the positive incentive channel has zero intensity. Then each wavelength representation μ eventually controls a negative incentive proportional to $G_{\lambda^-}(\mu)$.

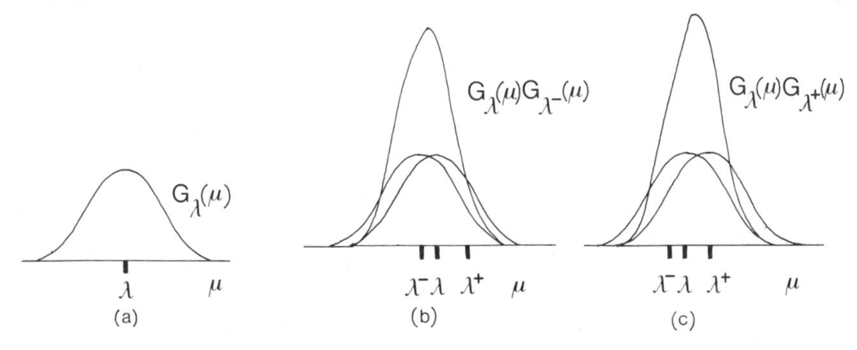

FIGURE 6. (a) An input with wavelength λ excites representations of other wavelengths μ with an intensity $G_\lambda(\mu)$. (b) Each wavelength representation μ gains control of a signal to the negative incentive channel whose strength is proportional to $G_{\lambda^-}(\mu)$. Then an input with wavelength λ excites a signal of strength proportional to $G_\lambda(\mu)G_{\lambda^-}(\mu)$ from every representation μ. (c) Each wavelength representation μ gains control of a signal to the positive incentive channel whose strength is proportional to $G_{\lambda^+}(\mu)$. Then an input with wavelength λ excites a signal of strength proportional to $G_\lambda(\mu)G_{\lambda^+}(\mu)$ from every representation μ.

At this point, let me pause to outline a general direction for parametric studies. It is not generally true that a dipole's on-reaction (when λ^+ is on) and its rebound off-reaction (when λ^- is on) have equal size. For example, suppose that we consider the rebound that is generated by offset of a cue (e.g. λ^+) which has elicited an input of intensity J to the dipole. Given a fixed value of J, the ratio of rebound size to on-reaction size is often an increasing function of the arousal level I (§5). In other words, the relative off-reaction to cue offset can get bigger if a higher arousal level is maintained. When this happens, the negative incentive properties of λ^- will be strengthened relative to the positive incentive properties of λ^+. We will see how this asymmetry can cause larger peak shift and behavioral contrast effects as we continue our argument below. Similarly, even if all cues stay on, a more unexpected event can cause a larger increment in I, which can cause a larger negative rebound, and thus more peak shift and behavioral contrast. Both these effects are modified by inverted U properties at underaroused and overaroused I levels (§7).

Having made these qualifications, I will continue analysing the case where the dipole on-reaction and off-reaction have equal strength. What happens when a wavelength λ is presented on extinction trials? First, λ activates each wavelength representation μ with intensity $G_\lambda(\mu)$. Due to prior learning, each μ controls a positive incentive pathway with strength proportional to $G_{\lambda^+}(\mu)$ and a negative incentive pathway with strength proportional to $G_{\lambda^-}(\mu)$. Thus the signal from μ to the positive incentive channel is proportional to $G_\lambda(\mu)G_{\lambda^+}(\mu)$, and the signal from μ to the negative incentive channel is proportional to $G_\lambda(\mu)G_{\lambda^-}(\mu)$. To get the *total* positive incentive signal, we sum over all μ to get

$$I^+(\lambda) = \int G_\lambda(\mu)G_{\lambda^+}(\mu)\, d\mu. \tag{45}$$

(Figure 6b). To get the *total* negative incentive signal, we sum over all μ to get

$$I^-(\lambda) = \int G_\lambda(\mu)G_{\lambda^-}(\mu)\, d\mu. \tag{46}$$

(Figure 6c). In general, the total inputs $I^+(\lambda)$ and $I^-(\lambda)$ are influenced by the dipole's arousal level before the competition acts to compute the net incentive. For simplicity, let us ignore the arousal level, and just suppose that the net positive incentive after dipole interaction is

$$N^+(\lambda) = \begin{cases} I^+(\lambda) - I^-(\lambda) & \text{if } I^+(\lambda) > I^-(\lambda), \\ 0 & \text{if } I^+(\lambda) \leq I^-(\lambda). \end{cases} \tag{47}$$

To understand behavioral contrast, we compare the effects of $I^+(\lambda)$ and of $N^+(\lambda)$ on STM competition. That is, we study how disconfirmed expectancies and/or the offset of positive incentive cues on learning trials can alter the STM activities of the wavelength representations that control performance on extinction trials.

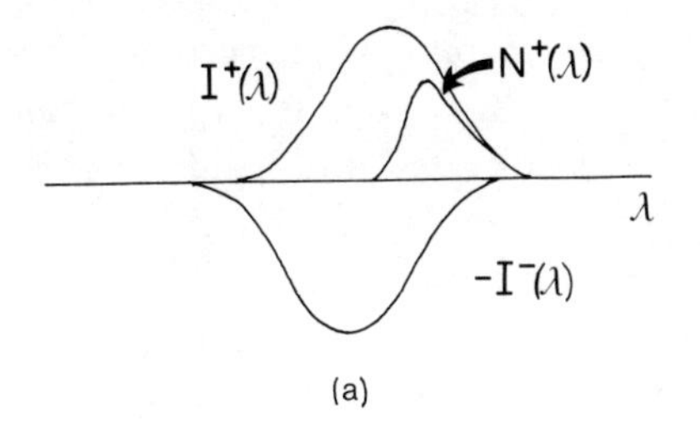

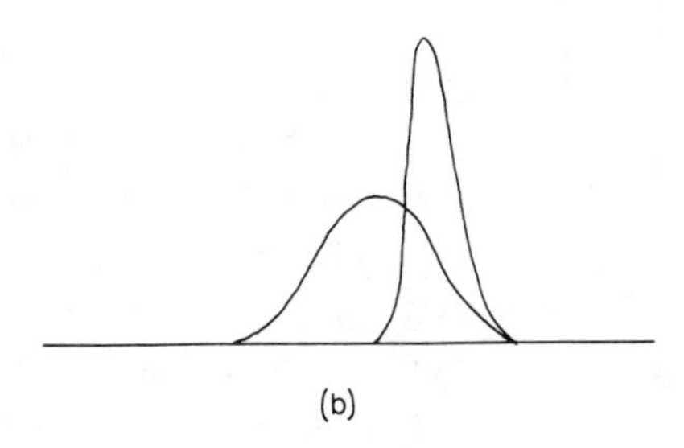

FIGURE 7. (a) The net gradient $N^+(\lambda)$ is lower than $I^+(\lambda)$ and is peak shifted away from λ^-. (b) The normalized net gradient of $N^+(\lambda)$ is higher than the normalized gradient of $I^+(\lambda)$.

The peak of $N^+(\lambda)$ is shifted away from λ^- because $N^+(\lambda)$ is the difference of $I^+(\lambda)$ and $I^-(\lambda)$ (Figure 7a). Also the height of the $N^+(\lambda)$ peak is less than the height of the $I^+(\lambda)$ peak because $N^+(\lambda)$ is the difference of $I^+(\lambda)$ and $I^-(\lambda)$. What then produces the *higher* peak-shifted peak that characterizes behavioral contrast? My answer is: STM normalization. Because gradient $N^+(\lambda)$ is the difference of $I^+(\lambda)$ and $I^-(\lambda)$, the $N^+(\lambda)$ gradient is *narrower* than the $I^+(\lambda)$ gradient. The normalization property says that STM competition tries to generate the same area under every gradient that it sees. In particular, STM competition will try to generate the same area in response to $I^+(\lambda)$ as it does in response to $N^+(\lambda)$. Since $N^+(\lambda)$ is narrower than $I^+(\lambda)$, after normalization takes place, the stored gradient of $N^+(\lambda)$ is higher than the stored gradient of $I^+(\lambda)$ (Figure 7b). This is behavioral contrast. In summary, STM competition bootstraps the lower but narrower net peak $N^+(\lambda)$ into a higher STM-stored peak than the STM-stored peak of $I^+(\lambda)$.

14. The Hinson and Staddon theory of schedule interactions. In their paper on positive and negative contrast effects within reinforcement experiments, Hinson and Staddon [**1978**] studied bar-pressing for milk reinforcement in rats. Their apparatus consisted of a Skinner box with a running wheel at one end and, at the other end, a response bar, a dipper feeder, and two stimulus lights. The sessions were forty minutes long and consisted of two alternating, one minute components each signalled by one or the other stimulus light.

After preliminary bar-press shaping, with access to the wheel blocked, the rats were broken into two groups. In condition 1 (prediscrimination), rats 1 and 2 had access to the wheel, whereas rats 3 and 4 did not. All rats were exposed to a multiple variable-interval 60 seconds, variable interval 60 seconds (mult VI 60 VI 60) schedule for milk reinforcement (Reynolds [**1968**, p. 90]). In other words, each stimulus light was turned on for one minute during which the rats' lever-presses were rewarded with milk on a VI schedule, after which the other stimulus light was turned on for one minute during which the rats' lever-presses were again rewarded with milk on a VI schedule. An interval schedule is one during which a certain amount of time must elapse irrespective of the animal's performance before it is rewarded. A VI schedule varies the length of this time interval across learning trials. Rats 1 and 2 could, therefore, either lever-press or wheel-turn at all times, whereas rats 3 and 4 could not wheel-turn. In condition 2 (discrimination), the schedule was altered to a multiple variable-interval 60 seconds, extinction (mult VI 60 Ext) schedule. During this phase, milk reinforcement was withheld when the second stimulus light was turned on. To balance experimental design, condition 3 reinstated the mult VI 60 VI 60 of condition 1. After responding stabilized in condition 3, conditions 1 and 2 were repeated, with animals interchanged, thereby exposing each rat to all four conditions.

Hinson and Staddon's experimental design explicitly manipulates the availability of wheel-turning, a response which can compete with lever-pressing. These authors suggest that an animal's terminal responses, or responses related to reinforcement (lever-pressing) compete for available time with the complementary class of interim responses, or responses not related to reinforcement (wheel-turning). They study how the frequency of lever-pressing and wheel-turning change when the extinction condition is imposed. During extinction, the VI schedule is maintained when the first light is on (unchanged component), but is withheld when the second light is on (changed component).

Hinson and Staddon argue that "In the first (prediscrimination) condition, . . . interim activities compete with terminal responses . . . leading to an intermediate level of terminal responding in both. In the second (discrimination) condition, however, there will be no terminal responding in the changed component, as a result of the absence of reinforcement; hence interim responding is free to increase. With this reallocation of interim activity into the changed component, the level of interim activity in the unchanged component is likely to decrease, reducing its inhibitory effect on the measured (terminal) response. This results in disinhibition of terminal responses in the unchanged component, thus producing positive behaviorial contrast. A similar, symmetrical account can be offered for negative contrast" (p. 432).

Hinson and Staddon show that the amount of positive contrast, as expressed by the increase in lever-pressing during the unchanged component (first light) of the discrimination condition is less with no wheel-available. They attribute this difference to "the degree of suppression of bar-pressing by the introduction of the wheel" (p. 433) during the prediscrimination trials. "The large positive contrast effect in the wheel-available condition was due to a lower rate in the unchanged component in the prediscrimination phase rather than to a higher rate in the discrimination phase" (p. 433).

Hinson and Staddon offer their theory as an alternative to several other theoretical positions. They criticize the "additivity theory" for offering no straightforward account of negative contrast and for finding difficulty explaining how positive contrast can be obtained with responses that are not induced by food-related stimuli. The additivity theory claims that terminal responses occur in the presence of stimuli that predict food and that the stimulus (e.g., the first light) in the unchanged component of a VI schedule elicits contrast when it is made more predictive by abolishing food delivery in the changed component.

Hinson and Staddon also criticize "relative value" theory, which claims that positive contrast is due to a "change for the worse" in the changed component during extinction trials. They claim that relative value theory cannot explain their data since "the change from VI to Ext in the changed component will entail a smaller reduction in the value of that component with the wheel available than without" (p. 434). They also claim (p. 434, footnote 15) that "the lack of contrast effects when reinforcement duration (as opposed to frequency) is varied is again inconsistent with a relative-value interpretation." By contrast, Hinson and Staddon view their emphasis on competition as a "commonsensical mechanism that may contribute to contrast effects, which seems to have escaped attention despite its simplicity" (p. 432).

I will argue that the Hinson and Staddon theory also falls into difficulties. A way out of these difficulties will be suggested which explains the Hinson and Staddon data and supports suitably refined "additivity theory" and "relative value theory" constructs. Before making explicit theoretical suggestions, I will place this discussion in a broader conceptual and historical framework.

Two problems faced by many psychological theories, even a relatively good theory like Hinson and Staddon's theory, are solipsism and the tendency to throw out the baby with the bathwater. Solipsism is fostered when authors cannot recognize their personal intuitions in the writings of other authors. For example, the dynamics of competition in discrimination learning generally, and behavioral contrast in particular, had been rigorously studied before Hinson and Staddon offered their "commonsensical" approach (Grossberg [**1975**]). Perhaps Hinson and Staddon overlooked these prior studies because competition is just one of the mechanisms that these studies invoke to explain schedule interactions. In particular, Hinson and Staddon say nothing about drive and reinforcement mechanisms which, I contend, permit wheel-turning to increase during the changed component of the discrimination sessions, and which thereupon permit wheel-turning to decrease during the unchanged component of the discrimination sessions. In a theory which says nothing about reinforcement mechanisms, what prevents us from expecting that frustrating lever-pressing during the changed component of the extinction sessions enhances the incentive value of wheel-turning, and that this heightened incentive thereupon generalizes to the unchanged component of the extinction sessions to yield more wheel-turning there too, instead of less wheel-turning? The "relative value theory" which Hinson and Staddon criticize at least attempts to define a reinforcement rule. The "additivity theory" at least attempts to identify the importance of predictable events and, conversely, of unexpected events in the control of reinforced behavior. Hinson and Staddon throw out the baby with the bathwater by ignoring reinforcement and expectancy mechanisms to place competition in center stage.

Such errors of overemphasis can lead us to reject ideas which contain as much truth, and falsehood (!), as the ideas which we embrace in their stead. For example, it is simply false that relative value theory is necessarily inconsistent with "the lack of contrast effects when reinforcement duration (as opposed to frequency) is varied." Antagonistic rebounds are driven by the onset or offset of events, not by their duration. Moreover, whereas all of Hinson and Staddon's rats lever-press less during the prediscrimination phase if a wheel is available than if it is not, three out of four of these rats lever-press *more* during the unchanged component of the discrimination phase if a wheel is available than if it is not. These data are paradoxical if all that is happening is that the wheel is competing with the lever press during the unchanged component of the discrimination phase, albeit less than it does during the changed component of the discrimination phase.

Such problems of detail are not the only ones faced by the Hinson and Staddon theory. To explain an animal's behavior as Hinson and Staddon do in terms of "limitations of time on the animal's ability to engage in different activities" (p. 432) does not address the basic question: what factors influence the animal's behavior at *each* time? If the animal is executing a goal-oriented plan whose components are organized serially in time, then one can properly argue that there is not enough time to execute the whole plan, because at each instant of time the unfolding of the plan describes how the animal's behavior is being controlled. However, if an animal is engaged in a series of acts whose commands get reset as time proceeds, then an explanation of its behavior must isolate the factors that control the animal's behavior at each choice point in time. Had the Hinson and Staddon theory faced the problem of determining the factors which control behavior at each time, the inadequacy of saying that wheel-turning would be "likely to decrease" (p. 432) during the unchanged component because it was previously so vigorous during the changed component would have been apparent. The authors would have been forced to ask: what signal which controls wheel-turning during the unchanged component was altered by the higher level of wheel-turning during the changed component?

15. Explaining the several types of behavioral contrast in real-time. An interpretation of the Hinson and Staddon data in terms of network mechanisms is sketched below. Consider the animal's behavior during the prediscrimination phase when a wheel is available. The animal shuttles between lever and wheel, driven first by such generalized drive sources as exploratory motor arousal, and gradually building up a cognitive expectancy and a positive net incentive due to fluid delivery at the lever on the VI schedule. Then the animal's approach to the lever is also motivated by conditioned positive incentives due to fluid reward. At any choice point in time, the internal representations of the cues controlling observable behavior compete in STM, bolstered by the net incentives that feed back to them at that time. Superimposed on this balance of incentives and STM competitive effects at any time is a temporal oscillation in relative incentive strength due to changes in internal drive inputs such that fluid incentives decrease after drinking, and motor incentives decrease after wheel-turning. As the expectancy of reward develops, the animal can experience negative rebounds (frustration) after nonrewarded lever-presses due to the VI schedule, but experiences no frustration, and possibly some complementary positive rebounds, as it turns from the lever after nonreward and approaches the wheel.

At least two types of effects, apart from competitive effects, need to be discussed in comparing the wheel and no-wheel situation: changes of internal drive levels through time, and changes in conditioned reinforcer levels through time. The conditioned reinforcer comparison is based on the fact that rats lever-press less in the prediscrimination component when the wheel is available than when the wheel is not available. The internal drive level comparison is based on the increased rate of wheel-turning during the changed component that is caused by the conditioned reinforcer change.

The conditioned reinforcer change can be summarized as follows. Due to the VI reinforcement schedule, during the prediscrimination phase when behavior alternates between lever and wheel, the animal emits fewer lever-presses per reward than during the prediscrimination phase with no wheel present. Consequently, the animal can learn to expect more reward per lever-press than in the wheel-absent case. When extinction trials begin, the disconfirmation of this greater expectancy can yield larger negative rebounds, which after sampling by active cue representations would make the net positive incentives that motivate lever-pressing during the changed component smaller in the wheel-present case than in the wheel-absent case. Simultaneously, the frustration reduction due to wheel approach can be greater and can drive larger positive rebounds. Thus the "relative value" of wheel-turning, and interim activities generally, can become greater in the wheel-present case than in the wheel-absent case, thereby causing a large increase in wheel-turning rate during the changed component of the extinction trials. During the unchanged component, animals in the wheel-present case can experience larger positive rebounds when reward occurs because, having built up stronger expectancies of no-reward, disconfirming these expectancies can drive larger positive rebounds. Consequently, the net positive incentive sampled by lever-press cues can increase more in the wheel-present case than in the wheel-absent case. These two effects can qualitatively explain the larger relative increments in lever-pressing during the unchanged component and the larger relative, but not necessarily absolute, decrements in lever-pressing during the changed component.

The above argument notes how the change from VI to Ext in the changed component can entail a *larger* reduction in the value of that component with the wheel available than without, by contrast with the authors' claim (p. 434). This argument is made cautiously, however, because it depends on whether the animals cognitively group, or chunk, together whole sequences of lever-presses into individual commands during the prediscrimination phase. For example, although animals in the wheel-absent prediscrimination phase typically push the lever more than animals in the wheel-present phase, it is possible that animals in the former case group together lever-presses into bursts that are each controlled by a single cognitive command. Then a smaller number of lever-press commands than of lever-presses can occur between rewards in both the wheel-present and the wheel-absent cases. Because the wheel is at the opposite end of the Skinner box than the lever, it is likely that more commands are reset per reward in the no-wheel case than in the wheel-present case. Were this not the case, disconfirming the expectancy of reward in the two situations during extinction trials could have comparable rebound effects, other things equal. Thus an animal's cognitive abilities and strategies can influence its incentive changes through time, and these capabilities surely vary across species (Bitterman [**1972**]). Other influences of cognitive variables on the effects of reinforcement have been actively studied since the experiments of Buchwald [**1967**].

The above argument concerning rebound size is made with caution also because rebound size can depend on the net arousal level I within an animal's motivational dipoles according to an inverted U function (§7), and the arousal level might differ in the wheel and no-wheel cases. Also the relative rebound size can vary as a function of I (§§5 and 6). A more subtle influence on the net arousal level is due to changes in the cues that are simultaneously stored in STM. The net incentive after a rebound is sampled does not depend just on the difference of the total input strengths to the complementary dipole channels. Rather, the net incentive is described by a function that resembles the difference of the ratios of the total inputs to the complementary channels (§5). Adding a wheel to a Skinner box can decrease the ratio of the inputs that control lever-pressing, and thereby decrease the positive incentive which is initially controlled by lever-press cues. Such an effect can reduce the number of lever-presses, as occurs in the Hinson and Staddon data, by lowering the probability of choosing the lever at times when the lever and the wheel are both simultaneously stored in STM. Such an effect could also alter the rebound size if lever cues continued to be stored in STM at times when the lever-press was extinguished. This latter possibility is minimized in Hinson and Staddon's apparatus by placing the lever and the wheel at opposite ends of the Skinner box, but it could become an important factor were the lever and the wheel placed side-by-side in the Skinner box. Thus, by either of two effects, the very existence of competing responses can influence the amount of contrast, which is one of Hinson and Staddon's main points. Despite these cautions, however, Hinson and Staddon's claim that a smaller reduction in value *necessarily* follows the change from VI to Ext in the wheel-present case is not supported when the effects of expectancy disconfirmation on net incentive are taken into account.

Can we explain the reduced wheel-turning during the unchanged component using only these facts? To do so, we would need to argue that the difference between the positive incentives controlled by lever cues vs. wheel cues is greater during the unchanged component with wheel present than with wheel absent. This conclusion follows from our previous arguments if the wheel cues do not acquire significantly stronger conditioned reinforcer properties during the changed component. This possibility raises an issue that is worth studying in its own right: when sensory STM is reset after a negative rebound occurs, under what circumstances will newly stored cues benefit from a secondary positive rebound?

Another possible factor which can reduce wheel turning during the unchanged component is an internal drive change. In particular, turning the wheel a great deal during the changed component can reduce the motor drive input which supports the motivation for wheel-turning. The reduction of this drive input can facilitate lever-pressing during the unchanged component by enabling lever-press cues more easily to win the STM competition at behavioral choice points. This type of disinhibitory effect is compatible with Hinson and Staddon's viewpoint, but it is based on a drive input concept that Hinson and Staddon do not mention, and it is insufficient to explain their data without also invoking net incentive changes due to rebounds that occur subsequent to unexpected events. Whether such a temporal oscillation in drive input occurs can possibly be tested by varying the duration of the changed component, and looking for decreases in wheel-turning rates towards the end of the changed component intervals.

Finally it is worth emphasizing two more points: The contrast effects due to rebounds are not transient just because the rebounds themselves are transient; if the rebound is sampled by a conditioned reinforcer pathway, then the cues that control this pathway can read out of LTM a long-term contrast effect when the pattern of their net incentive motivational feedback is acted on by the STM normalization property. Also the type of contrast which Hinson and Staddon have considered differs from the contrast that occurs during a peak shift (§13). In Hinson and Staddon's data, an important factor is how different expectancies can develop with or without a wheel present, and can thereupon alter the net incentive controlled by a lever-press cue. In peak shift data, an important factor is how several related cues, after acquiring positive or negative incentive value, compete for storage in STM constrained by the STM normalization property. In the light of this discussion, it would appear that behavioral contrast is not a unitary behaviorial entity and cannot be explained by any single mechanism.

16. Concluding remarks: Real-time analysis of multicomponent behavioral mechanisms. The above considerations suggest an alternative to Hinson and Staddon's claim that "contrast results from changes in relative reinforcement rate" (p. 432). One might begin by saying that contrast can result from a complex interaction between at least four types of competition acting in different proportions, and two transmitter systems acting on different time scales. For example, cues whose generalization gradients are narrowed by incentive feedback due to dipole competition between on-reactions and off-rebounds can cause behavioral contrast when their narrowed gradients are heightened by an STM storage process, if these stored representations thereupon release motor commands that control observable behavior.

One's first reaction might be to balk at the complexity that such a multicomponent description seems to entail. Such basic components are ignored, however, at the peril of being continually driven into error and paradox. For example, greater expectancy mismatch and arousal sometimes, but not always, yield a larger contrast effect. If the ambient arousal level is too high, they do not because the rebound is reduced by the inverted U law which results from transmitter adaptation and dipole competition. If all cues remain active before and after the mismatch, they do not because all the cues control small net incentive due to dipole competition between their conditioned reinforcer signals. If the cues that are on before the mismatch and on after the mismatch are too dissimilar, they do not if they are separately normalized. Thus all the component processes can be differentially varied to amplify or to cancel each other as a function of experimental manipulations and internal parameter changes. More importantly, all the component processes are needed to build up and stabilize the behavioral codes which lead to behavioral contrast effects as a special property. Such considerations suggest that it is better to avoid all one sentence generalities, and to study the interplay of underlying behavioral mechanisms as they react in real-time to different experimental conditions.

REFERENCES

Berlyne, D. E., 1969. *The reward-value of indifferent stimulation*, Reinforcement and Behavior (J. T. Tapp, ed.), Academic Press, New York.

Bitterman, M. E., 1972. *Phyletic differences in learning*, Biological Boundaries of Learning (M. E. P. Seligman and J. L. Hager, eds.), Appleton-Century-Crofts, New York.

Bloomfield, T. M., 1969. *Behavioral contrast and the peak shift*, Animal Discrimination Learning (R. M. Gilbert and N. S. Sutherland, eds.), Academic Press, New York.

Bolles, R. C., 1967. *Theory of motivation*, Harper and Row, New York.

Buchwald, A. M., 1967. *Effects of immediate vs. delayed outcomes in associative learning*, J. Verbal Learning and Verbal Behavior **6**, 317–320.

Butcher, L. L., ed., 1978. *Cholinergic-monoaminergic interactions in the brain*, Academic Press, New York.

Cox, V. C., J. W. Kakolewiski and E. S. Valenstein, 1969. *Inhibition of eating and drinking following hypothalamic stimulation in the rat*, J. Comparative and Physiological Psychology **68**, 530–535.

Denny, M. R., 1971. *Relaxation theory and experiments*, Aversive Conditioning and Learning, (F. R. Brush, ed.), Academic Press, New York.

Dunham, P. J., 1971. *Punishment: Method and theory*, Psych. Rev. **78**, 58–70.

Estes, W. K., 1969. *Outine of a theory of punishment*, Punishment and Aversive Behavior (B. A. Campbell and R. M. Church, eds.), Appleton-Century-Crofts, New York.

Estes, W. K. and B. F. Skinner, 1941. *Some quantitative properties of anxiety*, J. Exper. Psych. **29**, 390–400.

Friedhoff, A. J., ed., 1975a. *Catecholamines and behavior*. I. *Basic neurobiology*, Plenum Press, New York.

______, ed., 1975b. *Catecholamines and behavior*. II. *Neuropsychopharmacology*, Plenum Press, New York.

Grossberg, S.,1964. *The theory of embedding fields with applications to psychology and neurophysiology*, Rockefeller Institute for Medical Research, New York.

______, 1968. *Some physiological and biochemical consequences of psychological postulates*, Proc. Nat. Acad. Sci. U.S.A. **60**, 758–765.

______, 1969. *On the production and release of chemical transmitters and related topics in cellular control*, J. Theoret. Biol. **22**, 323–364.

______, 1971. *On the dynamics of operant conditioning*, J. Theoret. Biol. **33**, 225–255.

______, 1972a. *A neural theory of punishment and avoidance*. I. *Qualitative theory*, Math. Biosci. **15**, 39–67.

______, 1972b. *A neural theory of punishment and avoidance*. II. *Quantitative theory*, Math. Biosci. **15**, 253–285.

______, 1973. *Contour enhancement, short-term memory, and constancies in reverberating neural networks*, Stud. Appl. Math. **52**, 217–257.

______, 1975. *A neural model of attention, reinforcement, and discrimination learning*, Internat. Rev. Neurobiology **18**, 263–327.

______, 1976a. *Adaptive pattern classification and universal recoding*. I. *Parallel development and coding of neural feature detectors*, Biol. Cybernet. **23**, 121–134.

______, 1976b. *Adaptive pattern classification and universal recoding*. II. *Feedback, expectation, olfaction, illusions*, Biol. Cybernet. **23**, 187–202.

______, 1981a. *Adaptive resonance in development, perception and cognition*, these PROCEEDINGS.

______, 1981b. *A psychophysiological theory of normal and abnormal motivated behavior* (submitted).

Hebb, D. O., 1955. *Drives and the CNS (conceptual nervous system)*, Psych. Rev. **62**, 243–254.

Hinson, J. M. and J. E. R. Staddon, 1978. *Behaviorial competition: A mechanism for schedule interactions*, Science **202**, 432–434.

Mackintosh, N. J., 1974. *The psychology of animal learning*, Academic Press, New York.

Maier, S. F., M. E. P. Seligman and R. L. Solomon, 1969. *Pavlovian fear conditioning and learned helplessness effects on escape and avoidance behavior of* (a) *the CS-US contingency and* (b) *the independence of the US and voluntary responding*, Punishment and Aversive Behavior (B. A. Campbell and R. M. Church, eds.), Appleton-Century-Crofts, New York.

Masterson, F. A., 1970. *Is termination of a warning signal an effective reward for the rat?* J. Comparative and Physiological Psychology **72**, 471–475.

McAllister, W. R. and D. E. McAllister, 1970. *Behavioral measurement of conditioned fear*, Aversive Conditioning and Learning (F. R. Brush, ed.), Academic Press, New York.

McAllister, W. R., D. E. McAllister and W. K. Douglass, 1971. *The inverse relationship between shock intensity and shuttlebox avoidance learning in rats*, J. Comparative and Physiological Psychology **74**, 426–433.

Novin, D., W. Wyrwicka and G. A. Bray, eds., 1976. *Hunger: Basic mechanisms and clinical implications*, Raven Press, New York.

Olds, J., 1977. *Drives and reinforcements: Behaviorial studies of hypothalamic functions*, Raven Press, New York.

Reichsman, F., ed., 1972. *Hunger and satiety in health and disease*, Karger, Basel, Switzerland.

Reynolds, G. S., 1968. *A primer of operant conditioning*, Scott, Foresman and Co., Glenview, Ill.

Seligman, M. E. P. and J. C. Johnston, 1973. *A cognitive theory of avoidance learning*, Contemporary Approaches to Conditioning and Learning (F. J. McGuigan and D. B. Lumsden, eds.), Winston and Sons, Washington, D. C.

Solomon, R. L., L. J. Kamin and L. C. Wynne, 1953. *Traumatic avoidance learning: The outcomes of several extinction procedures with dogs*, J. Abnormal and Social Psychology **48**, 291–302.

Stricker, E. M. and M. J. Zigmond, 1976. *Recovery of function after damage to central catecholamine-containing neurons: A neurochemical model for the lateral hypothalamic syndrome*, Progress in Psychobiology and Physiological Psychology (J. M. Sprague and A. N. Epstein, eds.), Academic Press, New York.

Thomas, D. R., F. Freeman, J. C. Svinicki, D. E. Scott Burr and J. Lyons, 1970. *Effects of extradimensional training on stimulus generalization*, J. Exper. Psych. monograph **83**, 1–21.

DEPARTMENT OF MATHEMATICS, BOSTON UNIVERSITY, BOSTON, MASSACHUSETTS 02215

SIAM-AMS Proceedings
Volume 13
1981

Sociobiological Variations on a Mendelian Theme

M. FRANK NORMAN

1. Introduction. My articles in this volume illustrate two aspects of my work as a psychologically oriented mathematician. The present article shows elementary mathematics in the service of psychology. The other article [**1**] shows elementary psychology in the service of mathematics. Both papers bear on simple genetic models: deterministic models in this paper and stochastic models in the other one.

In psychology one frequently encounters facile evolutionary explanations of contemporary behaviors and their neural substrates. It is often easy to convince oneself that a certain characteristic of a currently predominant genotype permitted individuals of that genotype to have more offspring than conspecifics, thus ensuring the success of the genotype. The (modest) interest in such exercises is partially predicated on the validity of the transition between individual reproductive success ("fitness") and long-term success of the genotype. This paper shows that such transitions are not always valid.

2. Fitness and survival. According to E. O. Wilson, "Hamilton's theorem on altruism consists merely of a more general restatement of the basic axiom that *genotypes increase in frequency if their relative fitness is greater*" [**2**, pp. 415–416, italics added]. W. D. Hamilton's theory of the evolution of altruism will be considered in §4. The present section relates to the italicized proposition, which is an unusually explicit statement of a dominant theme of the literature of Evolutionary Biology.

Consider the following example. Suppose that there are three interbreeding varieties of zebras in a certain region. Call these varieties a, b, and c. They differ in probability of survival from conception to reproductive age, perhaps because of different diets. The common term for this survival probability is *viability*. Assuming that all varieties are equally fertile, viability is proportional to, and can thus be identified with, the expected number of offspring of a newly

1980 *Mathematics Subject Classification*. Primary 92A10, 92A20, 92A25.

conceived zebra. This number is termed *fitness*, and denoted w_a, w_b, w_c for the three varieties. Suppose, for definiteness, that $w_a = 1/5$, $w_b = 3/5$, $w_c = 4/5$. Obviously, the relative frequency of c will increase from generation to generation, in accordance with Wilson's proposition, and this variety will eventually displace the other two, in accordance with the notion of "survival of the fittest."

This is "obvious," but it need not be true. As in most casual evolutionary arguments, I have said nothing about the genetic structure of my zebras. This is tantamount to assuming that genetic structure is irrelevant to the course of evolution. This is a dangerous assumption, to say the least.

Suppose, for example, that the differences between a, b, and c are controlled by one genetic locus with two alleles, A_1 and A_2, and that the genotypes of a, b, and c are $a = A_2A_2$, $b = A_1A_1$, $c = A_1A_2$. The fitnesses w_a, w_b, and w_c will henceforth be denoted w_{22}, w_{11}, and w_{12}. Make all of the standard simplifying assumptions: random mating, infinite population, and nonoverlapping generations. Let

p_n = proportion of A_1 genes among newly conceived individuals in the nth generation

and $q_n = 1 - p_n$. Then the proportions of A_1A_1, A_1A_2, and A_2A_2 are p_n^2, $2p_nq_n$, q_n^2 at conception (this is the Hardy-Weinberg law [3, p. 4]) and proportional to $w_{11}p_n^2$, $2w_{12}p_nq_n$, $w_{22}q_n^2$ among adults, from which it follows that

$$p_{n+1} = \frac{w_{11}p_n^2 + w_{12}p_nq_n}{w_{11}p_n^2 + 2w_{12}p_nq_n + w_{22}q_n^2}. \tag{2.1}$$

Since the heterozygote, $A_1A_2 = c$, has greater fitness than either homozygote, p_n converges to an internal equilibrium value,

$$p_\infty = \frac{w_{12} - w_{22}}{2w_{12} - w_{11} - w_{22}} = \frac{3}{4}.$$

The asymptotic genotype proportions, derived from the Hardy-Weinberg law, are given in Table 1.

TABLE 1
Fitness and asymptotic frequency

Variety	Genotype	Fitness	Frequency
a	A_2A_2	1/5	1/16
b	A_1A_1	3/5	9/16
c	A_1A_2	4/5	6/16

Clearly "survival of the fittest" does not apply here, if this phrase is understood to imply that the fittest variety displaces all competitors. In our example, all three varieties remain in appreciable frequencies. Moreover, the second fittest variety, b, is asymptotically more numerous than the fittest variety, c. Nor is it correct that the fittest variety always increases in frequency, as Wilson alleges. For it can be shown that, if $p_1 < p_\infty$, then p_n increases to $p_\infty = 3/4$. Hence $P(c) = 2p_n q_n$ decreases for all n sufficiently large that $p_n > \frac{1}{2}$.

It is sometimes claimed that the theory of evolution by natural selection is vacuous, since its mechanism is "survival of the fittest" and fitness is often defined as survival probability. This conundrum arises from an ambiguity in the word "survival": it is used to refer to both individuals and varieties. Our example shows that superior individual survival probability (fitness) does not guarantee superior asymptotic frequency of the corresponding variety. (For a fuller discussion of the claim that natural selection is tautological, together with a different resolution, see [**4**, Chapter 4].)

The notion that one may equate individual and variety survival is deeply ingrained in biology. This notion can lead to error because it ignores the subtleties of Mendelian genetics. Its persistence is perhaps partly explicable by the long delay between publication of *The Origin of Species* and the rediscovery and acceptance of Mendel's ideas. Moreover, the linkage between fitness and asymptotic prevalence is often as direct as anyone could wish. If

$$\min\{w_{11}, w_{22}\} \leqslant w_{12} < \max\{w_{11}, w_{22}\},$$

then the fittest genotype (either A_1A_1 or A_2A_2) displaces all others at asymptote. However, if $\min\{w_{11}, w_{22}\} < w_{12} = \max\{w_{11}, w_{22}\}$, then the simple fitness-prevalence linkage again fails, since A_1A_2 is completely displaced by a homozygote with the same fitness.

Our findings in this section apply with equal force to the evolution of gross anatomical structures and to the evolution of neural microstructures that influence behavior. Subsequent sections have a specifically behavioral focus. Variants of the example presented in this section will be used to expose weaknesses in two popular theories of the evolution of social behavior: Maynard Smith's theory of the evolution of behavior in conflicts between animals, and Hamilton's theory of the evolution of altruism.

3. Evolutionarily stable strategies. The central notion in Maynard Smith's theory [**5**] is the *evolutionarily stable strategy* or *ESS*. Roughly speaking, an ESS is a frequency distribution over alternative genetically controlled behavior patterns that is resistant to incursion by small mutant or migrant groups. Stated in this way, it is possible to apply the notion directly to our three types of zebras with different dietary preferences and corresponding fitnesses $w_a = 1/5$, $w_b = 3/5$, $w_c = 4/5$. From a naturalistic, genetically naive, viewpoint, it would appear that the ESS for zebras is for them all to be of type c. For $w_c > w_a$ and $w_c > w_b$, so mutants or migrants of types a and b would be at a disadvantage in a population of individuals of type c (so the argument goes). However, the "all c" strategy is not, in fact, stable in the ordinary, dynamic, sense, if, as before, type c zebras are A_1A_2 heterozygotes. For if all zebras were of type c in the nth generation, then, in the next generation, types a, b, and c would have frequencies 1/4, 1/4, and 1/2. Another way of putting it is that the "all c" strategy is not available to the population under natural conditions. It ignores the constraints of Mendelian segregation. Oster and Rocklin [**6**, p. 31] have also noted the genetic naiveté of Maynard Smith's theory.

It may reasonably be objected that the preceding paragraph slights ESS by generalizing it beyond the game-theoretic context in which it was formulated. This context will now be described. Suppose that a population consists of k behaviorally distinct varieties of a certain species, which we label $1, 2, \ldots, k$. The distinctive behaviors are displayed in certain interactions (e.g., territorial conflicts) with other members of the population. We suppose, for simplicity, that an individual has exactly one such interaction in his lifetime, and that this interaction determines his fitness. The fitness of an i that interacts with a j is $W(i,j)$. Consequently, the expected fitness of an i that interacts with a randomly chosen opponent is $W(i, Q) = \sum_{j=1}^{k} W(i,j)Q_j$, where Q_j is the probability that the other animal is a j. Finally, if both contestants are randomly chosen, the expected fitness of the first is

$$W(P, Q) = \sum_{i=1}^{k} \sum_{j=1}^{k} P_i W(i,j) Q_j.$$

Here one should think of choosing the two contestants independently from subpopulations with respective distributions P and Q.

The definition of an ESS is formulated in terms of the expected fitness function. For any P, Q, and $\varepsilon > 0$, let $P_\varepsilon = (1 - \varepsilon)P + \varepsilon Q$.

DEFINITION. A distribution, P, is an *ESS* if, for every $Q \neq P$,

$$W(P, P_\varepsilon) > W(Q, P_\varepsilon),$$

for all sufficiently small ε.

Here Q corresponds to a mutant or migrant subpopulation attempting, unsuccessfully, to "gain a foothold" against an established population with distribution P. This formulation of the definition of ESS follows [**7**].

Maynard Smith proposes that distributions occurring in nature should be evolutionarily stable in this sense. This is an appealing notion, but it may be inconsistent with certain models for the genetic substrate, as we can see by reconsidering our zebras. Thus $k = 3$, and varieties 1, 2, and 3 correspond to our earlier $a = A_2A_2$, $b = A_1A_1$, and $c = A_1A_2$. Moreover, we assume that the fitnesses given earlier apply here, regardless of the other animal's behavior. In other words, $W(1,j) = 1/5$, $W(2,j) = 3/5$, $W(3,j) = 4/5$, for $j = 1, 2, 3$. Thus we have simply embedded our earlier example in the present game-theoretic framework. As was noted previously, the distribution that evolves is

$$P_1 = 1/16, \quad P_2 = 9/16, \quad P_3 = 6/16.$$

This distribution is not an ESS. It is easy to show that the only ESS is our old friend, the distribution concentrated on the heterozygous variety, 3.

The assumption that fitness is independent of the other animal's behavior makes this a degenerate example of an "interaction," much less a "conflict." However, I see nothing to prevent such inconsistencies between ESS and genetics from arising in bona fide conflict situations. The gap between ESS and genetics is clearly revealed by the fact that the rationale for ESS, such as it is, applies with equal force to interspecific and intraspecific conflict, while only the latter situation imposes genetic constraints.

Our example shows that the ESS notion is incompatible with *one* possible specification of genetic substrate. Perhaps it is more interesting to ask whether there is *any* genetic substrate with which ESS is consistent. I was surprised to discover, quite recently, that the answer is "yes." The anomalies of the zebra example are traceable to the superior fitness of heterozygotes. This led me to consider a scheme that assigns intermediate fitness to heterozygotes.

Suppose that there are only two behavioral phenotypes, 1 and 2 (e.g., the "hawks" and "doves" discussed in [5]). Underlying these are three genotypes, A_1A_1, A_2A_2, and A_1A_2. All A_1A_1 homozygotes, and a proportion, γ, of the heterozygotes are assumed to have phenotype 1; the remaining heterozygotes and the A_2A_2 homozygotes have phenotype 2. As in §2, let p_n be the proportion of A_1 genes in the nth generation, and let P_n be the corresponding distribution over the two phenotypes. Then $P_{n,1} = p_n^2 + 2\gamma p_n q_n$. Genotypes A_2A_2, A_1A_1, and A_1A_2 have respective fitnesses $W(2, P_n)$, $W(1, P_n)$, and $\gamma W(1, P_n) + (1 - \gamma) W(2, P_n)$, and the trajectory of the system is determined by (2.1) with w_{22}, w_{11}, and w_{12} replaced by these values.

For this model, evolutionary stability and dynamic stability are synonymous. This can be seen from the following catalog of cases corresponding to different values of the parameters

$$\delta_1 = W(1, 1) - W(2, 1) \quad \text{and} \quad \delta_2 = W(2, 2) - W(1, 2).$$

Proofs are omitted. It is assumed that $0 < P_{1,1} < 1$, and P_∞ denotes the limit of P_n as $n \to \infty$.

(a) $\delta_1 = 0$ and $\delta_2 = 0$. Then there is no ESS, and $P_\infty = P_1$.

(b.1) $\delta_1 \geqslant 0$, $\delta_2 \leqslant 0$, and at least one of these inequalities is strict. Then the only ESS is $P_\infty = (1, 0)$.

(b.2) $\delta_1 \leqslant 0$, $\delta_2 \geqslant 0$, and at least one of these inequalities is strict. Then the only ESS is $P_\infty = (0, 1)$.

(c) $\delta_1 < 0$ and $\delta_2 < 0$. Then the only ESS is $P_\infty = (\alpha, 1 - \alpha)$, where $\alpha = \delta_2/(\delta_1 + \delta_2)$.

(d) $\delta_1 > 0$ and $\delta_2 > 0$. Then both (1, 0) and (0, 1) are ESSs. Moreover

$$P_\infty = \begin{cases} (1, 0) & \text{if } P_{1,1} > \alpha, \\ (\alpha, 1 - \alpha) & \text{if } P_{1,1} = \alpha, \\ (0, 1) & \text{if } P_{1,1} < \alpha. \end{cases}$$

This model can be generalized in an obvious way to more than two behavioral phenotypes. It remains to be seen whether the identity between evolutionary and dynamic stability holds for the generalization. Moreover, there is no reason to believe that such models accurately represent the genetic contribution to animals' behavior in conflict situations.

In this section we have seen that it is possible to find a genetic model compatible with Maynard Smith's theory, as well as a model that is not compatible. This leaves one uneasy about most of the literature on ESS, since this literature characteristically avoids explicit consideration of genetic substrate. Two exceptions are [7] and [**8**], which, however, concentrate on behaviorally uninteresting haploid organisms.

4. The evolution of altruism. Altruistic behavior involves personal sacrifice for the benefit of others. In the present context, we must think of loss of personal fitness in the process of increasing the fitness of others. If genes that promote altruism are shared by the beneficiaries of the altruism, these genes may increase in frequency. This mechanism could favor evolution of sacrifice for the benefit of relatives.

As one would expect, the necessity of considering both donors and recipients imposes a certain inherent complexity on models for the evolution of altruism. However, for the important case of parental altruism, it is possible to formulate a relatively simple model.

Bidding goodby to our zebras, let us consider an avian species where both parents make equally important contributions to nestlings' survival. Their activities in this regard (incubating eggs, feeding chicks, guarding the nest, etc.) endanger their personal survival and are thus altruistic. Suppose, for simplicity, that the entire complex of parental behavior is controlled by one locus with two alleles, A_1 and A_2 (our usual assumption). We now assume, however, that there are two parameters, u_{ij} and v_{ij}, associated with the genotype A_iA_j. The parameter u_{ij}, called *viability*, is the probability that an A_iA_j survives to reproductive age, given that he survives until he leaves the nest. The other parameter, v_{ij}, called *nurturance*, is a measure of the quality of parental care provided by A_iA_j. Specifically, $v_{ij}v_{km}$ is the probability that an offspring of A_iA_j and A_kA_m parents survives until he leaves the nest. The v parameters in models of this kind are usually described as fertilities [**9**], but it is understood that what I call nurturance is an important component of genetic fertility parameters [**10**, p. 51]. For my present illustrative purposes, I am essentially assuming that nurturance is the only component of fertility.

Let p_n be A_1 gene frequency among birds leaving the nest in the nth generation. It is not difficult to show that this quantity satisfies (2.1), if we take

$$w_{ij} = u_{ij}v_{ij} = \text{viability} \times \text{nurturance}. \qquad (4.1)$$

In this model, an altruistic individual is one with high nurturance and low viability, while a selfish individual has low nurturance and high viability. It is not too surprising that the course of evolution is controlled by the product of viability and nurturance, which I will call *composite fitness*.

To facilitate comparison with Hamilton's theory, let me rewrite (2.1) in the form

$$\Delta p_n = p_n(w_{1\cdot} - w_{\cdot\cdot})/w_{\cdot\cdot}, \tag{4.2}$$

where $\Delta p_n = p_{n+1} - p_n$, $w_{1\cdot} = w_{11}p_n + w_{12}q_n$, and $w_{\cdot\cdot} = w_{11}p_n^2 + 2w_{12}p_nq_n + w_{22}q_n^2$. The theory in Hamilton's basic paper [**11**] covers relatives of all kinds, though he says that it is especially appropriate for interactions between relatives of the same generation. Using an argument involving "certain lapses from mathematical rigour" [**11**, p. 2], Hamilton obtains the equation

$$\Delta p_n = p_n(R_{1\cdot} - R_{\cdot\cdot})/(R_{\cdot\cdot} + \delta S_{\cdot\cdot}). \tag{4.3}$$

(This is Hamilton's equation (2) [**11**, p. 6], with $i = 1$ and dot superscripts omitted.)

Hamilton draws attention to the similarity between (4.3) and its classical analog, (4.2). Apart from the term $\delta S_{\cdot\cdot}$, which we may ignore, the only difference between (4.2) and (4.3) is that, in place of composite fitness, w_{ij}, Hamilton has *inclusive fitness*, R_{ij}. This quantity is the sum of an individual's personal fitness, taking account of his altruistic or selfish acts, and the changes in relatives' fitnesses due to these acts. The latter changes are weighted by the appropriate coefficient of relationship, r, which is the expected proportion of genes of the donor and recipient that are identical by descent (that is, copies of the same gene in some recent ancestor). For parent and offspring, this coefficient is $\frac{1}{2}$, so

$$R_{ij} = 1 + x_{ij} + \tfrac{1}{2}y_{ij}, \tag{4.4}$$

where x_{ij} and y_{ij} are, respectively, the parent's and offsprings' changes in personal fitness due to the parent's altruistic or selfish behavior. In the absence of such behavior, we are assuming, for simplicity, that all genotypes have the same fitness, which we have taken to be 1 without loss of generality.

We now come to the main point of this section. According to Hamilton, "With classical selection a genotype may be regarded as positively selected if its fitness is above the average and as counter-selected if it is below." In the case of altruistic and selfish behavior, "the kind of selection may be considered determined by whether the inclusive fitness of a genotype is above or below average" [**11**, p. 14]. For easy reference below, I will call this *Hamilton's criterion*. We saw in §2 that it is incorrect in the classical case. Superior w_{ij} does not ensure that A_iA_j increases in frequency if $i \neq j$. The analogy between (4.2) and (4.3) leaves no doubt that Hamilton's criterion is also inconsistent with Hamilton's basic theory as expressed by (4.3). Perhaps this is what Hamilton had in mind when, in a later paper [**12**, p. 196], he described this criterion as "approximate."

Given the conflict between Hamilton's criterion and his basic theory, it is of considerable interest that his most frequently cited result is derived from the criterion. I am referring to the rule relating k to $1/r$ [**11**, p. 16], where k is the ratio of recipients' and donor's changes in fitness, $k = y_{ij}/x_{ij}$. In cases of selfishness and altruism, x_{ij} and y_{ij} have different signs, so $k < 0$. For parent-offspring interaction, (4.4) yields

$$R_{ij} = 1 + (1 + k/2)x_{ij}. \tag{4.5}$$

For other relationships, the comparable equation is

$$R_{ij} = 1 + (1 + rk)x_{ij}.$$

Thus, according to Hamilton's criterion, natural selection should tend to favor an altruistic genotype ($x_{ij} < 0$) if $|k| > 1/r$. Similarly, a selfish genotype ($x_{ij} > 0$) should be favored if the reverse inequality holds.

These calculations with inclusive fitness, R, can be imitated with composite fitness, w. Recall that u_{ij} and $v_{ij}v_{km}$ are in units of offspring survival probability; hence v_{ij}^2, not v_{ij}, is comparable to u_{ij}. This suggests rewriting (4.1) in the form $w = u(v^2)^{1/2}$ or

$$\log w = \log u + \tfrac{1}{2} \log v^2. \tag{4.6}$$

(Here and below we omit subscripts.) Only ratios of $\underline{u}$s, $\underline{v}$s, and $\underline{w}$s are relevant to (4.2); hence these parameters can be rescaled so that a neutral condition (no selfishness of altruism) corresponds to $u = 1$, $v = 1$, and thus $w = 1$. Then the neutral values of the logarithms in (4.6) are 0, and these quantities are fully comparable to $R - 1$, x, and y in (4.4). Taking

$$k = \log v^2/\log u, \tag{4.7}$$

we obtain

$$\log w = (1 + k/2)\log u, \tag{4.8}$$

the analog of (4.5). Altruistic ($u < 1$, $v > 1$) or selfish ($u > 1$, $v < 1$) genotypes satisfy Hamilton's criterion for positive selection (large w) depending on whether $|k| > 2$ or $|k| < 2$, just as in the inclusive fitness formulation.

To see that this rule need not correctly predict the course of evolution, we return once again to the example described in Table 1 and assume that A_1A_1 is neither altruistic nor selfish. Rescaling to achieve $w_{11} = 1$, we obtain $w_{22} = 1/3$ and $w_{12} = 4/3$. As a consequence of (4.7) and (4.8), $u = w^{2/(k+2)}$ and $v = w^{k/(k+2)}$. For $k = -4$, these formulas yield the values given in Table 2. According to

TABLE 2

Viability, nurturance, and asymptotic frequency for $k = -4$

Genotype	u	v	w	Frequency
A_2A_2 (selfish)	3	1/9	1/3	1/16
A_1A_1 (neutral)	1	1	1	9/16
A_1A_2 (altruistic)	3/4	16/9	4/3	6/16

Hamilton's rule, A_1A_2 altruists should prosper, since $|k| > 2$, but, in fact, selection favors the neutral genotype, A_1A_1. Table 3 has been obtained by reversing entries in the u and v columns of Table 2. This transformation preserves w but changes k to $k' = 4/k = -1$.

TABLE 3
Viability, nurturance, and asymptotic frequency for $k = -1$

Genotype	u	v	w	Frequency
A_2A_2 (altruistic)	1/9	3	1/3	1/16
A_1A_1 (neutral)	1	1	1	9/16
A_1A_2 (selfish)	16/9	3/4	4/3	6/16

Since $|k| < 2$, the selfish genotype should be favored, but, again, it is asymptotically less prevalent than the neutral genotype. For further consideration of "$|k| > 1/r$," in the context of other models, see **[13]** and **[14]**.

5. Genes and genotypes. Fitnesses of genes can be defined by averaging over genotypes in which they occur. In the classical case, the fitnesses of A_1 and A_2 are

$$w_{1\cdot} = w_{11}p_n + w_{12}q_n \quad \text{and} \quad w_{2\cdot} = w_{12}p_n + w_{22}q_n.$$

Although (4.2) and (4.3) do not imply positive selection of genotypes with above average fitness, they do imply positive selection of genes with above average fitness. Thus discussions of evolution in terms of genes rather than genotypes avoid the pitfall which this paper has been concerned.

Many discussions of evolution (e.g., parts of [**15**]) are, in fact, cast in terms of genes. A limitation of this approach is that the fitness of a gene is not as intuitive as the fitness of a genotype. It is frequency dependent, and, if the heterozygote is most (or least) fit, even the ordering of $w_{1\cdot}$ and $w_{2\cdot}$ is different for large and small p_n. Discussions that ignore this frequency dependence implicitly (and perhaps unconsciously) rule out heterozygote superiority, which is the basis for the example considered repeatedly in this paper.

REFERENCES

1. M. F. Norman, *A "psychological" proof that certain Markov semigroups preserve differentiability*, these PROCEEDINGS.
2. E. O. Wilson, *Sociobiology: the new synthesis*, Harvard Univ. Press, Cambridge, Mass., 1975.
3. W. J. Ewens, *Mathematical population genetics*, Springer-Verlag, New York, 1979.
4. S. J. Gould, *Ever since Darwin: reflections in natural history*, Norton, New York, 1977.
5. J. Maynard Smith, *Evolution and the theory of games*, Amer. Sci. **64** (1976), 41–45.
6. G. F. Oster and S. M. Rocklin, *Optimization models in evolutionary biology*, Some Mathematical Questions in Biology, vol. 10 (S.A. Levin, ed.), Amer. Math. Soc., Providence, R. I., 1979.
7. P. D. Taylor and L. B. Jonker, *Evolutionarily stable strategies and game dynamics*, Math. Biosci. **40** (1978), 145–156.
8. J. Hofbauer, P. Schuster and K. Sigmund, *A note on evolutionary stable strategies and game dynamics*, J. Theoret. Biol. **81** (1979), 609–612.
9. W. F. Bodmer, *Differential fertility in population genetics models*, Genetics **51** (1965), 411–424.
10. T. Nagylaki, *Selection in one- and two-locus systems*, Springer-Verlag, New York, 1977.

11. W. D. Hamilton, *The genetical evolution of social behavior*. I, J. Theoret. Biol. **7** (1964), 1–16.

12. ______, *Altruism and related phenomena, mainly in social insects*, Ann. Rev. Ecology and Systematics **3** (1972), 193–232.

13. E. L. Charnov, *An elementary treatment of the genetical theory of kin-selection*, J. Theoret. Biol. **66** (1977), 541–550.

14. L. L. Cavalli-Sforza and M. W. Feldman, *Darwinian selection and "altruism,"* Theoret. Pop. Biol. **14** (1978), 268–280.

15. R. Dawkins, *The selfish gene*, Oxford, New York, 1976.

DEPARTMENT OF PSYCHOLOGY, UNIVERSITY OF PENNSYLVANIA, PHILADELPHIA, PENNSYLVANIA 19104

SIAM-AMS Proceedings
Volume 13
1981

A "Psychological" Proof that Certain Markov Semigroups Preserve Differentiability

M. FRANK NORMAN

1. Introduction. Unlike the preceding paper [1], the present one is directed toward a purely mathematical result. This result is linked to genetics because it can play a role in analyzing genetic models. It is linked to psychology because my proof makes essential use of a psychological model. However, the result itself is a fairly general theorem about a well-studied class of mathematical objects. The objects I have in mind are the semigroups of operators associated with certain diffusions.

For our purposes, a diffusion, $X(t)$, is a real-valued strong Markov process with continuous sample paths. (See [2, Chapter 2] for background information and terminology regarding diffusions.) We assume that the process is defined for all $t \geqslant 0$ and that its state space, $I = [r_0, r_1]$, is closed and bounded. A central role in the description and analysis of a finite Markov chain, X_n, is played by its transition matrix, P, with components

$$p_{xy} = P(X_{n+1} = y | X_n = x).$$

Associated with this matrix is a transformation of functions (regarded as column vectors) defined by matrix multiplication,

$$Pf(x) = \sum_y p_{xy} f(y),$$

or, equivalently, $Pf(x) = E(f(X_{n+1}) | X_n = x)$. Both the transition matrix and the transition operator can be generalized to diffusions, but the transition operator, defined by

$$T_t f(x) = E(f(X_{t+s}) | X_s = x),$$

is far more tractable and thus occupies a more prominent place in the theory of diffusions.

1980 *Mathematics Subject Classification.* Primary 47D07, 35K15, 35K65; Secondary 60F05, 60J60, 60J70.

These operators form a semigroup, i.e., $T_tT_u = T_{t+u}$ and T_0 is the identity operator. Moreover, T_t is linear, positive ($T_tf \geq 0$ if $f \geq 0$), conservative ($T_t1 = 1$), and contractive ($\|T_tf\|_0 \leq \|f\|_0$, where $\|f\|_0 = \sup_{x \in I}|f(x)|$). Although T_t is well defined for bounded measurable functions, it is advantageous to restrict it to continuous functions ($f \in C^0$). For most diffusions, if $f \in C^0$ then $T_tf \in C^0$ and T_tf is continuous in t with respect to the supremum norm. The latter property is called strong continuity of T_t. A strongly continuous semigroup of positive linear contractions mapping C^0 into C^0 is called a Fellerian semigroup (after W. Feller). The theory of such semigroups is presented in [**3**, Chapters I and II].

Diffusions are highly amenable to mathematical analysis, so there is considerable interest in using them as approximations for stochastic processes that are more recalcitrant. These processes often have discrete time scales. A prime example is the finite population version of the infinite population genetic model considered in [**1**]. Let $X_{K,n}$ be the proportion of A_1 genes in the nth generation for a population of size K, and suppose that selection differentials (and mutation rates, if nonzero) are of the order of magnitude of K^{-1}. Then $X_{K,n}$ can be approximated by $X(n/K)$ for a certain diffusion, $X(t)$. More precisely, if $f \in C^0$, then

$$E\big(f(X_{K,n})|X_{K,0} = x\big) - T_{n/K}f(x) \to 0$$

as $K \to \infty$, uniformly over x and $n \leq KC$, for any $C < \infty$ [**4**, §18.1].

General theorems of this type [**5**] can be proved very simply by an argument growing out of Khintchine's work [**6**, §1 of Chapter 3]. I will refer to this argument as Khintchine's, although modern versions are simpler than the original in certain respects. An essential ingredient of this approach is a suitable bound on x-derivatives of T_tf. If f is infinitely differentiable, then, typically, T_tf is too, and $|T_tf|_k \leq \exp(\lambda_k t)|f|_k$ for all $k \geq 1$ and $t \geq 0$, where λ_k is a constant, $|f|_k = \sum_{j=1}^k \|f^{(j)}\|_0$, and $f^{(j)}$ is the jth derivative of f. To prove diffusion approximation theorems, this bound is needed only for $k = 3$. However, for a broad class of diffusions, it holds for all positive integers. Our main objective in this paper is to establish such bounds by means of an argument that uses mathematical learning models.

2. Statement of results. We shall first attempt to motivate the conditions that define the class of semigroups to which our results apply. These semigroups correspond to diffusions, $X(t)$, that satisfy equations of the form

$$E(\Delta_\tau X(t)|X(t) = x) = \tau b(x) + o(\tau), \tag{2.1}$$

$$E\big((\Delta_\tau X(t))^2|X(t) = x\big) = \tau a(x) + o(\tau), \tag{2.2}$$

$$E\big(|\Delta_\tau X(t)|^3|X(t) = x\big) = o(\tau), \tag{2.3}$$

where $\Delta_\tau X(t) = X(t + \tau) - X(t)$. The drift and diffusion functions, b and a, are continuous, and $o(\tau)$ is uniform over I, i.e.,

$$\lim_{\tau \to 0} \|\tau^{-1}o(\tau)\|_0 = 0. \tag{2.4}$$

It follows from (2.1) and (2.2) that $b(r_0) \geqslant 0$, $b(r_1) \leqslant 0$, and $a(x) \geqslant 0$ for all $x \in I$. We shall assume that $a(x) > 0$ in the interior of I. However, applying $E(Y^2)^2 \leqslant E(|Y|)E(|Y|^3)$ to $Y = \Delta_\tau X(t)$, we see that $a(r_0) = a(r_1) = 0$.

A Fellerian semigroup is characterized by its generator, Γ, defined by

$$\Gamma f = \lim_{t \to 0} t^{-1}(T_t f - f).$$

Functions for which this limit exists (with respect to $\|\cdot\|_0$) constitute the domain, $\mathfrak{D}(\Gamma)$, of Γ. Let C^k be the collection of real-valued continuous functions on I with k derivatives continuous throughout I (even at endpoints). It follows easily from (2.1)–(2.4) that

$$\mathfrak{D}(\Gamma) \supset C^2 \tag{2.5}$$

and

$$\Gamma f(x) = 2^{-1}a(x)f''(x) + b(x)f'(x) \quad \text{for } f \in C^2. \tag{2.6}$$

These conditions provide the required characterization of the semigroups to be treated.

The following function spaces figure prominently in our results.

$$D^0 = \text{essentially bounded functions on } I,$$

and, for $k \geqslant 1$,

$$D^k = \{f \in C^{k-1}: f^{(k-1)} \text{ is absolutely continuous, and } f^{(k)} \in D^0\}.$$

Since I is compact, $C^k \subset D^k$. For $f \in D^0$, let $\|f\|_0$ = essential supremum of $|f(x)|$, and, for $f \in D^k$, $k \geqslant 1$, let $|f|_k = \sum_{j=1}^k \|f^{(j)}\|_0$ and $\|f\|_k = \sum_{j=0}^k \|f^{(j)}\|_0$. For all $k \geqslant 0$, D^k is a Banach space with respect to the norm $\|\cdot\|_k$.

THEOREM 1. *Suppose that* $a, b \in D^m$ *for some* $m \geqslant 2$, $a(r_0) = 0$, $a(r_1) = 0$, $a(x) > 0$ *for* $r_0 < x < r_1$, $b(r_0) \geqslant 0$, *and* $b(r_1) \leqslant 0$. *Then there is one and only one conservative Fellerian semigroup,* T_t, *whose generator,* Γ, *satisfies* (2.5) *and* (2.6). *Moreover, for any* $1 \leqslant k \leqslant m$ *and* $t \geqslant 0$, T_t *maps* D^k *into* D^k *and*

$$|T_t f|_k \leqslant \exp(\lambda_k t)|f|_k, \tag{2.7}$$

where

$$\lambda_k = \max_{1 \leqslant j \leqslant k} \sum_{i=j}^{k} \|d_{ij}\|_0 \tag{2.8}$$

and

$$d_{ij} = \binom{i}{j-1} b^{(i-j+1)} + \frac{1}{2}\binom{i}{j-2} a^{(i-j+2)} \qquad (d_{i1} = b^{(i)}). \tag{2.9}$$

Since $a(r_i) = 0$, the elliptic operator (2.6) is degenerate. A typical example of a diffusion function to which Theorem 1 applies is $a(x) = 2^{-1}x(1 - x)$. This is the diffusion function for the diffusion approximation to the genetic model mentioned in §1.

Since $\|T_t f\|_0 \leq \|f\|_0$, the seminorm, $|\cdot|_k$, can be replaced by the norm, $\|\cdot\|_k$, in (2.8), so that $\|T_t\|_k \leq \exp(\lambda_k t)$.

Theorem 1 is similar to a result of Ethier [7, Theorem 1], in which C^k takes the place of our D^k. Ethier's proof, involving a judicious combination of semigroup and martingale methods, is quite different from ours. Unlike our result, Ethier's is not restricted to finite intervals I. In order to apply our result to infinite intervals, it is necessary to make a preliminary spatial transformation of the type used in the proof of Lemma 6 in §7.

Readers with backgrounds in diffusion theory may be struck by the fact that (2.5) and (2.6) (or the equivalent conditions (2.1)–(2.4)) uniquely determine a semigroup (or diffusion) without reference to boundary conditions. This is worth considering in more detail. We begin with a review of Feller's boundary classification. This classification depends on the scale function, p, and speed function, m, defined by

$$p(x) = \int_r^x e^{-B(y)}\,dy \quad \text{and} \quad m(x) = \int_r^x \frac{2}{a(y)} e^{B(y)}\,dy,$$

where $B(x) = \int_r^x 2b(y)/a(y)\,dy$ and $r \in (r_0, r_1)$. Let

$$u(z) = \int_r^z m(x)\,dp(x) \quad \text{and} \quad v(z) = \int_r^z p(x)\,dm(x).$$

The classification of r_i is given by Table 1.

TABLE 1. Boundary classification

$u(r_i)$	$v(r_i)$	Boundary type
$< \infty$	$< \infty$	regular
$< \infty$	$= \infty$	exit
$= \infty$	$< \infty$	entrance
$= \infty$	$= \infty$	natural

Boundary conditions enter into the complete description of $\mathfrak{D}(\Gamma)$, which is as follows. Let $\mathfrak{D}$ be the subset of C^0 consisting of functions, f, with two continuous derivatives on (r_0, r_1) and for which

$$(d/dm)(d/dp)f(x) = 2^{-1}a(x)f''(x) + b(x)f'(x) \tag{2.10}$$

has finite limits at r_0 and r_1. Let

$$\mathfrak{D}_i = \begin{cases} \mathfrak{D}, & \text{if } r_i \text{ is entrance or natural,} \\ \{f \in \mathfrak{D} : (d/dp)f(r_i) = 0\}, & \text{if } r_i \text{ is regular,} \\ \{f \in \mathfrak{D} : (d/dm)(d/dp)f(r_i) = 0\}, & \text{if } r_i \text{ is exit.} \end{cases} \tag{2.11}$$

Then

$$\mathfrak{D}(\Gamma) = \mathfrak{D}_0 \cap \mathfrak{D}_1. \tag{2.12}$$

The interesting point is that, if r_i is regular, the boundary condition given above is not the only possibility for a conservative Fellerian semigroup. For example, any condition of the form

$$(-1)^{i+1}c_i(d/dp)f(r_i) + d_i(d/dm)(d/dp)f(r_i) = 0,$$

where $c_i \geqslant 0$, $d_i \geqslant 0$, and $c_i + d_i > 0$ is admissible. The diffusions obtained by boundary conditions other than those in (2.11) will satisfy (2.1)–(2.3) for each $x \in (r_0, r_1)$, but they will not satisfy the uniformity condition (2.4) or the closely related semigroup condition (2.5). Thus these conditions serve in lieu of boundary conditions in our approach. The semigroup thereby determined is subject to the bound (2.7), which, in the light of Khintchine's work, marks this semigroup as the appropriate diffusion approximation for a wide variety of processes.

If r_i is regular or exit, it is not obvious that the boundary conditions given by (2.11) are satisfied by all C^2 functions, as (2.5) requires. This fact follows easily from Theorem 2, which contains more information than is needed for our present purposes.

THEOREM 2. *Suppose that* $b \in D^1$, $a \in D^2$, $a(r_0) = 0$, $a(r_1) = 0$, $a(x) > 0$ *for* $r_0 < x < r_1$, $b(r_0) \geqslant 0$, *and* $b(r_1) \leqslant 0$. *Then the classification of* r_i *is determined by* $\alpha_i = (-1)^i a'(r_i)$ *and* $\beta_i = (-1)^i b(r_i)$ *in accordance with Table* 2.

TABLE 2. Boundary classification in terms of α_i and β_i

α_i	β_i	Boundary type
$= 0$	$=0$	natural
$= 0$	>0	entrance
> 0	$\geqslant \alpha_i/2$	entrance
> 0	$\in (0, \alpha_i/2)$	regular
> 0	$=0$	exit

This result is of independent interest, since the criteria for various boundary types are much easier to check than those of Table 1. Of course the criteria of Table 1 apply to speed and scale functions other than those considered here.

3. The role of learning models. In this section we introduce relevant learning models and sketch the proof of Theorem 1. Details are given in §§4–6. The proof of Theorem 2 is in §7.

Suppose that a rat has repeated trials in which he starts at the bottom of a T-shaped alley ("T-maze") and moves to the end of the right or left arm. Food is present in both arms on every trial. Let X_n be the rat's probability of turning right on the nth trial. According to the classical linear model for such an experiment [**8**, p. 115],

$$\Delta X_n = \begin{cases} \theta(1 - X_n) & \text{if he turns right,} \\ -\theta X_n & \text{if he turns left,} \end{cases}$$

where θ is a "learning rate parameter" $(0 < \theta < 1)$. In other words,

$$\Delta X_n = \begin{cases} \theta(1 - X_n) & \text{with probability } X_n, \\ -\theta X_n & \text{with probability } 1 - X_n. \end{cases}$$

This equation defines a discrete parameter Markov process with continuous state space [0, 1]. The genesis of the proof of Theorem 1 was my attempt **[9]** to obtain a diffusion approximation for this learning model and others like it.

It quickly emerged (see [**4**, Chapter 9]) that the techniques of **[9]** were equally applicable to processes satisfying equations of the form

$$\Delta X_n = \begin{cases} \theta\sigma_1(X_n) + \tau b(X_n) & \text{with probability } q(X_n), \\ -\theta\sigma_0(X_n) + \tau b(X_n) & \text{with probability } 1 - q(X_n), \end{cases} \tag{3.1}$$

where $\Delta X_n = X_{n+1} - X_n$, $\tau = \theta^2$,

$$q(x) = \sigma_0(x)/\left(\sigma_0(x) + \sigma_1(x)\right) \quad \text{for } r_0 < x < r_1, \tag{3.2}$$

$$\sigma_i(r_i) = 0, \qquad \sigma_i(x) > 0 \quad \text{for } r_0 < x < r_1, \tag{3.3}$$

$$q \text{ is nondecreasing,} \tag{3.4}$$

and

$$\sigma_i, b, q \in C^\infty. \tag{3.5}$$

To identify the appropriate diffusion approximation, $X(t)$, for discrete parameter processes, one usually considers moments of ΔX_n. In the present case,

$$E(\Delta X_n | X_n = x) = \tau b(x), \tag{3.6}$$

$$E\big((\Delta X_n)^2 | X_n = x\big) = \tau\sigma_0(x)\sigma_1(x) + O(\tau^{3/2}), \tag{3.7}$$

$$E\big(|\Delta X_n|^3 | X_n = x\big) = O(\tau^{3/2}), \tag{3.8}$$

where the O's are uniform over x. If $X(n\tau)$ is to approximate X_n, then $\Delta_\tau X(t)$ in (2.1)–(2.3) is comparable to ΔX_n in (3.6)–(3.8). Comparing these two sets of equations, we see that the putative limiting process is the one for which

$$a = \sigma_0\sigma_1. \tag{3.9}$$

The proof of Theorem 1 is based on the process (3.1) in a way that I will describe momentarily. I will refer to this process as a learning model, although few special cases beyond the linear model have received much attention from psychologists and, unbeknown to psychologists, mathematicians had studied such processes earlier **[10]**.

Our sketch of the proof of Theorem 1 begins with an alteration of viewpoint. Instead of starting with a learning model and looking for a diffusion to approximate it, we now start with a diffusion and seek an approximating learning model. Suppose initially that $b \in C^\infty$ and a is a polynomial (in addition to the other assumptions of Theorem 1). Let speed and scale functions, m and p, be defined accordingly. Feller's theory [**3**, Chapter II] assures us that the operator (2.10), restricted to the space $\mathfrak{D}(\Gamma)$ defined by (2.12), generates a conservative Fellerian semigroup, T_t. In view of Theorem 2, (2.5) and (2.6) are satisfied, and (2.1) – (2.4) follow. The question now arises: Can one choose the functions σ_1 and σ_0 in (3.1) so that (3.9) holds (hence (3.7) is analogous to (2.2)), and so that (3.3), (3.4), and (3.5) also hold, where q is defined by (3.2)? If a is a polynomial, as we are presently assuming, such σ_0 and σ_1 always exist. For if the zero of a at r_i has order z_i, we can take

$$\sigma_i(x) = |x - r_i|^{z_i}(A(x))^{1/2}, \tag{3.10}$$

where $A(x)$ is the polynomial

$$A(x) = a(x)/\left[(x - r_0)^{z_0}(r_1 - x)^{z_1}\right]. \tag{3.11}$$

Since $A(x) > 0$ for all $r_0 \leqslant x \leqslant r_1$, $(A(x))^{1/2} \in C^\infty$, and (3.5) holds, as do (3.3) and (3.4).

Now comes a crucial point. Khintchine's diffusion approximation argument, given in detail in the next section, presupposes two rather distinct circumstances. First, it presupposes corresponding expressions for moments of ΔX_n and $\Delta_\tau X(t)$. Second, it presupposes *either* a bound like (2.7) for $k = 3$ *or* an analogous bound for the learning model. We are trying to prove (2.7), so this bound is certainly not available to us at this juncture. However, a corresponding bound for the learning model is not difficult to establish. Let U^n be the discrete parameter semigroup for the learning model, $U^nf(x) = E(f(X_n)|X_0 = x)$. Then it is obvious that U^1 and thus U^n maps C^∞ into C^∞, and we shall show that, for each $k \geqslant 1$, there is a constant, γ_k, such that $|U^nf|_k \leqslant \exp(\gamma_k n\tau)|f|_k$. Khintchine's argument then yields $\|U^nf - T_tf\|_0 \to 0$ as $\tau \to 0$ and $n\tau \to t$, for $f \in C^\infty$. It follows that, if $f \in C^\infty$, then $T_tf \in C^\infty$ and

$$|T_tf|_k \leqslant \exp(\gamma_k t)|f|_k, \tag{3.12}$$

for all $k \geqslant 1$. These are the main conclusions of §4.

The constants, γ_k, of (3.12) depend on derivatives of b and the "artificial" functions σ_i and q, rather than b and a. This deficiency is corrected in §5, where a maximum principle is applied to $(\partial/\partial x)^kT_tf$ to derive (2.7) for infinitely differentiable f and b and for polynomial a. In §6, (2.7) is extended to $a, b \in D^m$ by taking the limit along a sequence $T_{n,t}$ of semigroups defined in terms of suitable polynomial approximations, a_n and b_n, to a and b. A simpler approximation argument allows us to pass from $f \in C^\infty$ to $f \in D^k$.

Uniqueness of T_t is established by the following observations. The argument in §4 (for polynomial a and infinitely differentiable b) actually shows that any Fellerian semigroup, T_t, that satisfies (2.5) and (2.6) also satisfies $U^nf \to T_tf$ as $\tau \to 0$ and $n\tau \to t$. Thus (2.5) and (2.6) uniquely determine T_t for polynomial a and $b \in C^\infty$. Similarly, the argument in §6 (for $a, b \in D^m$) derives $T_{n,t}f \to T_tf$ from (2.5) and (2.6). Since $T_{n,t}$ is uniquely determined, T_t is too.

The argument in §4 is similar to that of [**4**, Chapter 9], while our use of the maximum principle in §5 was suggested by [**11**]. The latter paper, by the way, gives a much better bound than (2.7) for the special case of linear b and quadratic a.

Throughout the remainder of the paper we shall assume, without loss of generality, that $r_0 = 0$ and $r_1 = 1$.

4. Proof of Theorem 1. Differentiability. In this section we assume that a is a polynomial and $b \in C^\infty$, and define σ_i and q by (3.10), (3.11), and (3.2). Of course $q(i)$ is defined so that $q \in C^0$. Then (3.3), (3.4), (3.5), and (3.9) are satisfied. Let $s_i = (-1)^{i+1}$, $x_i = x + \theta s_i \sigma_i + \tau b$, $q_1 = q$, and $q_0 = 1 - q$, so that the learning model has the form

$$X_{n+1} = (X_n)_i \quad \text{with probability } q_i(X_n). \tag{4.1}$$

Since $b(0) \geqslant 0$ and $b(1) \leqslant 0$, it follows that $x_i \geqslant 0$ for $x = 0$, and $x_i \leqslant 1$ for $x = 1$. For θ sufficiently small, $dx_i/dx \geqslant 0$, hence x_i maps $I = [0, 1]$ into I and the learning model is a well-defined Markov process in I. Its transition operator is

$$Uf(x) = \sum f(x_i) q_i(x), \tag{4.2}$$

where unindexed summations in this section are over $i = 0$ and $i = 1$. If $f \in C^\infty$, then $Uf \in C^\infty$, hence $U^n f \in C^\infty$.

Suppose now that T_t is any Fellerian semigroup satisfying (2.5) and (2.6). We shall use Khintchine's argument to show that, for any $f \in C^\infty$ and $K < \infty$,

$$\max_{0 \leqslant n \leqslant K/\tau} \|T_{n\tau} f - U^n f\|_0 \to 0 \tag{4.3}$$

as $\tau \to 0$. (This, of course, ensures the uniqueness of T_t.) Let $H_j = T_{(n-j)\tau} U^j f$. Then

$$\begin{aligned} T_{n\tau} f - U^n f = H_0 - H_n &= \sum_{j=0}^{n-1} (H_j - H_{j+1}) \\ &= \sum_{j=0}^{n-1} T_{(n-j-1)\tau} (T_\tau - U) U^j f. \end{aligned}$$

Therefore

$$\|T_{n\tau} f - U^n f\|_0 \leqslant \sum_{j=0}^{n-1} \|(T_\tau - U) U^j f\|_0. \tag{4.4}$$

If $g \in C^\infty$, a third-order Taylor expansion in conjuction with (3.6)–(3.9) yields $Ug = g + \tau \Gamma g + |g|_3 o(\tau)$. Similarly, in view of (2.1) – (2.4), $T_\tau g = g + \tau \Gamma g + |g|_3 o(\tau)$, where, in both cases, $o(\tau)$ is uniform over x and g. Thus

$$\|(T_\tau - U) g\|_0 \leqslant |g|_3 \tau \varepsilon_\tau,$$

where $\varepsilon_\tau \to 0$ as $\tau \to 0$, and (4.4) implies that

$$\|T_{n\tau} f - U^n f\|_0 \leqslant \tau \varepsilon_\tau \sum_{j=0}^{n-1} |U^j f|_3. \tag{4.5}$$

Suppose now that there were a bound of the form

$$|Uf|_k \leqslant \exp(\gamma_k \tau) |f|_k \tag{4.6}$$

for $k \geqslant 1$ and $0 < \theta < \rho_k$, where $\gamma_k \geqslant 0$. Then

$$|U^j f|_k \leqslant \exp(\gamma_k j\tau)|f|_k. \tag{4.7}$$

Taking $k = 3$ and substituting into (4.5), we obtain

$$\|T_{n\tau} f - U^n f\|_0 \leqslant \varepsilon_\tau |f|_3 \int_0^{n\tau} \exp(\gamma_3 t)\, dt \tag{4.8}$$

from which (4.3) follows.

As a consequence of (4.3) and strong continuity of T_t, $\|U^n f - T_t f\|_0 \to 0$ as $\tau \to 0$, and $n\tau \to t$. In conjunction with (4.7), this implies that $T_t f \in C^\infty$ and

$$|T_t f|_k \leqslant \exp(\gamma_k t)|f|_k \tag{4.9}$$

for all $t \geqslant 0$, $k \geqslant 1$, and $f \in C^\infty$.

It remains only to establish (4.6). This inequality follows directly from the expression

$$\begin{aligned}(d/dx)^j(f(x_i)q_i) &= f^{(j)}(x_i)(1 + j\theta s_i\sigma_i')q_i + \|f^{(j)}\|_0\omega_i j\theta s_i\sigma_i q_i' \\ &\quad + \sum_{n=0}^{j-1}\binom{j}{n} f^{(n)}(x)q_i^{(j-n)} + O(|f|_j\tau) \\ &\quad + \sum_{n=1}^{j-1} f^{(n)}(x)(\theta s_i\sigma_i q_i)^{(j-n+1)},\end{aligned} \tag{4.10}$$

where $|\omega_i| \leqslant 1$, valid for $j \geqslant 1$. The main points of the derivation of (4.10) are apparent in the case $j = 2$, to which we shall restrict our attention.

Clearly

$$(d/dx)^2(f(x_i)q_i) = \left[f''(x_i)(x_i')^2 + f'(x_i)x_i''\right]q_i + 2f'(x_i)x_i'q_i' + f(x_i)q_i''. \tag{4.11}$$

But

$$\begin{aligned}(x_i')^2 &= 1 + 2\theta s_i\sigma_i' + O(\tau),\\ f'(x_i)x_i'' &= f'(x)\theta s_i\sigma_i'' + O(|f|_2\tau),\\ f'(x_i)x_i' &= f'(x)x_i' + \|f''\|_0\omega_i(x_i - x)x_i'\\ &= f'(x)(1 + \theta s_i\sigma_i') + \|f''\|_0\omega_i\theta s_i\sigma_i + O(|f|_2\tau),\end{aligned}$$

and

$$f(x_i) = f(x) + f'(x)\theta s_i\sigma_i + O(|f|_2\tau).$$

When these expressions are substituted into (4.11), the case $j = 2$ of (4.10) is obtained.

Note that $\sum q_i = 1$ and

$$\sum s_i\sigma_i q_i = 0; \tag{4.12}$$

hence, in both cases, corresponding sums of derivatives are zero. Thus summation of (4.10) over i yields

$$\begin{aligned}(Uf)^{(j)}(x) &= \sum f^{(j)}(x_i)(1 + j\theta s_i\sigma_i')q_i\\ &\quad + \sum \|f^{(j)}\|_0\omega_i j\theta s_i\sigma_i q_i' + O(|f|_j\tau).\end{aligned}$$

Now $s_i q_i' \geqslant 0$ and, for θ sufficiently small, $1 + j\theta s_i \sigma_i' \geqslant 0$; hence

$$|(Uf)^{(j)}(x)| \leqslant \|f^{(j)}\|_0 \sum (1 + j\theta s_i \sigma_i') q_i$$
$$+ \|f^{(j)}\|_0 \sum j\theta s_i \sigma_i q_i' + K_j |f|_j \tau$$
$$= \|f^{(j)}\|_0 + K_j |f|_j \tau,$$

in view of (4.12). Thus $\|(Uf)^{(j)}\|_0 \leqslant \|f^{(j)}\|_0 + K_j |f|_j \tau$. Summing over $1 \leqslant j \leqslant k$, we obtain $|Uf|_k \leqslant (1 + \gamma_k \tau)|f|_k$, from which (4.6) follows.

5. Proof of Theorem 1. Better bound for $(\partial/\partial t)^k T_t f$. Throughout this section we continue to assume that a is a polynomial and b and f are infinitely differentiable. Clearly

$$T_t f - f = \int_0^t \Gamma T_u f \, du \tag{5.1}$$

and

$$\Gamma T_u f = T_u \Gamma f. \tag{5.2}$$

Since $\Gamma f \in C^\infty$, it follows that $T_u \Gamma f \in C^\infty$ and (4.9) applies to Γf. Thus integration and $(\partial/\partial x)^i$ can be interchanged on the right in (5.1), so that

$$(\partial/\partial x)^i T_t f - f^{(i)} = \int_0^t (\partial/\partial x)^i \Gamma T_u f \, du. \tag{5.3}$$

The integrand is bounded, so $(\partial/\partial x)^i T_t f$ is continuous in t (with respect to $\|\cdot\|_0$) for $f \in C^\infty$. Applying this observation to Γf, we see that the integrand in (5.3) is continuous, from which it follows that $(\partial/\partial x)^i T_t f$ is t-differentiable, and

$$(\partial/\partial t)(\partial/\partial x)^i T_t f = (\partial/\partial x)^i \Gamma T_t f. \tag{5.4}$$

Let $w_i(t, x) = (\partial/\partial x)^i T_t f(x)$, and let

$$\Gamma_i = 2^{-1} a (\partial/\partial x)^2 + b_i (\partial/\partial x),$$

where $b_i = b + (i/2)a'$. Then (5.4) can be rewritten in the form

$$(\partial/\partial t) w_i = \Gamma_i w_i + h_i, \tag{5.5}$$

where $h_i = \sum_{j=1}^i d_{ij} w_j$ and d_{ij} is as in the statement of Theorem 1.

We shall shortly use (5.5) and a maximum principle to derive the bound

$$\|w_i\|_{0,A} \leqslant \|f^{(i)}\|_0 + A\|h_i\|_{0,A}, \tag{5.6}$$

for all $A > 0$ and $i \geqslant 1$, where $\|w\|_{0,A} = \sup_{0 \leqslant t \leqslant A, x \in I} |w(t, x)|$. Let us accept (5.6) for the moment and pursue its implications. Estimating h_i in an obvious way, summing over $1 \leqslant i \leqslant k$, and then interchanging i and j summation in the final term on the right, we find that

$$\sum_{i=1}^k \|w_i\|_{0,A} \leqslant |f|_k + A\lambda_k \sum_{j=1}^k \|w_j\|_{0,A}.$$

Thus, if $A < 1/\lambda_k$,

$$\sum_{i=1}^k \|(\partial/\partial x)^i T_t f\|_{0,A} \leqslant |f|_k / (1 - A\lambda_k).$$

Consequently $|T_A f|_k \leqslant |f|_k/(1 - A\lambda_k)$, and, iterating,

$$|T_{nA} f|_k \leqslant |f|_k/(1 - A\lambda_k)^n.$$

Taking $A = t/n$ and letting $n \to \infty$ we obtain (2.7).

It remains only to derive (5.6). As a consequence of (5.2), (5.4), and (4.9), both $(\partial/\partial x)w_i$ and $(\partial/\partial t)w_i$ are bounded throughout $[0, A] \times I$; hence w_i is continuous, as is h_i. Moreover $w_i(t, \cdot) \in C^2$ and b_i, like $b_0 = b$, is continuous with $b_i(0) \geqslant 0$ and $b_i(1) \leqslant 0$. Henceforth we suppress the irrelevant i subscript.

For any ζ and ξ, let $W = e^{-\zeta t}w - \xi t$. Then (5.5) implies that

$$(\partial/\partial t)W = \Gamma W - \zeta W + e^{-\zeta t}h - \xi(1 + \zeta t).$$

If $\zeta > 0$ and

$$\xi = \|h\|_{0,A}, \tag{5.7}$$

then

$$(\partial/\partial t)W \leqslant \Gamma W - \zeta W, \tag{5.8}$$

for $0 \leqslant t \leqslant A$.

Since W is continuous, it attains its maximum over $[0, A] \times I$ at some point (t_0, x_0). If $0 < x_0 < 1$, then $(\partial/\partial x)W(t_0, x_0) = 0$, so

$$\Gamma W(t_0, x_0) = 2^{-1}a(x_0)(\partial/\partial x)^2 W(t_0, x_0) \leqslant 0.$$

If $x_0 = 0$ or 1, then $\Gamma W(t_0, x_0) = b(x_0)(\partial/\partial x)W(t_0, x_0) \leqslant 0$. Thus, in any case, $\Gamma W(t_0, x_0) \leqslant 0$.

If $t_0 > 0$, then $(\partial/\partial t)W(t_0, x_0) \geqslant 0$, so (5.8) yields $W(t_0, x_0) \leqslant 0$. Moreover $W(0, x_0) \leqslant \|f^{(i)}\|_0$. Hence, regardless of the value of t_0, $W(t_0, x_0) \leqslant \|f^{(i)}\|_0$ or

$$e^{-\zeta t}w(t, x) \leqslant \|f^{(i)}\|_0 + \xi A,$$

for $t \leqslant A$. Letting $\zeta \to 0$, this yields

$$w(t, x) \leqslant \|f^{(i)}\|_0 + \xi A.$$

Applying this inequality to $-w$ in place of w and recalling (5.7), we obtain (5.6).

6. Proof of Theorem 1. From C^∞ to D^m. Suppose that $m \geqslant 2$ and that $a \in D^m$ and $b \in D^m$ satisfy the hypotheses of Theorem 1, as do polynomials a_n and b_n. Let T_t and $T_{n,t}$ be corresponding semigroups whose generators, Γ and Γ_n, satisfy (2.5) and (2.6). Since a_n and b_n satisfy the hypotheses of §§4 and 5, $T_{n,t}$ is uniquely determined and, if $f \in C^\infty$, then $T_{n,t}f \in C^\infty$ and

$$|T_{n,t}f|_k \leqslant \exp(\lambda_{n,k}t)|f|_k \tag{6.1}$$

for all $k \geqslant 1$. Of course, $\lambda_{n,k}$ is defined by (2.8) and (2.9) with a_n and b_n in place of a and b.

If $f \in C^\infty$, then $T_{n,t}f \in \mathfrak{D}(\Gamma)$, so $T_{t-u}T_{n,u}f$ can be differentiated with respect to u. Its derivative is

$$(d/du)T_{t-u}T_{n,u}f = T_{t-u}(\Gamma_n - \Gamma)T_{n,u}f;$$

hence $T_t f - T_{n,t} f = \int_0^t T_{t-u}(\Gamma - \Gamma_n) T_{n,u} f\, du$. Consequently,

$$\|T_t f - T_{n,t} f\|_0 \leqslant \int_0^t \|(\Gamma - \Gamma_n) T_{n,u} f\|_0\, du$$

$$\leqslant \max\{2^{-1}\|a - a_n\|_0, \|b - b_n\|_0\} |f|_2 \int_0^t \exp(\lambda_{n,2} u)\, du. \quad (6.2)$$

(This inequality is analogous to (4.8).)

We shall show momentarily that the polynomials a_n and b_n can be chosen so that

$$\lim_{n\to\infty} \|a_n - a\|_0 = 0, \quad (6.3)$$

$$\lim_{n\to\infty} \|b_n - b\|_0 = 0, \quad (6.4)$$

and

$$\lim_{n\to\infty} \lambda_{n,k} = \lambda_k, \qquad 1 \leqslant k \leqslant m. \quad (6.5)$$

Since $m \geqslant 2$, it follows from (6.2) that

$$\lim_{n\to\infty} \|T_t f - T_{n,t} f\|_0 = 0. \quad (6.6)$$

Thus T_t is uniquely determined by (2.5) and (2.6). Furthermore, in the light of the lemma that follows, (6.1), (6.5), and (6.6) imply that, if $f \in C^\infty$, then $T_t f \in D^m$ and (2.7) holds for $1 \leqslant k \leqslant m$.

LEMMA 1. *If* $k \geqslant 1$, $g_n \in D^k$, $\lim_{n\to\infty} \|g_n - g\|_0 = 0$ *and* $\liminf_{n\to\infty} |g_n|_k < \infty$, *then* $g \in D^k$ *and*

$$|g|_k \leqslant \liminf_{n\to\infty} |g_n|_k.$$

PROOF. We may assume, without loss of generality, that $|g_n|_k \leqslant K < \infty$. Then the sequences $\{g_n^{(j)}\}$ are bounded and equicontinuous for $1 \leqslant j < k$, so a straightforward compactness argument yields $g \in C^{k-1}$ and $\|g_n - g\|_{k-1} \to 0$. Let

$$L(h) = \sup_{x\neq y} |h(x) - h(y)|/|x - y|.$$

Then $L(h) < \infty$ if and only if h is absolutely continuous and $\|h'\|_0 < \infty$. Moreover, in this case, $L(h) = \|h'\|_0$. Since $g_n^{(k-1)}(x) \to g^{(k-1)}(x)$ for all x, $L(g^{(k-1)}) \leqslant \liminf_{n\to\infty} L(g_n^{(k-1)})$ and the conclusions of the lemma follow immediately.

To extend (2.7) from $f \in C^\infty$ to $f \in D^k$, $1 \leqslant k \leqslant m$, assume that $f \in D^k$, and consider the Taylor expansion

$$f(x) = \sum_{j=0}^{k-1} f^{(j)}(0)\frac{x^j}{j!} + \frac{1}{(k-1)!}\int_0^x (x-y)^{k-1} f^{(k)}(y)\, dy. \quad (6.7)$$

Since $f^{(k)} \in D^0$, there is a sequence, p_n, of polynomials such that $p_n \to f^{(k)}$ almost everywhere and $\|p_n\|_0 \to \|f^{(k)}\|_0$, from which it follows that

$$\lim_{n\to\infty} \int_0^1 |p_n(x) - f^{(k)}(x)|\, dx = 0.$$

Replacing $f^{(k)}$ in (6.7) by p_n, we obtain a sequence, $\bar{f}_n$, of polynomials such that $\|\bar{f}_n - f\|_{k-1} \to 0$ and $\bar{f}_n^{(k)} = p_n$, so $\|\bar{f}_n^{(k)}\|_0 \to \|f^{(k)}\|_0$. Consequently $|\bar{f}_n|_k \to |f|_k$. Now apply (2.7) to $\bar{f}_n$. Since $\lim_{n\to\infty}\|T_t\bar{f}_n - T_tf\|_0 = 0$, Lemma 1 implies that $T_tf \in D^k$ and (2.7) holds if $f \in D^k$.

To complete the proof of Theorem 1, it remains only to show that there are polynomials a_n and b_n satisfying (6.3)–(6.5) in addition to the hypotheses of Theorem 1. Let $\bar{b}_n$ be a sequence of polynomials constructed as in the last paragraph for $k = m$, and let $b_n(x) = \bar{b}_n(x) + (b(1) - \bar{b}_n(1))x$. Then

$$\lim_{n\to\infty} \|b_n - b\|_{m-1} = 0, \tag{6.8}$$

$$\lim_{n\to\infty} \|b_n^{(m)}\|_0 = \|b^{(m)}\|_0, \tag{6.9}$$

and $\bar{b}_n(i) = b(i)$, so $\bar{b}_n(0) \geqslant 0$ and $\bar{b}_n(1) \leqslant 0$.

The construction of a_n is similar, but certain refinements are required in order to satisfy (6.5) for $k = m$ and $a_n(x) > 0$ for $0 < x < 1$. The only coefficient d_{ij}, $1 \leqslant j \leqslant i \leqslant m$, that involves $a^{(m)}$ is

$$d_{m2} = mb^{(m-1)} + 2^{-1}a^{(m)}.$$

Let $d^*_{n,m2}$ be a sequence of polynomials such that $d^*_{n,m2} \to d_{m2}$ almost everywhere and $\|d^*_{n,m2}\|_0 \to \|d_{m2}\|_0$. Let p_n be the polynomial defined by $d^*_{n,m2} = mb_n^{(m-1)} + 2^{-1}p_n$. Then $p_n \to a^{(m)}$ almost everywhere and

$$\int_0^1 |p_n(x) - a^{(m)}(x)|\, dx \to 0.$$

Let $\bar{a}_n$ be polynomial approximations to a constructed as in the penultimate paragraph, let $\tilde{a}_n(x) = \bar{a}_n(x) - \bar{a}_n(1)x$, and, finally, let

$$a_n(x) = \tilde{a}_n(x) + (16/3)\|a - \tilde{a}_n\|_1 x(1 - x).$$

Then $a_n(0) = a_n(1) = 0$, $a_n(x) \geqslant a(x)$ so that $a_n(x) > 0$ for all $0 < x < 1$, $\lim_{n\to\infty}\|a_n - a\|_{m-1} = 0$, and $\lim_{n\to\infty}\|d_{n,m2}\|_0 = \|d_{m2}\|_0$. The latter equations, together with (6.8) and (6.9), readily yield

$$\lim_{n\to\infty} \|d_{n,ij}\|_0 = \|d_{ij}\|_0$$

for all $1 \leqslant j \leqslant i \leqslant m$, from which (6.5) follows easily.

This completes the proof of Theorem 1.

7. Proof of Theorem 2. We shall restrict our attention to the upper boundary, $r_1 = 1$. Since $a \in D^2$ and $b \in D^1$, we have at our disposal the Taylor expansions $b(x) = -\beta + O(1 - x)$ and

$$a(x) = \alpha(1 - x) + O\big((1 - x)^2\big), \tag{7.1}$$

where $\beta = \beta_1$ and $\alpha = \alpha_1$ are nonnegative.

LEMMA 2. *If $\beta > 0$, then $v(1) < \infty$.*

PROOF. Choose $c \in (r, 1)$ sufficiently large that $b(x) < -\beta/2$ for $x > c$. Then for $1 > z > c$,

$$(\beta/2)\int_c^z p(x)\,dm(x) \leqslant -\int_c^z p(x)b(x)\,dm(x) = -\int_c^z p(x)\,de^{B(x)}$$
$$= p(c)e^{B(c)} - p(z)e^{B(z)} + \int_c^z e^{B(x)}\,dp(x)$$
$$\leqslant p(c)e^{B(c)} + (1-c).$$

Thus $v(1) = v(1^-) < \infty$, as claimed.

LEMMA 3. *If $\beta > 0$ and $\alpha = 0$, then 1 is entrance.*

PROOF. Since $2/a(x) = p'(x)m'(x)$, the Schwarz inequality yields

$$\int_r^1 (2/a(x))^{1/2}\,dx \leqslant \left(\int_r^1 p'(x)\,dx \int_r^1 m'(x)\,dx\right)^{1/2} = (p(1)m(1))^{1/2}.$$

If $\alpha = 0$, it follows from (7.1) that $p(1) = \infty$ or $m(1) = \infty$, so $u(1) = \infty$ or $v(1) = \infty$. In view of Lemma 2, $v(1) < \infty$, so $u(1) = \infty$ and 1 is entrance.

LEMMA 4. *If $\beta > 0$, then 1 is entrance if $0 < \alpha \leqslant 2\beta$ and regular if $\alpha > 2\beta$.*

PROOF. If $\alpha > 0$, then

$$\frac{b(x)}{a(x)} = \frac{-\beta}{\alpha(1-x)} + O(1)$$

for $r \leqslant x < 1$. Hence $B(x) = 2(\beta/\alpha)\ln(1-x) + O(1)$ and $p'(x) = (1-x)^{-2(\beta/\alpha)}e^{O(1)}$. If $2\beta \geqslant \alpha$ then $p(1) = \infty$, $u(1) = \infty$, and 1 is entrance by Lemma 2. If $2\beta < \alpha$, then $p(1) < \infty$. But $m(1) < \infty$ by Lemma 2, so 1 is regular.

LEMMA 5. *If $\beta = 0$ and $\alpha > 0$, then 1 is exit.*

PROOF. Under these conditions $B(x) = O(1)$; hence $m(x) = O(-\ln(1-x))$ for $r \leqslant x < 1$, and $u(1) < \infty$. However, for any $c \in (r, 1)$, $v(1) \geqslant K\int_c^1 (a(x))^{-1}\,dx$ for some $K > 0$, so $v(1) = \infty$.

LEMMA 6. *If $\beta = 0$ and $\alpha = 0$, then 1 is natural.*

PROOF. We shall transform (0, 1) onto $(0, \infty)$ in order to apply a result of Ethier [7, Lemma 3]. Let $h(x) = x/(1-x)$, $\tilde{p}(y) = p(h^{-1}(y))$, and $\tilde{m}(y) = m(h^{-1}(y))$. Then

$$\int_r^1 m\,dp = \int_{h(r)}^{\infty} \tilde{m}\,d\tilde{p},$$

and similarly for $\int_r^1 p\,dm$, so the boundary classification of ∞ for $\tilde{p}$ and $\tilde{m}$ is the same as the classification of 1 for p and m. It is easy to check that the coefficients $\tilde{a}$ and $\tilde{b}$ corresponding to $\tilde{p}$ and $\tilde{m}$ are

$$\tilde{a}(y) = a(x)(1+y)^4$$

and

$$\tilde{b}(y) = a(x)(1 + y)^3 + b(x)(1 + y)^2,$$

where $x = h^{-1}(y) = y/(1 + y)$. If $\alpha = 0$ and $\beta = 0$, then $\tilde{a}(y) = O(y^2)$ and $\tilde{b}(y) = O(y)$ for $y \geq 1$. Hence Ethier's lemma implies that ∞ is natural for $\tilde{p}$ and $\tilde{m}$, so that 1 is natural for p and m.

This completes the proof of Theorem 2.

REFERENCES

1. M. F. Norman, *Sociobiological variations on a Mendelian theme*, these PROCEEDINGS.

2. D. Freedman, *Brownian motion and diffusion*, Holden-Day, San Francisco, Calif., 1971.

3. P. Mandl, *Analytical treatment of one-dimensional Markov processes*, Springer-Verlag, New York, 1968.

4. M. F. Norman, *Markov processes and learning models*, Academic Press, New York, 1972.

5. ______, *Diffusion approximation of non-Markovian processes*, Ann. Probab. **3** (1975), 358–364.

6. A. Khintchine, *Asymptotische Gesetze der Wahrscheinlichkeitsrechnung*, Chelsea, New York, 1948.

7. S. N. Ethier, *Differentiability preserving properties of Markov semigroups associated with one-dimensional diffusions*, Z. Wahrsch. Verw. Gebiete **45** (1978), 225–238.

8. R. R. Bush and F. Mosteller, *Stochastic models for learning*, Wiley, New York, 1955.

9. M. F. Norman, *Slow learning with small drift in two-absorbing-barrier models*, J. Math. Psych. **8** (1971), 1–21.

10. O. Onicescu and G. Mihoc, *Sur les chaînes de variables statistiques*, Bull. Soc. Math. France **59** (1935), 174–192.

11. S. N. Ethier and M. F. Norman, *Error estimate for the diffusion approximation of the Wright-Fisher model*, Proc. Nat. Acad. Sci. U.S.A. **74** (1977), 5096–5098.

DEPARTMENT OF PSYCHOLOGY, UNIVERSITY OF PENNSYLVANIA, PHILADELPHIA, PENNSYLVANIA 19104

SIAM-AMS Proceedings
Volume **13**
1981

Axiomatic Measurement Theory[1]

R. DUNCAN LUCE AND LOUIS NARENS

INTRODUCTION

Everyone is aware that measurement is a cornerstone of science, one that in some cases is highly controversial. Much complex technology underlies the refined measurement of certain physical quantities, some of which can be estimated to surprisingly large numbers of significant figures; one of the more elaborate businesses spawned by the social sciences, a business that affects all of our lives, attempts to measure intellectual ability and/or achievement; and elaborate computer programs are widely used to provide numerical representations (and simplifications), e.g., by factor analysis and multidimensional scaling, of complexes of data. Behind all of this activity is a belief, often sustained by a mixture of intuition and successful–if ill understood–procedures, that certain bodies of data can be represented in some fashion by numbers and their relations to each other. The goal of the semiphilosophical, semimathematical field of our title is to lay bare the types of empirical structures that admit such numerical representations.

The reason for the term "axiomatic" in the title is that this is how the structures involved are described. The task is to isolate axioms that, on the one hand, are empirically and/or philosophically acceptable for at least one important scientific interpretation of the primitives and that, on the other hand, permit us to prove mathematically that the structure is closely similar (usually, isomorphic or homomorphic) to some numerical structure. Ultimately, one aims for a finite collection of different classes of structures that span all the scientifically interesting cases.

At present, our knowledge appears quite adequate for the better developed parts of classical physics. It is interesting to note that many influential writers

1980 *Mathematics Subject Classification.* Primary 06F05; Secondary 06F25, 92A20, 92A90.

[1]The research for this paper was partially supported by a grant (ITS-79-24019-1) from the National Science Foundation.

considered that classical physics had achieved an adequate measurement-theoretic underpinning by 1920, but that view simply was wrong and a tolerably adequate theory was only forged in the past few years and it is still being improved. It is considerably less clear that existing theory is adequate for some of the classically intractable concepts, such as those having to do with turbulence and hardness. In relativistic physics things are much worse, and it is doubtful if existing theory is suitable for such variables as relativistic velocity. More generally, any bounded variable, including probability and sensory concepts such as loudness and brightness, raises tricky problems that are not yet fully understood. Further, it is quite clear that the existing theories are not suited to the measurement questions that arise in quantum mechanics, and to date little effort has been spent on these questions.

If the field remains incompletely developed for physics, it is anyone's guess as to how adequate it will prove to be for the biological, behavioral, and social sciences. These fields, especially the latter two, have struggled for years with problems of measurement, and although much is "measured", the underlying conceptual basis is still poorly understood. In fact, a major motive for much of the work on axiomatic measurement since 1947, when the work of von Neumann and Morgenstern [**1944**, **1947**, **1953**] on utility theory became widely known, has been to clarify just what the measurement options are. There was a time–Campbell [**1920**], Cohen and Nagel [**1934**]–when measurement was said to be limited to those structures isomorphic to the additive reals and things that could be "derived" from them. This position was asserted, and asserted strongly, despite the fact that the ring of real numbers clearly played a role in the representation of physical measures–witness the additivity of length, mass, time, and the like, and the product of powers of measures to form other measures, as reflected in the units of physical measurement. Today, we know of several classes of structures with vastly richer numerical representations than the additive group of the reals, which nonetheless still plays a highly central role, and we hope that behavioral and social scientists will find useful some of the recently developed generalizations. There exists a small group of theoretical and empirical scientists who are working on the interplay of these kinds of measurement concepts with data, but it probably will take a considerable time before we have any clear sense of just how applicable this kind of "applicable mathematics" is outside of physics.

At this juncture, two quite different types of programs are needed. The one is to recast, to simplify, and to inject these ideas into the mainstream of the behavioral and social sciences, just as was done with statistics over the past 50 years. This has begun on many fronts, including texts, expository articles, empirical methods, computer programs, and the like. It will have to continue for a long time, just as it has with statistics, and it will have to be shown to make a difference. The other is to enlist the help of the mathematical community to enlarge our understanding of the relevant structures. It is not, however, just a problem of mathematical generalization and more powerful proof techniques–although both are surely needed–but of developments that are sensitive to the possible empirical interpretations that can be given to the primitives. It is, after all, applicable mathematics, and more is required of the mathematician than just mathematical skills. It is all too easy to lose sight of or to underestimate the added sensitivity that is needed if the work is to remain scientifically deep as it becomes mathematically deeper.

The body of this paper summarizes a number of the main results and approaches that have been taken. Some history and references are provided, but it is far from a scholarly survey. It is, rather, a highlighting of what we think is most important–always an idiosyncratic criterion–and is intended more as an overview and an invitation to dig more deeply and to contribute than it is a precise account of the whole field. A number of books provide more detail and depth. The most elementary and the one with the greatest number of illustrative social science examples is Roberts [**1979**]. The earliest and most compact is Pfanzagl [**1968**, **1971**]. The most comprehensive, provided one takes into account the projected second volume, is Krantz et al. [**1971**], [in preparation]. The most advanced is the nearly completed one by Narens [in preparation].

I. Structures with One Operation

All measurement rests upon having a qualitative ordering $\succsim$ of the set X of objects. It is well known (Krantz, et al. [**1971**, §2.1]) that an order preserving numerical representation exists if and only if $\langle X, \succsim \rangle$ is a total order with a finite or countable order dense subset. Moreover, any two such representations are related by a monotonically strictly increasing function. Such so-called ordinal scales are far too weak to be useful for measurement: concepts such as the derivative of a quantity are not invariant under admissible changes in the representation. In order for the representation to be firmer, it is necessary that the numerical measures preserve structure in addition to but related to the order. In practice, this has meant one of four things: either a single operation is included which is represented by some numerical operation, often addition; or the ordered elements are themselves structured in the sense that X is the Cartesian product of two or more sets, which structure is represented by some numerical operation, often multiplication; or X is a Cartesian product of the form $A \times A$ and the representation is in some geometric space with distance preserving the ordering relation; or there is even more structure such as two operations or a Cartesian product and an operation on one of the components, which is represented by two (or more) numerical operations. We deal with the first two cases, which are closely related, in the first major part of the paper, and the latter in the second major part. The third case, the geometric one, is not covered in this paper; see Beals and Krantz [**1967**], Beals, Krantz and Tversky [**1968**], Tversky and Krantz [**1970**], and Krantz et al. [in preparation].

1. Extensive structures.

DEFINITION 1. Let $\bigcirc$ be a partial binary operation on the nonempty set X (i.e., a function from a subset of $X \times X$ into X), $\succcurlyeq$ be a total ordering on X, R a subset of Re, and $\odot$ a partial operation on R. Then φ is said to be a $\odot$-*representation* for the structure $\mathfrak{X} = \langle X, \succcurlyeq, \bigcirc \rangle$ if and only if φ is an isomorphic imbedding of $\mathfrak{X}$ into the structure $\langle R, \geqslant, \odot \rangle$. If φ is a $\odot$-representation and $\odot$ is $+$, then φ is said to be an *additive representation*.

Partial operations, rather than operations (which we will often call *closed* operations to distinguish them clearly from partial operations) play a critical role in some measurement situations.

Often in measurement theory, structures of the form $\langle X, \succsim, \bigcirc \rangle$ are considered where $\succsim$ is a weak ordering (transitive and connected) rather than a total ordering (also asymmetric). The measurement theoretic results for such structures are almost identical to those of the totally ordered case. In this paper, the totally ordered case is often invoked (although not always) to simplify notation and some definitions.

The first serious results in the foundations of measurement go back to Helmholtz [**1887**] and Hölder [**1901**], who presented axiomatizations for additive physical attributes. These axiomatizations have been greatly refined by a number of researchers, and today find their most useful formulation in the following definition due to Krantz et al. [**1971**].

DEFINITION 2. Let X be a nonempty set, $\succcurlyeq$ a binary relation on X, and $\bigcirc$ a partial binary operation on X. The structure $\mathfrak{X} = \langle X, \succcurlyeq, \bigcirc \rangle$ is said to be an *extensive structure* if and only if the following eight axioms hold for all w, x, y, z in X:

1. *Total ordering.* $\succcurlyeq$ is a total ordering.

2. *Nontriviality.* There exist u, v in X such that $u \succ v$.

3. *Local definability.* If $x \bigcirc y$ is defined, $x \succcurlyeq w$, and $y \succcurlyeq z$, then $w \bigcirc z$ is defined.

4. *Monotonicity.* (1) If $x \bigcirc z$ and $y \bigcirc z$ are defined, then $x \succcurlyeq y$ iff $x \bigcirc z \succcurlyeq y \bigcirc z$, and

(2) if $z \bigcirc x$ and $z \bigcirc y$ are defined, then $x \succcurlyeq y$ iff $z \bigcirc x \succcurlyeq z \bigcirc y$.

5. *Restricted solvability.* If $x \succ y$, then there exists u such that $x \succ y \bigcirc u$.

6. *Positivity.* If $x \bigcirc y$ is defined, then $x \bigcirc y \succ x$ and $x \bigcirc y \succ y$.

7. *Archimedean.* There exists $n \in I^+$ such that either nx is not defined or $nx \succcurlyeq y$, where mx is inductively defined by $1x = x$, and if $(mx) \bigcirc x$ is defined, then $(m + 1)x = (mx) \bigcirc x$.

8. *Associativity.* If $x \bigcirc (y \bigcirc z)$ and $(x \bigcirc y) \bigcirc z$ are defined, then $x \bigcirc (y \bigcirc z) = (x \bigcirc y) \bigcirc z$.

If $\bigcirc$ is a closed operation, then $\mathfrak{X}$ is said to be a *closed* extensive structure.

The theoretical measurement of length is often taken as an example of an extensive structure. Let X be a set of (straight) measuring rods. For each x, y in X, let $x \succcurlyeq y$ stand for "the rod x is at least as long as the rod y," and let $x \bigcirc y$ be the rod that is obtained by abutting x to y along a straight edge. In theoretical physics, it is assumed that $\mathfrak{X} = \langle X, \succcurlyeq, \bigcirc \rangle$ is an extensive structure and $\bigcirc$ is a closed operation. Another, slightly more subtle application, is to probability theory. Here we assume Ω to be a nonempty set, Y an algebra of subsets of Ω, and $X = Y - \{\varnothing\}$. Let $\succsim$ be a binary relation on X, to be interpreted as the concept of "at least as likely as." Thus, axiomatically, we assume $\succsim$ to be a weak ordering. A natural partial operation to define on X is $\oplus$, where for all x, y in X,

$$x \oplus y = z \quad \text{if and only if} \quad x \cap y = \varnothing \text{ and } x \cup y = z.$$

Then $\mathfrak{X} = \langle X, \succsim, \oplus \rangle$ starts to resemble an extensive structure: $\succsim$ is a weak ordering and $\oplus$ is associative for the elements for which it is defined. Furthermore, monotonicity of $\oplus$ is a natural assumption to make, i.e., for all x, y, z, w in X such that $x \cap z = \varnothing$ and $y \cap w = \varnothing$ and $z \sim w$,

$$x \succsim y \quad \text{iff} \quad x \oplus z = x \cup z \succsim y \cup w = y \oplus w.$$

What is missing is that the partial operation $\oplus$ is not defined for sufficiently many pairs of events. This can be partially rectified by letting $\mathbf{X} = X/\sim$, $\boldsymbol{\succcurlyeq} = \succsim/\sim$, and defining $\boldsymbol{\oplus}$ by: for each A, B, C in $\mathbf{X}$, $A \boldsymbol{\oplus} B = C$ if and only if for some x in A, y in B, z in C, $x \oplus y = z$, and considering the totally ordered structure $\boldsymbol{\mathfrak{X}} = \langle \mathbf{X}, \boldsymbol{\succcurlyeq}, \boldsymbol{\oplus} \rangle$. $\boldsymbol{\mathfrak{X}}$ is very close to an extensive structure. Its primary lack is that local definability and Archimedean may not hold. However, rather plausible axioms in terms of the primitives $\succsim$ and $\cup$ can be given that guarantee that $\boldsymbol{\mathfrak{X}}$ is an extensive structure. Such an extensive structure $\boldsymbol{\mathfrak{X}}$ in this paper will be called a *qualitative probability structure*. The interested reader should consult Luce [**1965**] or Krantz et al. [**1971**, Chapter 5] or Fine [**1971a**], [**1971b**] for a detailed axiomatization. It should also be noted that it is inherent in the nature of probability, which has Ω as a maximal element, that $\oplus$ must be a partial, not a closed, operation.

The following theorem shows that extensive structures have a restricted set of additive representations, and this fact is widely used to justify and establish numerical scales of empirical variables.

THEOREM 1. *Suppose $\mathfrak{X} = \langle X, \succcurlyeq, \bigcirc \rangle$ is an extensive structure and $\langle X, \succcurlyeq \rangle$ does not have a maximal element. Then the following three statements are true*:

(i) *there exists an additive representation for $\mathfrak{X}$*;

(ii) *if φ and ψ are both additive representations of $\mathfrak{X}$, then for some r in* Re^+, $\varphi = r\psi$;

(iii) *$r\varphi$ is an additive representation for $\mathfrak{X}$ for each r in* Re^+ *and each additive representation φ of $\mathfrak{X}$.*

A proof of Theorem 1 is given in Chapters 2 and 3 of Krantz et al. [**1971**].

Theorem 1 can be used to show for the example of length presented above that positive numbers can be assigned to measuring rods so that rod x is at least as long as y if and only if the number assigned to x is $\geqslant$ the number assigned to y, and the number assigned to the rod resulting from abutting x to y is the sum of the numbers assigned to x and y. Furthermore, any other assignment with these properties is essentially the same: it differs by at most multiplication by a positive constant. It also follows by applying Theorem 1 to the probabilistic situation discussed above with $\mathfrak{X}$ being a qualitative probability structure that a unique, finitely additive probability representation exists, i.e., there exists a unique function P from the algebra of events Y of Ω such that

(1) $P(\Omega) = 1$ and $P(\varnothing) = 0$;

(2) $P(x \cup y) = P(x) + P(y)$ for all x, y in Y such that $x \cap y = 0$; and

(3) for each x, y in X, x is at least as likely as y (i.e., $x \succsim y$) if and only if $P(x) \geqslant P(y)$.

The axioms for extensive structures are sufficient for the existence of additive representations but not necessary. For the case of a closed operation, Roberts and Luce [**1968**] have given necessary and sufficient conditions for the existence of additive representations and showed a result like Theorem 1. (These results are presented in Krantz et al. [**1971**].)

2. Generalizations of extensive structures. A number of generalizations of extensive structures have appeared in the literature. A very brief description of some of these will be now given.

Structures with weakened forms of Axiom 1, total ordering, are considered in Narens [in preparation] and Holman [**1974**]. Narens considers the case where $\circ$ is a closed operation and $\succcurlyeq$ is a transitive and reflexive relation, and gives necessary and sufficient conditions for such structures to have an additive representation. Holman considers a case that has an equivalence relation instead of an ordering relation. By considerably strengthening the Archimedean axiom, he shows a theorem analogous to Theorem 1.

Falmagne [**1971**], [**1975**] considers structures which have additive representations, but in which local definability (Axiom 3) is weakened so that arbitrarily small elements need not exist. "Arbitrarily small" here means arbitrarily small in terms of some additive representation rather than in terms of the ordering relation $\succcurlyeq$, i.e., in terms of some additive representation of $\mathfrak{X}$ assuming values arbitrarily close to 0. Falmagne's axiomatization yields a theorem analogous to Theorem 1.

Structures without Archimedean axioms are considered in Narens [**1974a**], [**1974b**], [in preparation]. In general, such structures do not have additive representations in the reals. However, Narens shows that they have additive representations in certain structures richer than the reals, namely the nonstandard reals and structures that resemble lexicographically ordered vector spaces. Skala [**1975**] has collected together various results about nonarchimedean measurement.

Perhaps the most important generalization of extensive structures comes from deleting Axiom 8, associativity. This structure, which was first considered in

Narens and Luce [**1976**], has surprisingly strong measurement theoretic properties.

DEFINITION 3. $\mathcal{X} = \langle X, \succcurlyeq, \circ \rangle$ is said to be a *positive concatenation structure* if and only if $\mathcal{X}$ satisfies all the axioms for an extensive structure except possibly Axiom 8, associativity.

Narens and Luce [**1976**] showed that positive concatenation structures have $\odot$-representations for some $\odot$, and that such representations have strong uniqueness properties. Cohen and Narens [**1979**] gave a slightly different version of uniqueness for these structures, and their version is given in statement (ii) of the following theorem.

THEOREM 2. *Suppose* $\mathcal{X} = \langle X, \succcurlyeq, \circ \rangle$ *is a positive concatenation structure. Then the following two statements are true*:

(i) $\mathcal{X}$ *has a* $\odot$*-representation for some* $\odot$;

(ii) *if* φ *and* ψ *are* $\odot$*-representations for* $\mathcal{X}$ *such that* $\varphi(X) = \psi(X)$ *and if for some* x *in* X, $\varphi(x) = \psi(x)$, *then* $\varphi = \psi$.

There are some scientifically important concatenation operations, such as temperature and averaging, that do not satisfy positivity of Definition 2. In these cases Axioms 6, 7, and 8 do not hold, but they can be replaced by another Archimedean axiom and the following property (called intern):

$$\text{if } x \succ y, \text{ then } x \succ x \circ y \succ y.$$

Such concatenation structures are called *intensive*. Some work on a special case, satisfying a property called bisymmetry, was reported by Pfanzagl [**1959a**], [**1959b**] (see Krantz et al. [**1971**, §6.9]). Narens and Luce [**1976**] showed that a broad class of intensive structures is closely related to positive concatenation structures.

3. Conjoint structures. Structures of the form $\langle X, \succsim, \circ \rangle$ where $\succsim$ is an ordering and $\circ$ is a partial operation naturally arise in physical science with $\succsim$ and $\circ$ being directly observable relations on physical variables. The corresponding situation of directly observable concatenation operations happens rarely in the behavioral sciences. Still they play an important, indirect role as follows: A prevalent type of structure both in the physical and behavioral sciences is a directly observable ordering on a Cartesian product–e.g., the ordering by energy over mass-velocity pairs or by loudness over energy pairs to the two ears. Krantz [**1964**] first showed how extensive structures arise in the simplest such cases, and Narens and Luce [**1976**] showed that more general ordered structures are often transformable into positive concatenation structures. In these cases, the concatenation operation $\circ$ results from the interaction of objects in different components of the Cartesian product. The following definition describes two types of such structures.

DEFINITION 4. $\mathcal{C} = \langle X \times P, \succsim, ab \rangle$ is said to be a *conjoint structure solvable with respect to the element ab* if and only if ab is in $X \times P$ and the following seven axioms hold:

1. *Weak ordering.* $\succsim$ is a weak ordering on $X \times P$.

2. *Nontriviality.* There exists xp in $X \times P$ such that $xp \succ ab$.

3. *Density.* For each xp, yp in $X \times P$, if $xp \succsim yp$, then for some z in X, $xp \succ zp \succ yp$.

4. *Solvability* (*with respect to ab*). For each xp in $X \times P$, there exist z and q such that $xp \sim zb$ and $xb \sim aq$.

5. *Archimedean.* For each x, y in X such that $xb \succ ab$, there exists n in I^+ such that $(nx)b \succ yb$, where nx is defined inductively as follows: $1x = x$, and if nx is defined and s is such that $xb \sim as$, then $(n+1)x$ is some u such that $ub \sim (nx)s$.

6. *Independence.* For each x, y in X and p, q in P, (i) if $xs \succsim ys$ for some s, then $xp \succsim yp$; and (ii) if $wp \succsim wq$ for some w, then $xp \succsim xq$.

From Axiom 6, independence, it easily follows that the relations $\succsim_X$ and $\succsim_P$ defined on X and P respectively by: for each x, y in X and each p, q in P,

$$x \succsim_X y \text{ iff for some } s, \quad xs \succsim ys,$$

and

$$p \succsim_P q \text{ iff for some } w, wp \succsim wq,$$

are weak orderings on X and P respectively. Once again, to simplify notation and some definitions, we will assume the following axiom.

7. *Component total ordering.* $\succsim_X$ and $\succsim_P$ are total orderings, which will be written as $\geqslant_X$ and $\geqslant_P$.

If in addition to Axioms 1–7 above, ab is the minimal element of $X \times P$ (i.e., $xp \succsim ab$ for all xp in $X \times P$), then $\mathcal{C}$ is said to be a conjoint structure *solvable with respect to a minimal element ab.*

Let $\mathcal{C} = \langle X \times P, \succsim \rangle$ be a conjoint structure solvable with respect to a *minimal element ab* (i.e., $xp \succsim ab$ for all xp in $X \times P$). We will now sketch the construction of Narens and Luce **[1976]** which shows how to code $\mathcal{C}$ as a positive concatenation structure; it generalizes the proof for the additive case given in Holman **[1971]**. By solvability and component total ordering, let ξ: $X \times P \to X$ and σ: $X \to P$ be defined as the unique solutions to the following equations for each xp in $X \times P$,

$$xp \sim \xi(xp)b \quad \text{and} \quad xb \sim a\sigma(x).$$

Let $X^+ = \{x | x \in X \text{ and } x \succ_X a\}$ and for each x, y in X^+, let $x \bigcirc y = \xi[x\sigma(y)]$. Then it follows from results in Narens and Luce **[1976]** that $\mathcal{X}^+ = \langle X^+, \geqslant_X, \bigcirc \rangle$ is a positive concatenation structure. Note that for each x, y, u, v in X^+, $x \bigcirc y \geqslant_X u \bigcirc v$ iff $\xi[x\sigma(y)] \geqslant_X \xi[u\sigma(v)]$ iff $x\sigma(y) \succsim u\sigma(v)$.

For $\bigcirc$ to be associative, the following condition on $\mathcal{C}$ is necessary and sufficient.

The Thompsen condition. For each x, y, z in X and each p, q, r in P, if $xp \sim yq$ and $yr \sim zp$, then $xr \sim zq$.

THEOREM 3. *Suppose* $\mathcal{C}$ *is a solvable conjoint structure with a minimal element* ab *and satisfies the Thompsen condition. Then* $\bigcirc$ *is associative and there exist functions* $\varphi: X \to \mathrm{Re}^+ \cup \{0\}$, $\psi: P \to \mathrm{Re}^+ \cup \{0\}$ *such that*

(i) $\varphi(a) = \psi(b) = 0$,

(ii) *for each* xp, yq *in* $X \times P$,

$$xp \succsim yq \text{ iff } \varphi(x) + \psi(p) \geqslant \varphi(y) + \psi(q), \tag{1}$$

and

(iii) *if* φ', ψ' *is another pair of functions on* X *and* P *respectively that satisfy* (i) *and* (ii) *above, then for some* r *in* Re^+, $\varphi' = r\varphi$ *and* $\psi' = r\psi$.

The interested reader should consult Narens and Luce [**1976**] for the proof of Theorem 3 and details of the above construction. (Structures $\mathcal{C}$ that satisfy Equation 1 for some φ, ψ are called *additive conjoint structures*.)

The solvability condition in Definition 3 requires $\bigcirc$ to be a closed operation. A weaker form of solvability that yields partial operations is considered in Narens and Luce [**1976**], and a weaker form for conjoint structures satisfying the Thompsen condition is considered in Luce [**1966**] and Chapter 6 of Krantz et al. [**1971**].

4. Uniqueness of positive concatenation structures. The uniqueness of representations of positive concatenation structures, as given in statement (ii) of Theorem 2, takes a form different from that of extensive structures, as given in statements (ii) and (iii) of Theorem 1. The two kinds of uniqueness are equally "unique" in the sense that a value at one point determines the representation. However, Theorem 1 also tells how any two representations are related to each other, whereas Theorem 2 does not.

To clarify that question for a positive concatenation structure $\mathcal{X} = \langle X, \succcurlyeq, \bigcirc\rangle$, we consider the automorphism group of $\mathcal{X}$. Suppose φ is a $\odot$-representation and α is an automorphism of $\mathcal{X}$. It is easy to verify that $\varphi\alpha$ is a $\odot$-representation of $\mathcal{X}$. Furthermore, each $\odot$-representation ψ of $\mathcal{X}$ such that $\psi(X) = \varphi(X)$ is of this form since it easily follows that $\varphi^{-1}\psi$ is an automorphism of $\mathcal{X}$. Thus to understand how $\odot$-representations of $\mathcal{X}$ with the same range are related, it is sufficient to understand how automorphisms are related.

Cohen and Narens [**1979**] showed that the group of automorphisms $\langle A, *\rangle$ of a positive concatenation structure $\mathcal{X} = \langle X, \succcurlyeq, \bigcirc\rangle$ has a natural ordering $\succcurlyeq$ defined on it by: for each α, β in A,

$$\alpha \succcurlyeq \beta \text{ iff for some } x \text{ in } X, \alpha(x) \succcurlyeq \beta(x).$$

They showed that the structure $\mathcal{G} = \langle A, \succcurlyeq, *\rangle$ is an Archimedean, totally ordered group. It is well known that such groups are of the following three types. Let ι be the identity automorphism, and $A^+ = \langle \alpha \in A | \alpha \succ \iota\rangle$. Then $\mathcal{G}$ is *trivial* if $A^+ = \varnothing$, *discrete* if A^+ has a least element, and *dense* if A^+ has no least element. There are positive concatenation structures with automorphism groups of each type: Consider $\langle \mathrm{Re}^+, \geqslant, \oplus\rangle$ and its group of automorphisms $\mathcal{G}$. In each of the following choices for $\oplus$, the structure is a positive concatenation structure.

1. $x \oplus y = x + y$. $\oplus$ is commutative and associative. $\mathcal{G}$ is dense and consists of multiplication by every positive real.

2. $x \oplus y = x + y + x^{1/2}y^{1/2}$. $\oplus$ is commutative and nonassociative. $\mathcal{G}$ is dense and consists of multiplication by every positive real.

3. $x \oplus y = x + y + x^{1/3}y^{2/3}$. $\oplus$ is noncommutative and nonassociative. $\mathcal{G}$ is dense and consists of multiplication by every positive real.

4. $x \oplus y = x + y = (xy)^{1/2}[2 + \sin(\frac{1}{2}\log xy)]$. $\oplus$ is commutative and nonassociative. Cohen and Narens [**1979**] showed $\mathcal{G}$ to be discrete and consist of multiplication by $e^{2\pi n}$, $n = 0, 1, 2, \ldots$.

5. $x \oplus y = x + y + x^2y^2$. $\oplus$ is commutative and nonassociative. Cohen and Narens [**1979**] showed $\mathcal{G}$ to be trivial.

All of the above structures are Dedekind complete (every nonempty bounded subset has a least upper bound). Narens and Luce [**1976**], Cohen and Narens [**1979**] and Narens [**1981**] investigate conditions under which positive concatenation structures are extendable to Dedekind complete ones. The arguments are much more complicated and subtle than those familiar from the associative case, and several results are established concerning what sort of measurement-theoretic properties are inherited by such Dedekind completions. The interested reader should consult the above papers. Throughout the rest of this part we consider only the Dedekind complete case with a dense automorphism group.

5. Homogeneous structures.

DEFINITION 5. Let $\mathcal{X} = \langle X, R_1, R_2, \ldots \rangle$ be a relational structure (i.e., X is a nonempty set and $R_1, R_2, \ldots$ are relations and/or functions on X). Then $\mathcal{X}$ is said to be *homogeneous* if and only if for each x, y in X there exists an automorphism α of $\mathcal{X}$ such that $\alpha(x) = y$.

Cohen and Narens [**1979**] showed the following theorem.

THEOREM 4. *Suppose $\mathcal{X} = \langle X, \succcurlyeq, \bigcirc \rangle$ is a Dedekind complete positive concatenation structure. Then the following three conditions are equivalent*:

(i) *$\mathcal{X}$ is homogeneous*;

(ii) *$\mathcal{X}$ has a dense automorphism group*;

(iii) *for each n in I^+, $n(x \bigcirc y) = (nx) \bigcirc (ny)$.*

DEFINITION 6. Dedekind complete positive concatenation structures that satisfy one of the conditions of Theorem 4 are called *fundamental unit structures*.

Condition (iii) in Theorem 4 is of particular interest since it formulates in the language of the first order predicate calculus what is meant by $\mathcal{X}$ being homogeneous despite the fact that the concept of "automorphism" is a higher order concept, not formulable in the first order predicate calculus. In an extensive structure, condition (iii) of Theorem 4 follows from associativity, and it amounts to an interesting generalization of associativity, as the following theorem shows.

THEOREM 5. *Suppose* $\mathfrak{X} = \langle X, \succcurlyeq, \bigcirc \rangle$ *is a fundamental unit structure. Then there exists a* $\odot$*-representation* φ *of* $\mathfrak{X}$ *such that the following three statements are true*:

(1) *There exists* $f\colon X \to \mathrm{Re}^+$ *such that, for all* x, y *in* X, $\varphi(x) \odot \varphi(y) = \varphi(y) f[\varphi(x)/\varphi(y)]$;

(2) *for each* $\odot$*-representation* ψ *of* $\mathfrak{X}$, *there exists* r *in* Re^+ *such that* $\psi = r\varphi$; *and*

(3) *for each* r *in* Re^+, $r\varphi$ *is a* $\odot$*-representation of* $\mathfrak{X}$.

DEFINITION 7. $\odot$-representations φ of fundamental unit structures $\mathfrak{X}$ that satisfy statements (1), (2) and (3) of Theorem 5 are called *unit representations* of $\mathfrak{X}$.

Properties of fundamental unit structures are thoroughly explored in Cohen and Narens [**1979**], and we know a great deal about this class of structures. In particular the form of the function f in statement (1) of Theorem 5 is highly constrained.

The method of establishing Theorem 5 extends to other kinds of structures, and Narens [**1981**] has exploited this fact to show that ratio scalability–uniqueness of the representation up to multiplication by positive reals–holds in a variety of structures. The following theorem, which generalizes Theorem 5, is an instance of this approach.

THEOREM 6. *Suppose* X *is a nonempty set,* $\succcurlyeq$ *is a binary relation on* X, *and* $\mathfrak{X} = \langle X, \succcurlyeq, R_1, R_2, \ldots \rangle$ *is a relational structure that has the following four properties*:

(1) $\langle X, \succcurlyeq \rangle$ *is a totally ordered and Dedekind complete.*

(2) $\mathfrak{X}$ *is homogeneous* (*Definition* 4).

(3) *The group of automorphisms of* $\mathfrak{X}$ *is commutative.*

(4) $\langle X, \succcurlyeq \rangle$ *is dense in the sense that if* x, y *are in* X *and* $x \succ y$, *then there is a* z *in* $\mathfrak{X}$ *with* $x \succ z \succ y$.

Then there exists a structure $\mathfrak{N} = \langle \mathrm{Re}^+, \geqslant, S_1, \ldots, S_n, \ldots \rangle$ *that is isomorphic to* $\mathfrak{X}$ *and is such that for all isomorphisms* φ, ψ *of* $\mathfrak{X}$ *onto* $\mathfrak{N}$, (i) *there exists* r *in* Re^+ *such that* $\psi = r\varphi$, *and* (ii) *for each* s *in* Re^+, $s\varphi$ *is an isomorphism of* $\mathfrak{X}$ *onto* $\mathfrak{N}$.

II. MORE THAN ONE OPERATION

6. Introduction. It is easy enough to speak of studying general relational structures of the form $\langle A, \succsim, R_2, \ldots, R_n \rangle$, where $\succsim$ is a weak (or quasi) order. However, until recently (Narens [**1981**]), little measurement research has been done on the general case, and a good deal of attention has been concentrated on structures with two operations (or the equivalent thereof) that in one way or another can be mapped into addition and multiplication.

Perhaps the most natural example for mathematicians is the concept of a ring that can be mapped into subrings of $\langle \mathrm{Re}, \geqslant, +, \cdot \rangle$, and a generalization of this will be considered in §7 with applications to two different measurement problems: qualitative conditional probability and polynomial conjoint measurement.

To a physicist, it probably seems more natural to consider an ordering on the Cartesian product of two (or more) distinct sets, which under certain assumptions induces an operation on each component, together with an explicit operation on either the product itself or on one of the components. Here the explicit operation is represented by addition and the Cartesian product by multiplication. A typical example is the measure of kinetic energy, $\frac{1}{2}mv^2$, where mass and velocity are the component attributes, and mass, at least, possesses a concatenation operation. Much the same sort of representation occurs when we think of the probability measure over independent repetitions of an identical experiment (as in a random sample). We take up the latter in §8 and the former in §9.

§10 is devoted to the philosophical topic of meaningful statements in measurement contexts and, in particular, the relationship between this topic and the concept of dimensionally invariant laws in physics.

To a social scientist or a statistician, still another role for addition and multiplication comes to mind, namely, in the computation of expected values or, more generally, weighted averages. Much work along this line has centered on the specific economic problem of subjective expected utility theory, but the formalism can be interpreted more generally. This we take up in §11.

7. Semirings. In formulating the concept of a semiring in a form suitable for use in measurement theory, it is again necessary to work with partial operations.

DEFINITION 8. Suppose A is a set, $\succsim$ a binary relation on A, $\bigcirc$ and $*$ partial binary operations with domains $B^{\bigcirc}$ and $B^* \subseteq A \times A$. Then $\langle A, \succsim, B^{\bigcirc}, B^*, \bigcirc, *\rangle$ is said to be a *positive, regular Archimedean ordered local semiring* if and only if

1. $\langle A, \succsim, B^{\bigcirc}, \bigcirc\rangle$ is an extensive structure (Definition 2).
2. $\langle A, \succsim, B^*, *\rangle$ satisfies Axioms 1–4 and 8 of Definition 2.
3. (i) If $(b, c) \in B^{\bigcirc}$, $(a, b\bigcirc c) \in B^*$ then (a, b), $(a, c) \in B^*$ and $(a * b)\bigcirc(a * c) \in B^{\bigcirc}$ and $a*(b\bigcirc c) = (a * b)\bigcirc(a * c)$.
   (ii) The right distributive analogue of (i).
4. For $a \in A$, there exist $b, c \in A$ such that $(b, c) \in B^{\bigcirc}$ and $(a, b\bigcirc c) \in B^*$.

(The notion of an Archimedean ordered semiring presented here is a little more restricted than the one presented in Chapter 2 of Krantz et al. which does not assume that $\langle A, \succsim, B^{\bigcirc}, \bigcirc\rangle$ satisfies positivity as given in Definition 2 and assumes a weaker form of restrictive solvability than given in Definition 2.)

This generalizes the concept of an Archimedean ordered ring in which both operations are closed, $\langle A, \succsim, \bigcirc\rangle$ is an Archimedean ordered group, $\langle A, \bigcirc, *\rangle$ is a ring with zero element e, and if $a \succ e$, $b \succ c$, then $a * b \succ a * c$ and $b * a \succ c * a$. The following theorem generalizes the classic result that any

Archimedean ordered ring is uniquely isomorphic to a subring of $\langle \mathrm{Re}, \geqslant, +, \cdot \rangle$ (see Krantz et al. [**1971**, 2.27]).

THEOREM 7. *Suppose* $\langle A, \succsim, B^{\circ}, B^{*}, \circ, * \rangle$ *is a positive, regular Archimedean ordered local semiring. Then there is a unique homomorphism of* $\langle A, \succsim, B^{\circ}, B^{*}, \circ, * \rangle$ *into* $\langle \mathrm{Re}^{+}, \geqslant, \mathrm{Re}^{+2}, \mathrm{Re}^{+2}, +, \cdot \rangle$.

To date, this has been used in the proof of two measurement theorems. The first arises in the study of qualitative conditional probability.

Let $\mathcal{E}$ be an algebra of subsets of X, $\mathcal{N} \subset \mathcal{E}$, and $\succsim$ a relation on $\mathcal{E} \times (\mathcal{E} - \mathcal{N})$. The interpretation of $(A, B) \succsim (C, D)$ is that event A given event B is at least as probable as event C given event D. The following theorem can be shown.

THEOREM 8. *Suppose* $\langle X, \mathcal{E}, \mathcal{N}, \succsim \rangle$ *satisfies the following eight axioms: for all* $A, A' \in \mathcal{E}$, $B, B', C, C' \in \mathcal{E} - \mathcal{N}$,

1. $\succsim$ *is a weak order.*
2. $X \in \mathcal{E} - \mathcal{N}$ *and* $A \in \mathcal{N}$ *iff* $(A, X) \sim (\varnothing, X)$.
3. $(X, X) \sim (C, C)$ *and* $(X, X) \succsim (A, B)$.
4. $(A, B) \sim (A \cap B, B)$.
5. *Suppose* $A \cap B = A' \cap B' = \varnothing$. *If* $(A, C) \succsim (A', C')$ *and* $(B, C) \succsim (B', C')$, *then* $(A \cup B, C) \succsim (A' \cup B', C)$; *and if* $\succ$ *holds in either antecedent, it holds in the conclusion.*
6. *Suppose* $A \subseteq B \subseteq C$ *and* $A' \subseteq B' \subseteq C$. *If* $(B, C) \succsim (A', B')$ *and* $(A, B) \succsim (B', C')$, *then* $(A, C) \succsim (A', C')$.
7. (*Archimedean*) *Every standard sequence is finite, where* $\{A_i\}$ *is a standard sequence iff* $A_i \in \mathcal{E} - \mathcal{N}$, $A_{i+i} \supseteq A_i$, *and* $(X, X) \succ (A_i, A_{i+i}) \sim (A_1, A_2)$.
8. (*Solvability*) *If* $(A, B) \succsim (A', C)$, *there exists* $A'' \in \mathcal{E}$ *such that* $A' \cap C \subseteq A''$ *and* $(A, B) \sim (A'', C)$.

There then exists a unique real-valued function P on $\mathcal{E}$ *such that for all* $A, A' \in \mathcal{E}$, $B, B' \in \mathcal{E} - \mathcal{N}$,

(i) $\langle X, \mathcal{E}, P \rangle$ *is a finitely additive probability space.*
(ii) $N \in \mathcal{N}$ *iff* $P(N) = 0$.
(iii) $(A, B) \succsim (A', B')$ *iff* $P(A \cap B)/P(B) \geqslant P(A' \cap B')/P(B')$.

The proof of Theorem 8 is given in Krantz et al. [**1971**, §5.6], and involves defining $\succsim'$ on $\mathcal{E}$ by $(A, X) \succsim (A', X)$ which with the union of disjoint sets as $\circ$ leads to an Archimedean ordered local ordered semigroup on $\mathcal{E}/\sim'$. Define $*$ on $\mathcal{E}/\sim'$ by $[A] * [B] = [C]$ iff $(A, X) \sim (C, B)$. Then one can show the conditions of Theorem 7 hold, from which the representation follows readily.

The other application of Theorem 7, which will be described in less detail, concerns the generalization of additive conjoint measurement to representation by simple polynomials (see Krantz et al. [**1971**, Chapter 7]). For $n = 3$, these are

$x + y + z$, xyz, $x(y + z)$ and $xy + z$ and all permutations of the symbols. More generally, a polynomial on n factors is *simple* if and only if, for some m, $0 < m < n$, it is the sum or the product of a simple polynomial on a set of m factors and another simple polynomial on the remaining $n - m$ factors. In some cases, such as $x(y + z)$, Theorem 7 is used to arrive at the representation from axioms somewhat like, but more complex than, those for additive conjoint measurement.

8. Independence in qualitative probability. In §1, the reduction of qualitative probability to extensive structures required the assumption of solvability conditions. This is also true for the reduction just given of conditional probability to local semirings.

The major idea to overcome the assumption of strong solvability conditions has been to incorporate, in one way or another, independence of events as a primitive of the structure. This is, of course, contrary to the spirit of Kolmogorov's [**1933**] axiomatization of numerical probability in which independence is a defined concept [A and B are independent iff $P(A \cap B) = P(A)P(B)$]. However, since there are qualitative probability structures with nonunique representations, it is clear that, in general, independence cannot be defined in terms of the qualitative ordering of the events. Moreover, in scientific practice, the independence of many events is assumed on considerations such as physical isolation, and not just through the above numerical definition, and this kind of postulated independence is widely used in arriving at estimates of probabilities from relative frequencies.

Two directions for incorporating independence have been tried. The first, and the one that initially seems more straightforward, is simply to add a binary relation $\perp$ on $\mathcal{E}$ to the structure $\langle X, \mathcal{E}, \succsim \rangle$, where X is a nonempty set, $\mathcal{E}$ an algebra of subsets on X, and $\succsim$ the "at least as likely as" ordering on $\mathcal{E}$, and then to search for axioms sufficient to yield a unique probability measure P on $\mathcal{E}$ such that P preserves the ordering $\succsim$ and if $A \perp B$, then $P(A \cap B) = P(A)P(B)$. At first one might hope for a local ring to be involved, but that hope is dashed when one realizes that the relation of independence is really quite irregular; in particular, it utterly fails the property that if $A \perp B$ and $A \succsim A'$ and $B \succsim B'$, then $A' \perp B'$, which is part of the ring definition. No one has yet seen how to make effective use of the extra independence primitive without imposing strong structural conditions that are unacceptable to many researchers (see Domotor [**1970**], and Krantz et al. [**1971**, §5.8]).

The second and far more successful approach is to model qualitatively the idea of an indefinite number of independent repetitions of an experiment. This is, after all, what lies qualitatively beneath the idea of independent repetitions of identical random variables, the resulting central limit theorem, and our standard procedure for estimating probabilities from data. True, there are experiments that cannot be repeated, and those await a better theory. For those that can be repeated, it is really quite surprising how strong the results are–in essence, they amount to the joint qualitative space being rich enough that the solvability conditions are met and so Theorem 1 is applicable, and this leads to a unique probability measure that is multiplicative over the repetitions. This, and more, were worked out in Kaplan's [**1971**] important master's thesis which was exposited in Kaplan and Fine [**1977**]. We summarize one of the key results here.

Suppose $\langle X, \mathcal{E}, \succsim \rangle$ is a nontrivial (there exists $A \in \mathcal{E}$ with $X \succ A \succ \varnothing$) qualitative probability structure, and $\langle X_i, \mathcal{E}_i, \succsim_i \rangle$, $i = 1, 2, \ldots,$ are isomorphic copies of it. Let $X^* = \times_{i=1}^{\infty} X_i$. For any $A \in \mathcal{E}$, let A^i denote that subset of elements in X^* such that the ith component falls in the isomorphic image of A. For any $\alpha \subseteq \{1, 2, \ldots\}$, denote by $\mathcal{E}(\alpha)$ the σ-algebra generated by A^i for all $A \in \mathcal{E}$ and all $i \in \alpha$. Let $\mathcal{E}^* = \mathcal{E}(\{1, 2, \ldots\})$. Finally, let $\succsim^*$ be a binary relation on $\mathcal{E}^*$.

DEFINITION 9. $\langle X^*, \mathcal{E}^*, \succsim^* \rangle$ is an *infinite, independent, identically distributed product space* for $\langle X, \mathcal{E}, \succsim \rangle$ if and only if the following three conditions hold for all $A, B \in \mathcal{E}$:

1. $A^i \succsim^* B^i$ iff $A \succsim B$.
2. Suppose $\alpha, \beta, \gamma, \delta \subseteq \{1, 2, \ldots\}$, $\alpha \cap \beta = \gamma \cap \delta = \varnothing$, $A \in \mathcal{E}(\alpha)$, $B \in \mathcal{E}(\beta)$, $C \in \mathcal{E}(\gamma)$, and $D \in \mathcal{E}(\delta)$. Then $A \succsim^* C$ and $B \succsim D$ implies $A \cap B \succ^* C \cap D$, and if $\succ^*$ holds in either antecedent, then it holds in the conclusion.
3. $A^i \sim^* A^j$ *for all* $i, j \in \{1, 2, \ldots\}$.

THEOREM 9. *Suppose* $\langle X^*, \mathcal{E}^*, \succsim^* \rangle$ *is an infinite, independent, identically distributed product space for* $\langle X, \mathcal{E}, \succsim \rangle$ *with the property that* $\succsim^*$ *is monotonely continuous. Then there exists a unique countably additive probability measure* P^* *on* $\mathcal{E}^*$ *that preserves the ordering* $\succsim^*$ *and, for* $i \neq j$,

$$P(A^i \cap B^j) = P(A^i)P(B^j).$$

The proof rests on showing that $\langle X, \mathcal{E}^*, \succsim \rangle$ has no atoms and then invoking Villegas' [**1964**] theorem (for that case see Krantz et al. [**1971**, p. 216]). The multiplicative property then falls out moderately easily.

Luce and Narens [**1978**] present axioms for $\succsim_2$ on $\mathcal{E} \times \mathcal{E}$ that are sufficient to insure a finitely additive probability representation with the multiplicative representation of $\succsim_2$. They also prove that if $\succsim_n$ are orderings on $\mathcal{E}^n$, $n = 1, 2, \ldots,$ which all can be represented multiplicatively by a probability measure, then that measure is necessarily unique.

9. Distributive triples and dimensional analysis. Next we turn to the physicist's simultaneous use of addition and multiplication, where the former represents the combining operation within a dimension and the latter the combining operation between dimensions, as in expressions such as $\frac{1}{2}mv^2$. This is of considerable importance since, as is well known, the interaction of the major dimensions of classical physics is as products of powers of additive scales. This structure is reflected in the nature of the units of physical scales and it underlies the surprisingly powerful method of dimensional analysis. Although a variety of axiomatizations exist for the mathematical structure representing the dimension of physics, the most satisfactory is Whitney's **[1968]**, which is reproduced in §10.2 of Krantz et al. **[1971]**. Within that framework, one formulates the idea of a similarity transformation, imposes the requirement that any numerical law of physics be invariant under those transformations, which is referred to as the property of *dimensional invariance*, and then proves Buckingham's **[1914]** π-theorem establishing that any such law is some function of the maximal number of independent dimensionless quantities that can be formed from the variables involved in the law (see Theorem 10.4 of Krantz et al. **[1971**, p. 466**]**).

From the point of view of a measurement theorist, two major questions must be considered. First, what is the qualitative nature of the interlock among dimensions that permits the representation of each dimension as the product of powers of a limited number of dimensions each of which has an associative operation? That is the topic of this section. Second, why should physical laws be dimensionally invariant? That is the topic of the next section.

DEFINITION 10. Let $\succsim$ be a reflexive relation on $A \times P$, and suppose $\bigcirc_A$ is a closed binary operation on A. Then $\bigcirc_A$ is said to be *distributive* if and only if for all $a, b, c, d \in A$, $p, q \in P$, whenever $(a, p) \sim (c, q)$ and $(b, p) \sim (d, q)$, then $(a \bigcirc_A b, p) \sim (c \bigcirc_A d, q)$.

As the following theorem of Narens **[1981]** shows, distributivity seems to be the key interlock exploited in the dimensional structures of physics.

THEOREM 10. *Suppose* $\langle A \times P, \succsim \rangle$ *is a conjoint structure that satisfies weak ordering, independence, and solvability with respect to ap for each ap in* $A \times P$ *(Definition* 3), *and suppose* $\bigcirc_A$ *is a closed operation on* A, *and that* $\mathcal{A} = \langle A, \succsim_A, \bigcirc_A \rangle$ *is a fundamental unit structure with unit representation* φ_A *(Definition* 7).

1. *If* $\bigcirc_A$ *is distributive, there exists* φ_P *on* P *such that, for all* $a, b \in A$, $p, q \in P$,

$$(a, p) \succsim (b, q) \quad \text{iff} \quad \varphi_A(a)\varphi_P(p) \geqslant \varphi_A(b)\varphi_P(q).$$

2. $\bigcirc_A$ *is distributive if and only if, for all automorphisms* θ *of* $\mathcal{A}$ *and all* $a, b \in A, p, q \in P$,

$$(a, p) \sim (b, q) \quad \text{iff} \quad (\theta(a), p) \sim (\theta(b), q).$$

3. *If, in addition* $\bigcirc_P$ *is a closed operation on* P *such that* $\langle P, \succsim_P, \bigcirc_P \rangle$ *is a fundamental unit structure with unit representation* φ_P, *then there is some constant* α *such that, for every* $\beta > 0$, $\varphi_A^{\beta}\varphi_P^{\alpha\beta}$ *preserves the ordering relation* $\succsim$ *on* $A \times P$.

Parts 1 and 3 of Theorem 10 were established for associative operations in Narens and Luce **[1976]**. Proofs of Theorem 10 can be found in Narens **[1981]**, Krantz et al. [in preparation], and Narens [in preparation].

In order to develop the usual numerical representation of a large number of interlocked physical dimensions, it is necessary to postulate an adequate density of distributive triples of the sort described in parts 1 and 4 of Theorem 10. When done properly–the details are given in Luce [**1978**] and Krantz et al. [in preparation]–one is able to show the existence of a finite basis in fundamental unit structures such that all other scales are represented as products of powers of these scales.

From the point of view of a nonphysical scientist, the major interest in these results is that they show exactly what is involved in adding a new dimension to those already discovered by physics. Whether or not such a fundamental dimension will be discovered in the biological realm is conjectural. What is not conjectural is the qualitative conditions that such a new dimension would have to exhibit in order to become a part of the existing structure of dimensions.

10. Qualitative and quantitative meaningfulness. Once qualitative information is recast numerically–when measurement is possible–one needs to consider carefully just which numerical statements do and do not correspond to something meaningful in the underlying structure. As a trivial example, if only order is preserved, there is nothing in the qualitative structure corresponding to $x + y = z$. This problem has been explored with some care, but a complete consensus as to its solution does not yet exist. The source of the uncertainty centers on exactly what is meant by saying that a relation is "meaningful" in a relational structure that is characterized axiomatically. Intuitively, one would like to say that a relation is meaningful if and only if it can be defined in terms of the primitives of the structure, but there is at this time no appropriate formal definition of "defined" that is useful in general measurement contexts.

This inability to define directly these kinds of relevant concepts by some procedure of formal logic forces one to consider some sort of indirect procedure. The one exploited by measurement theorists rests on the idea that if a relation is definable in terms of the defining relations of a structure, then adding it to the structure does not further restrict the structure from the point of view of its measurement theoretic properties. Looking just at a structure, this implies that (1) it should remain invariant under the same transformations of the structure into itself that leave invariant the defining relations of the structure, i.e., the *endomorphisms* of the structure. But it also implies that (2) it does not alter the homomorphisms into numerical structures, which in turn suggests that (3) the homomorphic images of the relation should be invariant under the endomorphisms of the representing structure. However, it is by no means obvious which of these three conditions is the strongest or when they agree. We examine this more carefully.

Suppose $\mathcal{A} = \langle A, S_1, S_2, \ldots, S_n \rangle$ is a relational structure and $\mathcal{R} = \langle R, T_1, T_2, \ldots, T_n \rangle$ is a numerical relational structure with T_i being of the same order as S_i, $i = 1, 2, \ldots, n$. Suppose there is at least one homomorphism

φ from $\mathcal{A}$ into $\mathcal{R}$, i.e., for each $i = 1, 2, \ldots, n$, $(a_1, a_2, \ldots, a_{k_i}) \in S_i$ iff $(\varphi(a_1), \varphi(a_2), \ldots, \varphi(a_{k_i})) \in T_i$. Let $\Phi(\mathcal{A}, \mathcal{R})$ denote the set of all homomorphisms. For any relation T of order k on R and any $\varphi \in \varphi(\mathcal{A}, \mathcal{R})$ define

$$S(\varphi, T) = \{(a_1, \ldots, a_k) | (\varphi(a_1), \ldots, \varphi(a_k)) \in T\}.$$

DEFINITION 11. A relation T on R is *$\mathcal{A}$-reference invariant relative* to $\mathcal{R}$ if and only if, for all $\varphi, \psi \in \Phi(\mathcal{A}, \mathcal{R})$, $S(\varphi, T) = S(\psi, T)$, in which case the common value is denoted $S(T)$. A relation S on A that is $\mathcal{A}$-reference invariant relative to $\mathcal{A}$ is said to be *structure invariant*.

Put another way, S is structure invariant if and only if it is invariant under the endomorphisms of $\mathcal{A}$. The term "reference invariant" is due to Adams, Fagot and Robinson [**1965**], and the general definition, to Pfanzagl [**1968**, **1971**].

In order to avoid pointless distinctions, let us suppose that $\mathcal{A}$ is *irreducible* in the sense that all homomorphisms are one-to-one.

The following theorem is shown in Krantz et al [in preparation].

THEOREM 11. *Suppose $\mathcal{A}$ is irreducible and $\Phi(\mathcal{A}, \mathcal{R}) \neq \varnothing$. Let T be a relation of order k on R. If T is $\mathcal{A}$-reference invariant relative to $\mathcal{R}$, then $S(T)$ is structure invariant in $\mathcal{A}$. If, further, $T \subseteq \bigcup_{\varphi \in \Phi} \Phi(A^k)$, where A^k is the Cartesian product of A with itself k times, then T is structure invariant in $\mathcal{R}$.*

One can show that if a relation S on A is structure invariant in $\mathcal{A}$, it is not necessary for there to exist T on R such that T is $\mathcal{A}$-reference invariant relative to $\mathcal{R}$ and $S = S(T)$. Since, however, if such a T were to exist, the three notions of invariance would agree, it is thus interesting to know when this occurs. We formulate a simple, intuitively natural sufficient condition for such an occurrence.

Let $\varphi \in \Phi(\mathcal{A}, \mathcal{R})$ and let γ be a mapping of $\varphi(A)$ into R that leaves invariant the restriction of $\mathcal{R}$ to $\varphi(A)$; we speak of this as a *partial endomorphism*. It is not difficult to see that $\varphi' \in \Phi(\mathcal{A}, \mathcal{R})$ if and only if there is a partial endomorphism γ such that $\varphi' = \gamma\varphi$. The converse–given φ and γ, there exists the endomorphism α of $\mathcal{A}$ such that $\varphi\alpha = \gamma\varphi$–is not generally true. In case it is, we say $\mathcal{R}$ is *compatible* with $\mathcal{A}$. Krantz et al. [in preparation] show the following result:

THEOREM 12. *Suppose $\mathcal{A}$ is irreducible and $\mathcal{R}$ is compatible with $\mathcal{A}$.*

1. *A relation T on R is $\mathcal{A}$-reference invariant relative to $\mathcal{R}$ iff T is structure invariant in $\mathcal{R}$.*

2. *A relation S on A is structure invariant if and only if there exists a relation T on R such that T is $\mathcal{A}$-reference invariant relative to $\mathcal{R}$ and $S(T) = S$.*

Under the conditions of Theorem 12, it is widely believed that the common concept of invariance correctly captures the idea of a meaningful relation. When they do not agree, the strongest–reference invariance–is probably the appropriate one, if any are, for most measurement situations.

An easy to remember and often applicable sufficient condition for compatibility, and so for Theorem 12, is that for every $\varphi \in \Phi(\mathcal{A}, \mathcal{R})$, $\varphi(A) = R$; in this case all endomorphisms are automorphisms.

There have been two major applications of these ideas. The first and perhaps the most important for the behavioral and social sciences is to characterize the types of statistical hypotheses that are meaningful in different measurement structures. Indeed, the whole problem of meaningfulness was initiated when Stevens **[1946]**, **[1951]** pointed out that various familiar statistics are invariant under some groups of automorphisms that arise in well-known measurement systems but are not invariant under other groups. This observation and the structures he drew from it have led to a somewhat confused literature on the subject, but we need not go into that here.

The other application (Luce **[1978]**) is to the problem of dimensionally invariant laws in the structure of dimensional quantities. In essence, one proceeds as follows. The qualitative development based on distributive triples, outlined in the last section, is assumed, and it has a homomorphic representation in Whitney's structure which in turn is isomorphic to a multiplicative vector space over the reals. Both of these are then mapped in a quite natural way into relational structures, which it turns out are compatible. In that reformulation, dimensionally invariant laws correspond to relations that are structure invariant, and so by Theorem 12 they correspond exactly to qualitative relations that are structure invariant and hence to qualitatively meaningful ones. Thus, according to Luce, the answer to the question, "Why are physical laws dimensionally invariant?" is that this class of laws corresponds exactly to the class of all meaningful qualitative relations, provided we accept reference invariance or its equivalent in this context, structure invariance, as defining meaningfulness.

The concepts of meaningfulness discussed above are explored more fully in Narens **[1981]**, Krantz et al. [in preparation] and Narens [in preparation]. The relationship between dimensional invariance and meaningfulness is discussed in Luce **[1978]** and Krantz et al. [in preparation].

11. Averages and expected utility theory. Our final examples of a numerical representation involving both addition and multiplication are averages of the form $\Sigma w_i \varphi_i / \Sigma w_i$, $w_i \geqslant 0$. In economics, psychology, and statistics this representation occurs as the expectation of random variables, especially in the theories of subjective expected utility. In psychology Anderson **[1974a]**, **[1974b]** has also successfully applied averaging representations to category scale data from a wide variety of substantive areas.

The literature includes two somewhat different approaches which, fortunately, can be described in closely parallel terms. Let $\mathcal{C}$ be a set (of outcomes), X a set (sample space), and $\mathfrak{S}$ an algebra of subsets of X. One type of theory concerns a weak ordering of the set $\mathcal{Q}$ of all functions f: X into $\mathcal{C}$ subject to the restriction that, for all $c \in f(X)$, $f^{-1}(c) \in \mathfrak{S}$. Such functions are called "acts" because of the decision theory interpretation. The most important examples of such a theory is Savage's **[1954]** subjective (personalistic) expected utility.

The other type of theory concerns a weak order of a set $\mathcal{D}$ of functions of the form: for $A \in \mathcal{E}$, $A \neq \varnothing$, f_A: $A \to \mathcal{C}$ subject to the following two restrictions: for all $A, B \in \mathcal{E} - \{\varnothing\}, f_A, g_B \in \mathcal{D}$,

(i) if $A \cap B = \varnothing, f_A \cup g_B \in \mathcal{D}$;

(ii) if $A \supseteq B$, the restriction of f_A to B is $\in \mathcal{D}$.

These functions are called "conditional acts". Pfanzagl [**1959**], [**1967b**] examined a special case of such a theory, and Luce and Krantz [**1971**] studied the general case (also see Krantz et al. [**1971**, Chapter 8]).

The axioms, especially the structural and Archimedean ones, are sufficiently complex in both cases that we do not present them here. Rather we discuss the two most important necessary conditions, describe the line of proof, and state the representations obtained.

Both theories capture two essential facts about averages. The one is that on a *fixed* domain it looks just like measurement on additive conjoint structures. The other is that on disjoint domains, the average of two equivalent conditional acts is equivalent to them. To state these in the unconditional theory, one in essence has to define what is meant by a conditional act. By contrast, they are quite direct in the conditional theory, so we state them explicitly.

For all $A, B \in \mathcal{E} - \{\varnothing\}, A \cap B = \varnothing, f_A, f'_A, g_B \in \mathcal{D}$,

(i) $f_A \succsim f'_A$ iff $f_A \cup g_B \succsim f'_A \cup g_B$;

(ii) if $f_A \sim g_B$, then $f_A \cup g_B \sim f_A$.

The line of argument in the unconditional theory is to let $\succsim$ on $\mathcal{Q}$ induce an ordering on $\mathcal{E}$ and to assume an axiom adequate to prove the existence of a unique probability measure P that preserves the induced ordering. At that point the proof follows that of von Neumann and Morgenstern [**1947**, p. 617] leading to a real-valued function u on $\mathcal{C}$ such that, for all $f, g \in \mathcal{Q}, f \succsim g$ iff $E_P[u(f)] \geqslant E_P[u(g)]$, where E_P is the expectation operator with respect to P.

The line of argument in the conditional theory is to note that any partition of X into three or more nonnull subevents generates an additive conjoint structure. Using the results about uniqueness of representations for this structure, one is able to show simultaneously the existence of a real-valued function v on $\mathcal{D}$, unique up to positive linear transformations, and a unique probability measure P on $\mathcal{E}$ such that, for all $f_A, g_B \in \mathcal{D}$,

(i) $f_A \succsim g_B$ iff $v(f_A) \geqslant v(g_B)$, and

(ii) if $A \cap B = \varnothing$, then $v(f_A \cup g_B) = v(f_A)P(A|A \cup B) + v(g_B)P(B|A \cup B)$.

If, in addition, for each $c \in \mathcal{C}$ there exists $A \in \mathcal{E} - \{\varnothing\}$ and $c_A \in \mathcal{D}$ with $c_A(a) = c$ for $a \in A$, and if whenever $c_A, c_B \in \mathcal{D}$, $c_A \sim c_B$, then one can show there exists a real-valued function u on $\mathcal{C}$ such that $v(f_A) = E_P[u(f_A)]$.

For finite X, it has been shown how to map either expectation representation into the other (Krantz et al. [**1971**, §8.6.4]).

As theories of decision making, both suffer from problems of interpretation. The inclusion of all possible acts, including all constant ones, in the unconditional theory is highly unrealistic. The meaning of $f_A \cup g_B$ in the conditional theory is obscure since the choice of a conditional act appears to place $\cup$ under the control of the decision maker, whereas the axioms force a fixed probability on $\mathcal{E}$. For an extended discussion of these matters, see Balch [**1974**], Balch and Fishburn [**1974**], Krantz and Luce [**1974**], and Spohn [**1977**].

A general averaging model of the type assumed by Anderson [**1974a**], [**1974b**] arises from the conditional theory as follows. Let $X = \{1, 2, \ldots, n\}$, $\mathcal{E} = 2^X$, $\mathcal{C}_i$, $i \in X$, be sets and define

$$\mathfrak{D} = \{f_A | A \in \mathcal{E} - \{\varnothing\}, f_A(i) \in \mathcal{C}_i \text{ for } i \in A\}.$$

Observe that if $A \cap B = \varnothing$ and $f_A, g_B \in \mathfrak{D}$, then, automatically, $f_A \cup g_B \in \mathfrak{D}$. Assuming the axioms of the conditional theory, it follows readily that there exist nonnegative weights $w_i = P(\{i\})$ and a real-valued function φ_i on $\mathcal{C}_i$ such that

$$u(f_A) = \sum_{i \in A} w_i \varphi_i[f_A(i)] / \sum_{i \in A} w_i$$

preserves the ordering relation $\succsim$. For details, see Luce [**1981**].

Conclusions

Our understanding of positive concatenation and of conjoint structures is reasonably adequate when the automorphism group is dense and especially so for fundamental unit structures. In contrast, we know very little about the discrete and trivial cases. Undoubtedly, many of these are so irregular as to be of no conceivable scientific interest, but some are clearly of importance, witness the case of probability.

Our understanding of the interplay of solvable conjoint structures having at least one component that is a fundamental unit structure is adequate when they satisfy distributivity, as appears to be the case for classical physics. It is totally unsatisfactory when distributivity does not hold, as arises with velocity in relativistic physics. The problem evidences itself in our inability to relate closely the automorphisms of the conjoint structure and that of the positive concatenation one. Closely related to these problems of dimensional interlocks is the general conceptual issue of meaningfulness and how it should be defined in terms of automorphisms and/or endomorphisms and/or some other invariance concept. Judging by the importance of dimensional analysis, this issue is of rather more than just philosophical interest.

The study of structures with more than one operation, including the case just mentioned, is probably susceptible to considerable generalization, just as fundamental unit structures generalized extensive ones. At the moment, all of the theories lead only to polynomial or metric representations. Almost certainly, more powerful algebraic proof techniques will be required since the existing methods seem to be leading to ever more complex, not easily generalized proofs.

References

Adams, E. W., R. F. Fagot and R. E. Robinson, 1965. *A theory of appropriate statistics*, Psychometrika **30**, 99–127.

Anderson, N. H., 1974a. *Information integration theory: A brief survey*, Contemporary Developments in Mathematical Psychology. II (D. H. Krantz, R. C. Atkinson, R. D. Luce and P. Suppes, Eds.), Freeman, San Francisco, Calif., pp. 236–305.

______, 1974b. *Algebraic models in perception*, Handbook of Perception, II (E. C. Carterette and M. P. Friedman, Eds.), Academic Press, New York, pp. 215–298.

Balch, M., 1974. *On recent developments in subjective expected utility*, Essays on Economic Behavior under Uncertainty (M. S. Balch, D. L. McFadden and S. Y. Wu, Eds.), North-Holland, Amsterdam, pp. 45–55.

Balch, M. and P. C. Fishburn, 1974. *Subjective expected utility for conditional primitives*, Essays on Economic Behavior under Uncertainty (M. S. Balch, D. L. McFadden and S. Y. Wu, Eds.), North-Holland, Amsterdam, pp. 56–69.

Beals, R., D. H. Krantz and A. Tversky, 1968. *Foundations of multidimensional scaling*, Psych. Rev. **75**, 127–142.

Beals, R. and D. H. Krantz, 1967. *Metrics and geodesics induced by order relations*, Math. Z. **101**, 285–298.

Buckingham, E., 1914. *On physically similar systems: illustrations of the use of dimensional equations*, Phys. Rev. **4**, 345–376.

Campbell, N. R., 1920. *Physics: the elements*, Cambridge Univer. Press, New York. Reprinted as *Foundations of science: the philosophy and theory of measurement*, Dover, New York, 1957.

Cohen, M. R. and E. Nagel, 1934. *An introduction to logic and the scientific method*, Harcourt, Brace, New York.

Cohen, M. and L. Narens, 1979. *Fundamental unit structures: a theory of ratio scalability*, J. Math. Psych. **20**, 193–232.

Domotor, Z., 1970. *Qualitative information and entropy structures*, Information and Inference (J. Hintikka and P. Suppes, Eds.), Reidel, Dordrecht, pp. 184–194.

Falmagne, Jean-Claude, 1971. *Bounded versions of Hölder's theorem with application to extensive measurement*, J. Math. Psych. **8**, 495–503.

______, 1975. *A set of independent axioms for positive Hölder systems*, Philos. Sci. **421**, 137–151.

Fine, T., 1971a. *A note on the existence of quantitative probability*, Ann. Math. Statist. **42**, 1182–1186.

______, 1971b. *Theories of probability*, Academic Press, New York.

v. Helmholtz, H., 1887. *Zahlen und Messen erkenntnis-theoretisch betrachet*, Philosophische Aufsatze Eduard Zeller gewidmet, Leipzig; reprinted Gesammelte Abhandl., vol. 3, 1895, pp. 356–391; English transl., C. L. Bryan, Counting and measuring, van Nostrand, Princeton, N.J., 1930.

Hölder, O., 1901. *Die Axiome der Quantitat und die Lehre vom Mass.*, Ber. Verh. Kgl. Sachsis. Ges. Wiss. Leipzig, Math.-Phys. Classe **53**, 1–64.

Holman, E. W., 1969. *Strong and weak extensive measurement*, J. Math. Psych. **6**, 286–293.

______, 1971. *A note on conjoint measurement with restricted solvability*, J. Math. Psych. **8**, 489–494.

______, 1974. *Extensive measurement without an order relation*, Philos. Sci. **41**, 361–373.

Kaplan, M. A., 1971. *A characterization of independence in comparative probability*, M.S. thesis, Cornell Univ.

Kaplan, M. and T. L. Fine, 1977. *Joint orders in comparative probability*, Ann. Probab. **5**, 161–179.

Kolmogorov, A. N., 1933. *Grundbegriffe der Wahrscheinlichkeitsrechnung*, Springer, Berlin; English transl., N. Morrison, *Foundations of the theory of probability*, Chelsea, New York, 1956.

Krantz, D. H., 1964. *Conjoint measurement: the Luce-Tukey axiomatization and some extensions*, J. Math. Psych. **1**, 248–277.

Krantz, D. H. and R. D.Luce, 1974. *The interpretation of conditional expected-utility theories*, Essays on Economic Behavior under Uncertainty (M. S. Balch, D. L. McFadden and S. Y. Wu, Eds.), North-Holland, Amsterdam, pp. 70–73.

Krantz, D. H., R. D. Luce, P. Suppes and A. Tversky, 1971; in preparation. *Foundations of measurement*, Academic Press, New York, vol. I; vol. II.

Luce, R. D., 1965. *A "fundamental" axiomatization of multiplicative power relations among three variables*, Philos. Sci. **32**, 301–309.

______, 1966. *Two extensions of conjoint measurement*, J. Math. Psych. **3**, 348–370.

______, 1978. *Dimensionally invariant numerical laws correspond to meaningful qualitative relations*, Philos. Sci. **45**, 1–16.

______, 1981. *Axioms for the averaging and adding representations of functional measurement*, Math. Social Sci. **1**, 139–144.

Luce, R. D. and D. H. Krantz, 1971. *Conditional expected utility*, Econometrica **39**, 253–271.

Luce, R. D. and L. Narens, 1978. *Qualitative independence in probability theory*, Theory and Decision **9**, 225–239.

Narens, L., 1974a. *Minimal conditions for additive conjoint measurement and qualitative probability*, J. Math. Psych. **11**, 404–430.

______, 1974b. *Measurement without Archimedean axioms*, Philos. Sci. **41**, 374–393.

______, 1981. *A general theory of ratio scalability with remarks about the measurement-theoretic concept of meaningfulness*, Theory and Decision **13**, 1 – 70.

______, in preparation. *Abstract measurement theory*, M.I.T. Press, Cambridge, Mass.

Narens, L. and R. D. Luce, 1976. *The algebra of measurement*, J. Pure Appl. Algebra **8**, 197–233.

Pfanzagl, J., 1959a. *Die axiomatischen Grundlagen einer allgemeinen Theorie des Messens*, Schr. Stat. Inst. Univ. Wien. Neue Folge no. 1, Physica-Verlag, Wurzburg.

______, 1959b. *A general theory of measurement–applications to utility*, Naval Res. Logist. Quart. **6**, 283–294.

______, 1967a. *Subjective probability derived from the Morgenstern-von Neumann utility concept*, Essays in Mathematical Economics in Honor of Oskar Morgenstern (M. Shubik, Ed.), Princeton Univ. Press, Princeton, N.J., pp. 237–251.

______, 1967b. *Characterizations of conditional expectations*, Ann. Math. Statist. **38**, 415–421.

______, 1968, 1971. *Theory of measurement*, Wiley, New York.

Roberts, F. S., 1979. *Measurement theory*, Addison-Wesley, Reading, Mass.

Roberts, F. S. and R. D. Luce, 1968. *Axiomatic thermodynamics and extensive measurement*, Synthese **18**, 311–326.

Savage, L. J., 1954. *The foundations of statistics*, Wiley, New York.

Skala, H. J., 1975. *Non-archimedean utility theory*, Reidel, Dordrecht, Holland.

Spohn, W., 1977. *Where Luce and Krantz do really generalize Savage's decision model*, Erkenntnis **11**, 113–134.

Stevens, S. S., 1946. *On the theory of scales of measurement*, Science **103**, 677–680.

______, 1951. *Mathematics, measurement and psychophysics*, Handbook of Experimental Psychology (S. S. Stevens, Ed.), Wiley, New York, 1–49.

Tversky, A. and D. H. Krantz, 1970. *The dimensional representation and metric structure of similarity data*, J. Math. Psych. **7**, 572–596.

Villegas, C., 1964. *On qualitative probability σ-algebras*, Ann. Math. Stat. **38**, 1787–1796.

von Neuman, J. and O. Morgenstern, 1944, 1947, 1953. *Theory of games and economic behavior*, Princeton Univer. Press, Princeton, N.J.

Whitney, H., 1968. *The mathematics of physical quantities*. I. *Mathematical models for measurement*; II. *Quantity structures and dimensional analysis*, Amer. Math. Monthly **75**, 115–138; 227–256.

Department of Psychology and Social Relations, Harvard University, Cambridge, Massachusetts 02138

School of Social Science, University of California, Irvine, California 92664

SIAM-AMS Proceedings
Volume **13**
1981

Optimal Decision Rules for Some Common Psychophysical Paradigms

DAVID L. NOREEN[1]

Abstract. A decision rule specifies how information from a presented stimulus situation may be used to determine a course of action, or *response*. This paper discusses decision rules that are optimal from the standpoint of both a minimal-error criterion and a maximized-expected-value objective. In the first three sections of the paper, the basic decision problem is outlined and well-known results from the yes-no and two-alternative forced-choice paradigms are reviewed. In subsequent sections, the same basic techniques of analysis used in the earlier sections are applied to the "same-different" and "ABX" discrimination tasks to obtain new results. For important special cases of these two discrimination tasks, the optimal decision strategy involves an implicit classification or "categorization" process on the part of the observer. While such a decision strategy differs in a fundamental way from one in which the observer compares differences in sensations ("Sensory Difference Theory"), the predictions of averaged percentage correct scores by the optimal decision strategy and by Sensory Difference Theory are nevertheless often quite similar — for example, for data from a task involving the discrimination of tones differing in frequency, the two sets of predictions are within several percentage points of each other, and each set provides at least a first-order approximation to the obtained performance. The discussion section of the paper considers implications of these results and shows how the optimal decision rule analysis can be applied to a variety of other psychophysical paradigms.

1. Introduction. An experimenter can use a variety of different methods to investigate how an observer discriminates between two classes of stimuli. For instance, on any given trial he can select a single stimulus from one of the two classes, present it to the observer, and ask him to identify which class it represents (the "yes-no" or "two-stimulus complete identification" paradigm). Alternatively, he can present an instance of each of the two classes and ask the observer to specify which of the two possible presentation orders was used (the "two-alternative forced-choice" paradigm).

1980 *Mathematics Subject Classification.* Primary 92A27.

[1]The author wishes to thank Dirk Vorberg for helpful suggestions, Stephen Coffin and John Holmgren for comments on an earlier version of the manuscript, and Beverly Heravi for assistance with the production of the manuscript.

As a third possibility, he can ask the observer to indicate whether two presented stimuli represent instances of the same stimulus class or instances of different stimulus classes (the "same-different" discrimination paradigm). As yet another possibility, he can present an instance of each of the two classes as standards and then ask the observer to indicate whether a third, test stimulus represents a replication of the first-presented standard or a replication of the second-presented standard (the "ABX" discrimination paradigm).

While these represent only four of the many methods that can be used to study how well an observer can discriminate between two classes of stimuli, yet they have received a preponderate amount of attention in the experimental literature. These four paradigms, together with several less familiar but related tasks that will be mentioned briefly in the discussion section, comprise the basic set of paradigms that will be analyzed in this paper.

In our analysis, we will assume that in each of these various tasks the observer has received a sufficient number of practice trials to be familiar both with the basic structure of the experimental paradigm and with the nature of the two stimulus classes. However, despite this familiarity with the basic experimental procedure, his task is by no means an easy one, since the information that he receives on any trial from the presented stimulus situation is indeterminate — that is, it contains noise. Because of this noise, it is not possible for him to respond with perfect accuracy.

Decision rules. A decision rule is a device that takes the noisy stimulus information and uses it to specify a response. In other words, decision rules can be interpreted as mappings from the stimulus information space to the set of available response alternatives. Just as there are a number of possible such mappings, there often are a variety of plausible decision rules. The existence of more than one plausible decision rule, in turn, naturally leads one to examine the question of which decision rule is best, in the sense of optimizing some decision objective. A major goal of this paper consists of analyzing the properties associated with such optimal decision rules and determining the constraints that these properties impose upon measures of decision-making performance.

Organization of paper. The paper is divided into two major portions; the first consists of a tutorial review of some well-known results on optimal decision rules, and the second consists primarily of new results. The tutorial review begins in the remainder of this introduction, where the notion of an optimal decision rule is defined in a bit more detail, and continues through Sections 2 and 3 of the paper, where the yes-no and two-alternative forced-choice paradigms are discussed. The primary objective of this tutorial review is to set forth a number of theoretical results and establish some basic techniques of analysis that are used in the derivation of new results in the second major portion of the paper. In this second portion, Sections 4 and 5 treat the same-different and ABX discrimination paradigms, and Section 6 shows how the optimal decision rule analysis can be extended to a variety of additional psychophysical tasks.

Some basic notation. Throughout our discussion of optimal decision rules, we will use S_1 and S_2 to denote the two stimulus conditions that the observer must discriminate. In addition, we initially assume for illustrative purposes that there are n distinct sensations, or *sensory states* $x_1, x_2, \ldots, x_n$, that the observer can experience as a result of a presentation of stimulus condition S_1 or S_2. As shown in Figure 1, the conditional probability that sensory state x_j results given that stimulus condition S_i is presented is denoted $f(x_j|S_i)$, while the unconditional probability that S_i is presented, which is controlled by the experimenter, is denoted π_i. The problem confronting the trained observer involves using these differential probabilities with which the stimulus conditions and the various sensory states occur in order to optimally determine a response.

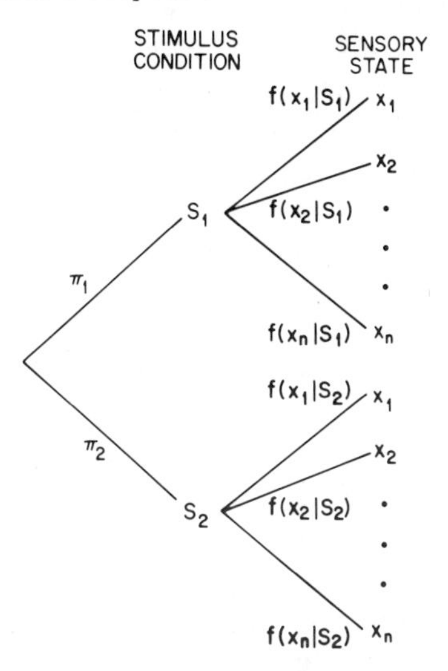

FIGURE 1. Structure of a discrimination task in which stimulus condition S_1 must be discriminated from stimulus condition S_2. The sensory states $x_1, x_2, \ldots, x_n$ represent all possible sensations that an observer can experience as a consequence of an S_1 or S_2 presentation, and $f(x_j|S_i)$ denotes the conditional probability that sensory state x_j occurs, given that stimulus condition S_i is presented. The experimenter controls the probabilities π_1 and $\pi_2 = 1 - \pi_1$, which determine the relative frequencies with which the two stimulus conditions appear.

Minimizing errors. Since the notion of optimality is meaningful only with respect to the attainment of a particular decision objective, it is necessary at this point to specify the decision objective that we wish to consider. One such objective that is often implicitly or explicitly stated in the experimental instructions is the minimization of errors — that is, the observer is often encouraged to make as few incorrect responses as possible.

To see how this decision objective of error minimization is satisfied, suppose that as the result of a stimulus presentation on a given trial the observer finds himself in sensory state x_j. Thus given the sensory information associated with this sensory state, he must determine which of the two stimulus conditions, S_1 or S_2, is more likely to have been presented. In other words, the problem is one of determining which is greater, $Pr(S_1|x_j)$ or $Pr(S_2|x_j)$, where these latter two quantities represent the so-called *posterior probabilities* that stimulus condition S_1 or S_2 was presented, given the event that the observer finds himself in sensory state x_j. By choosing as his response the stimulus condition that has the higher posterior probability of having been presented, the observer maximizes his probability of making a correct decision on the current trial, since the chosen stimulus condition will in the long run tend to produce the observed sensory state x_j with a relative frequency that is greater than that associated with the alternative stimulus condition.

To calculate the posterior probability that a particular stimulus condition, say S_1, was presented, given that sensory state x_j is experienced, one takes the total probability associated with reaching sensory state x_j by way of S_1, or $\pi_1 f(x_j|S_1)$, and divides it by the total probability associated with reaching sensory state x_j by way of either S_1 or S_2, which is given by $\pi_1 f(x_j|S_1) + \pi_2 f(x_j|S_2)$; this of course represents a simple application of Bayes' rule to the decision problem depicted in Figure 1. When an analogous ratio is formed to obtain the posterior probability associated with stimulus S_2, the result is the pair of equations

$$Pr(S_1|x_j) = \frac{\pi_1 f(x_j|S_1)}{\pi_1 f(x_j|S_1) + \pi_2 f(x_j|S_2)}, \tag{1a}$$

$$Pr(S_2|x_j) = \frac{\pi_2 f(x_j|S_2)}{\pi_1 f(x_j|S_1) + \pi_2 f(x_j|S_2)}. \tag{1b}$$

In the remainder of this paper, it will be convenient to drop our illustrative assumption that there are but n discrete sensory states and assume instead, somewhat more generally, that there is simply a decision variable X that represents the perceptual information conveyed by the presented stimulus. Thus the general results of the paper will not depend upon whether there is a discrete number of, or a continuum of, sensory states; they also will not depend on the exact dimensionality of the perceptual space in which the decision variable X resides. Rather, the only assumption that will be made is that for whatever unidimensional or multidimensional space in which X is defined, there exist probability density functions that specify the probability density associated with obtaining each point of the space as the observed value of the decision variable, given the particular stimulus conditions of the study.

In general, this extended domain of discourse does not lead to any complications in the form of the analysis. Instead of speaking of a particular sensory state x_j, we will speak of a particular observation value x that is assumed by the decision variable X. Expressions for the posterior probabilities of stimulus conditions S_1 and S_2 having been presented given that the decision variable X yields the observation value x on a particular trial may be obtained from Equations (1) by simply dropping the "j" subscripts from the equations, since an exactly analogous argument as that leading to Equations (1) in the previous case may be made for this more general case as well.

To summarize the argument that has been made thus far, an observer wishing to maximize the probability of making a correct response on a given trial will choose as his response the stimulus condition that has the higher posterior probability of having been presented, given the particular sensory state, or observation value, experienced on the particular trial. In other words, over the course of an experiment the observer will achieve a maximum number of correct responses — or, equivalently, a minimum number of errors — if he consistently follows

DECISION RULE 1. *Respond "R_1" iff $Pr(S_1|x) \geq Pr(S_2|x)$. Otherwise, respond "R_2".*

This Decision Rule will prove somewhat more useful if it is written in terms of the conditional probability density functions $f(x|S_1)$ and $f(x|S_2)$, which assign probability density values to the various outcomes assumed by the decision variable X. Since in Equations (1) the two posterior probabilities share the same denominators, it is clear that the inequality in Decision Rule 1 will hold whenever the numerator in Equation (1a) exceeds the numerator in Equation (1b). This fact, together with a rearrangement of terms, allows Decision Rule 1 to be rewritten in the alternative form

DECISION RULE 1′. *Respond "R_1" iff*

$$L(x) = \frac{f(x|S_1)}{f(x|S_2)} \geq \frac{\pi_2}{\pi_1}.$$

Otherwise, respond "R_2".

The quantity $L(x)$ that appears on the left-hand side of the inequality in this decision rule is the so-called *likelihood ratio* of the observation value x; the terminology reflects the fact that when $\pi_1 = \pi_2$ the value of $L(x)$ represents the relative likelihood, or odds-ratio, that x derives from stimulus condition S_1 rather than S_2. In words, Decision Rule 1′ specifies that an observer wishing to maximize the probability of making a correct response should make response "R_1" if and only if the likelihood ratio associated with the observation value x exceeds the inverse ratio of the presentation probabilities, π_2/π_1. Thus it is clear that the likelihood ratio plays a central role in specifying the optimal decision rule for an observer's discriminating S_1 from S_2; indeed, the remainder of this paper is largely devoted to deriving the form that the likelihood ratio takes for a number of experimental paradigms that are commonly employed in the study of psychophysical discrimination performance.

Before we proceed to these various paradigms, however, it is worthwhile to note that there are decision objectives other than maximizing the probability of a correct response for which a decision rule based on the likelihood ratio has optimal properties (cf. Green and Swets, 1974, pp. 18-25). For instance, if asymmetric costs and payoffs are associated with the various stimulus-response outcomes of the experiment, it may be to the observer's advantage to minimize the overall *cost* associated with making errors rather than the marginal *probability* of an error per se.

Minimizing costs. This alternative optimization problem can be formulated more explicitly by determining expected payoff values associated with the various responses, given that the observation value x is obtained on a particular trial. Thus, suppose V is a random variable representing the value of the payoff that the observer will receive on a particular trial, and let $E(V|R_j, x)$ be the expected value of this payoff given that he jointly has the observation value x and makes response "R_j". Then, if he wishes to maximize his expected payoff value on any given trial, a strategy which in the long run will minimize the overall cost of his errors, he will choose the response having the higher expected payoff value; this gives

DECISION RULE 2. *Respond "R_1" iff $E(V|R_1, x) \geq E(V|R_2, x)$. Otherwise, respond "$R_2$".*

To express the expected payoff in a more explicit form, let v_{ij} denote the payoff value that the observer receives when response "R_j" is made to a presentation of stimulus condition S_i. Then because the expected payoff value is a weighted combination of payoffs received when S_1 occurs and those received when S_2 occurs, we have for the expected payoff value associated with the observer making response "R_j" ($j = 1,2$)

$$E(V|R_1, x) = Pr(S_1|x)E(V|S_1, R_1) + Pr(S_2|x)E(V|S_2, R_1)$$

$$= \frac{\pi_1 f(x|S_1)v_{11} + \pi_2 f(x|S_2)v_{21}}{\pi_1 f(x|S_1) + \pi_2 f(x|S_2)}, \tag{2a}$$

$$E(V|R_2, x) = Pr(S_1|x)E(V|S_1, R_2) + Pr(S_2|x)E(V|S_2, R_2)$$

$$= \frac{\pi_1 f(x|S_1)v_{12} + \pi_2 f(x|S_2)v_{22}}{\pi_1 f(x|S_1) + \pi_2 f(x|S_2)}. \tag{2b}$$

When these expressions are substituted in Decision Rule 2 and the terms of the resulting equation are suitably rearranged, the result is an alternate form of the decision rule, given by

DECISION RULE 2′. *Respond "R_1" iff*

$$L(x) = \frac{f(x|S_1)}{f(x|S_2)} \geq \frac{\pi_2}{\pi_1} \frac{v_{22}-v_{21}}{v_{11}-v_{12}}.$$

Otherwise, respond "R_2".

In this form, Decision Rule 2′ is identical to Decision Rule 1′ except for a term called the *regret ratio* that appears on the right-hand side of the inequality in Decision Rule 2′, reflecting the ratio of the differences in correct and error payoffs for stimulus conditions S_2 and S_1, respectively. In other words, the optimal decision rule for maximizing the expected payoff value, like the optimal decision rule for maximizing the percentage of correct responses, utilizes a likelihood ratio decision criterion. The only effect on the optimal decision rule of incorporating payoffs that may be asymmetrically assigned to the various stimulus-response outcomes is to induce a shift in the numerical cut-off value which the likelihood ratio must exceed in order for the observer to emit response "R_1".

Thus, for the remainder of this paper we will use as our optimal decision rule for making response "R_1" the requirement that the likelihood ratio must exceed the inverse ratio of the presentation probabilities π_2/π_1, with it understood that this quantity must be adjusted to account for the effect of the regret ratio if there are asymmetric costs and payoffs incorporated into the experimental design. In essence, then, we will use "π_2/π_1" throughout the rest of the paper as a shorthand representation for the right-hand side of the inequality in Decision Rule 2′, which of course simplifies down to the inverse ratio of presentation probabilities when costs and payoffs are symmetric.

2. The yes-no paradigm. The yes-no paradigm, also known as the two-stimulus complete identification task, is the simplest of the various psychophysical paradigms that we will consider. In such a task, there are two basic types of stimuli, which may be labeled A stimuli and B stimuli. Stimulus condition S_1 consists of a single presentation of an A stimulus, while stimulus condition S_2 consists of a single presentation of a B stimulus.

Thus on any given trial of a yes-no task, the observer is required to report which of the two alternatives he perceived, $S_1 = \langle A \rangle$ or $S_2 = \langle B \rangle$. For example, in the standard yes-no signal detection task, $S_1 = \langle A \rangle$ is often a visual or auditory signal embedded in a background of noise, while $S_2 = \langle B \rangle$ is the noise background alone. The observer responds by indicating his choice of "R_1" or "R_2", where "R_1" represents the response "Yes" (the signal was present) while "R_2" represents the response "No" (there was merely a presentation of noise alone).

The optimal decision rule. In the most general case we will consider, the decision variable X is a random vector, whose various components correspond to the perceived attributes of the presented stimulus. On any given trial, this random vector assumes a particular observation value x, which can be represented as a point in the multidimensional attribute space. Associated with each such point x is a non-negative numerical value, $L(x)$, representing the relative likelihood (or *odds*) that the experienced configuration of stimulus component values would result from a presentation of $S_1 = \langle A \rangle$ rather than $S_2 = \langle B \rangle$ (this is discussed in more detail in Luce, 1963, pp. 108-113, and in Lappin, 1978). The optimal decision problem confronting the observer involves determining which response is more appropriate given a particular odds-value $L(x)$.

An important special case of this general framework occurs when the dimensionality of the sensory space assumes the value $r = 1$. That is, the decision variable X is often treated as a random variable that corresponds to some *unidimensional* measure of the sensory effect of the stimulus.

For example, it is often hypothesized that the observer in the signal detection task decides whether or not a signal was present by relying upon the strength of the sensory impression registered during the observation interval; this sensory strength is represented by a random variable X that takes on one of two distributions of values, depending upon the presented stimulus condition — a distribution with a relatively weak mean value in the case when noise alone is presented and a distribution with a somewhat stronger mean value in the case when a signal is embedded in the noise. As a second example, if the observer's task is one of identifying which of two tones is presented on a given trial, a high frequency tone or a tone of a slightly lower frequency, then the sensation left behind by the stimulus, as represented in the decision variable X, is often assumed to be defined along a unidimensional continuum that could be interpreted as "subjectively experienced frequency".

This assumption of unidimensionality is often accompanied by an assumption that the two distributions of "subjective frequency" or "sensory strength" corresponding to the two different stimulus conditions S_1 and S_2 are Gaussian distributions that share a common variance; Green and Swets (1974) may be consulted for a discussion of the appropriateness of these assumptions, along with a review of their usefulness in interpreting experimental data. The approach taken here will be to discuss optimal decision rules in terms of properties associated with likelihood ratios, so that the general results of the paper will not depend upon the Gaussian equal-variance assumptions. However, because so much empirical work has been done with these assumptions as a basis, and because they have proven so useful in practice, we will discuss the form the results take when they are appropriate, and specifically use them in applying the optimal decision rule analysis to some data on tone frequency discrimination collected by Creelman and Macmillan (1979).

In summary, the optimal decision rule for a yes-no task involves responding "R_1", or "yes" (a signal is present), if and only if the likelihood ratio of the observation value x exceeds a certain numerical value, or *criterion*, determined by the presentation probabilities and the various costs and payoffs (cf. Decision Rule 2′). In the next subsection we will briefly consider the form this likelihood ratio decision rule takes when the Gaussian, equal-variance assumptions (which we will henceforth collectively abbreviate as simply the "Gaussian case") are appropriate.

The Gaussian case. To illustrate the special assumptions that comprise the Gaussian case, consider once again the classical yes-no signal detection task. One of the basic premises underlying the special assumptions in this case is that a Gaussian distribution of sensory strength can be used to represent the sensory effect of a presentation of the noise background. A justification often given for this assumption is that the noise derives from a number of sources that combine additively to produce a Gaussian-distributed random variable by means of the Central Limit Theorem (however, see Laming, 1973, pp. 336-339, for a critical discussion of this justification). Embedding a signal in the noise is assumed to shift the noise distribution by a positive constant, reflecting the strength of the signal. This translation constant, measured in units of standard deviations, is the familiar d' of signal detection theory, although for simplicity of notation we will drop the prime and simply refer to it as the displacement constant, d, throughout the rest of the paper.[2]

Thus, the picture is as shown in Figure 2. Since the zero point of the scale is arbitrary, it has been placed at the location where the two densities cross, so that the mean of the signal plus noise (S_1) density is at $d/2$, while the mean of the noise alone (S_2) density is at $-d/2$.

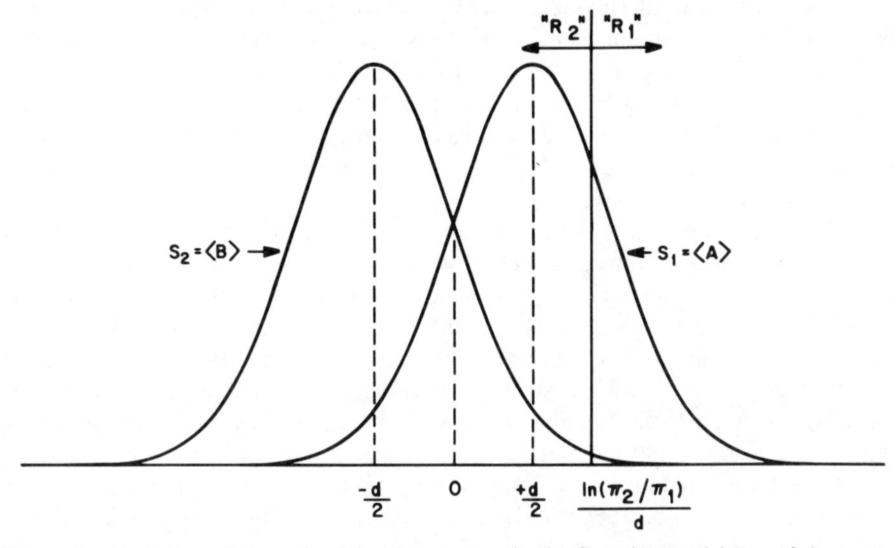

FIGURE 2. Illustration of the optimal decision strategy in the Gaussian special case of the yes-no paradigm. For all values of sensory strength greater than $(1/d)\ln(\pi_2/\pi_1)$, the observer should respond that an A was presented ("R_1"), while for all sensory strengths less than this criterion value, he should respond that a B was presented ("R_2"). In this particular example, π_2 exceeds π_1, and as a consequence the response criterion is biased in the positive direction to favor a larger percentage of "R_2" responses than "R_1" responses. In the commonly-encountered case in which π_1 and π_2 are equal, the criterion coincides with the zero-point of the scale, and the decision problem is symmetric with respect to the two stimuli.

The equation defining a Gaussian density with mean μ takes the form

$$f(x) = \frac{e^{-(x-\mu)^2/2}}{\sqrt{2\pi}}, \tag{3}$$

where the abscissa value x is measured in units of standard deviations, as is assumed in the definition of d'. As Green and Swets (1974, p. 60) note, a simple expression for the Gaussian likelihood ratio can be obtained by using this defining equation together with the appropriate values for the means of the S_1 and S_2 density functions to write

$$L(x) = \frac{f(x|S_1)}{f(x|S_2)} = \frac{e^{-(x-d/2)^2/2}}{e^{-(x+d/2)^2/2}} = e^{dx}. \tag{4}$$

[2]To be more specific, we will use d to signify the theoretical parameter that represents the distance between the Gaussian A and B distributions, scaled in units of their common standard deviation. On the other hand, we will use d' to signify the quantity that one estimates from psychophysical discrimination data, ordinarily by adding the normal-deviate z-scores that correspond to the percentage of hits and the percentage of correct rejections.

Thus the Gaussian likelihood ratio is a simple exponential function of the perceived sensory strength x. When we substitute this equation in Decision Rule 1′ and suitably rearrange the terms, we obtain the optimal decision rule for the Gaussian yes-no task in the form of

DECISION RULE 3. *Respond "R_1" iff $x \geq (1/d)\ln(\pi_2/\pi_1)$. Otherwise, respond "$R_2$".*

An illustration of this decision rule is contained in Figure 2. A cut-off, or criterion, has been placed along the sensory strength continuum at the point $x = (1/d)\ln(\pi_2/\pi_1)$. The optimal decision rule specifies that for values of sensory strength greater than this cut-off the observer should make response "R_1", reporting that a signal is present, while for values of sensory strength less than the cut-off, he should report that merely noise alone is present by making response "R_2".

Thus, in this Gaussian case — or, more generally, whenever there is a monotonic relation between likelihood ratio and sensory strength — the observer can perform optimally by making decisions simply on the basis of values of sensory strength, without having to explicitly calculate likelihood ratios. To be sure, we employed the likelihood ratio of Equation (4) in determining the level of sensory strength that acts as the optimal cut-off value in Decision Rule 3; however, a well-practiced observer could presumably arrive at this optimal cut-off value by a process of successive adjustments following errors, as is hypothesized in the adaptive, error-correction models of Kac (1962, 1969) and others for the initial training phase of the yes-no signal detection task.

That is, if the observer moved his criterion in the positive direction to increase his proportion of "R_2" responses every time he responded incorrectly to S_2, and likewise moved it in the negative direction after responding incorrectly to S_1, then after a suitable period of training the criterion would tend to be located near the optimal location specified in Decision Rule 3 (see Kac, 1962, for a more precise statement of the conditions for convergence). Thus, while we will continue to specify the general results for optimal decision rules in terms of likelihood ratios throughout the paper, it is worthwhile to keep in mind the fact that whenever a monotonic relation between sensory strength and likelihood ratio exists, the results could as well be specified in terms of sensory strengths, and in such cases the observer would not be required to explicitly calculate values of likelihood ratio per se.

3. The two-alternative forced-choice paradigm. The second experimental paradigm that we will review is the two-alternative forced-choice (2AFC) task. As shown in Figure 3, a single trial of this paradigm is comprised of two presentation intervals, I_1 and I_2; these two intervals may be either spatial intervals or temporal intervals depending upon the particular stimulus conditions of a given study, and they are complementary in the sense that one contains an A presentation while the other contains a B presentation. The observer's task is to specify which of the two intervals contains the A presentation, by responding "R_1" when it occurs in the first interval and "R_2" when it occurs in the second interval. The experimenter controls the probabilities π_1 and π_2, which determine the relative frequencies of occurrence of stimulus sequence $S_1 = <AB>$ versus $S_2 = <BA>$.

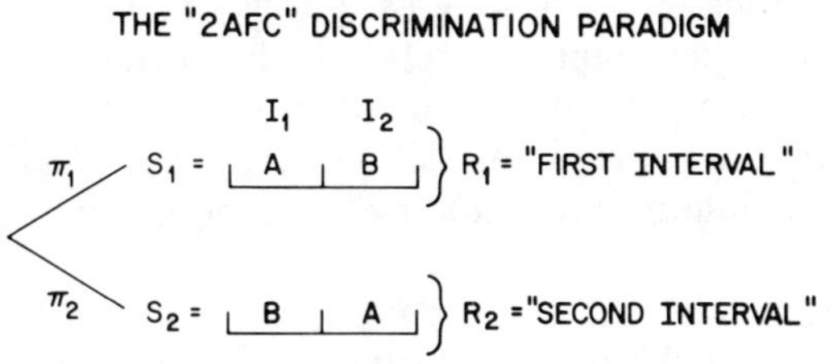

THE "2AFC" DISCRIMINATION PARADIGM

I_1 I_2

π_1 S_1 = A B } R_1 = "FIRST INTERVAL"

π_2 S_2 = B A } R_2 = "SECOND INTERVAL"

FIGURE 3. Structure of the two-alternative forced-choice paradigm. The observer's task is to specify the presentation order of a pair of complementary stimuli, labeled A and B, by responding "R_1" if the A presentation appears in the first presentation interval, I_1, and "R_2" if it appears in the second presentation interval, I_2. The experimenter controls the probabilities π_1 and $\pi_2 = 1 - \pi_1$, which determine the relative frequencies with which the two stimulus conditions appear.

In the standard analysis of this paradigm (cf. Green and Swets, 1974, pp. 64-69), a decision variable, X_1, is used to represent the perceptual information acquired from the first presentation interval, while a second decision variable, X_2, is used to represent the perceptual information acquired from the second presentation interval. Together, these two decision variables can be interpreted as the components of a decision vector $X = (X_1, X_2)$. On any given trial, the observed values x_1 and x_2 of the decision variables X_1 and X_2 comprise the observation vector $x = (x_1, x_2)$. The optimal decision problem confronting the observer involves determining which response is more appropriate given the particular observation vector x.

Two basic assumptions. There are two assumptions that are often made to facilitate analysis of the 2AFC paradigm (Green and Swets, 1974, p. 67); they will be utilized here both for the 2AFC task and for the other paradigms that will be considered later in this paper. These two assumptions take the form:

(A1) X_1 and X_2 are independent decision variables.

(A2) There is no effect of the presentation intervals per se.

Assumption (A1) simplifies analysis of the optimal decision strategy by allowing the joint density associated with the observation vector x to be written as a product of the densities associated with each of the components x_1 and x_2. Psychologically, (A1) asserts that the two presentation intervals are perceived independently of each other.

Assumption (A2), on the other hand, allows the decision variables X_1 and X_2 to be conditioned solely on whether an A or B stimulus was presented, without explicit account being taken of the specific interval in which the A or B presentation occurred. Thus, given an A presentation in the first interval, X_1 is assumed to have a distribution identical to that which would result for X_2, given an A presentation in the second interval. If the two intervals represent temporal periods in which the stimuli occur, as we will assume for illustrative purposes throughout the remainder of the paper, then the effect of making assumption (A2) is to rule out any role of temporal factors per se.

Discussion of assumptions. Lest these two assumptions be taken as unrealistically restrictive, a brief digression may be in order, to consider various ways in which they might be relaxed. For instance, it may be the case that for some types of stimuli and/or for particular experimental procedures there is necessarily a nonzero correlation between X_1 and X_2 due to the particular way the stimuli are presented by the experimenter or processed by the observer.[3]

While the presence of such a correlation would certainly be expected to complicate analysis of the optimal decision rule for the given paradigm, yet it would not be expected to render the analysis completely intractable in all cases. For instance, Green and Swets (1974, p. 67, Equation (3.8)) show that for one method of introducing a correlation of magnitude r between the two presentation intervals in the Gaussian 2AFC case, the only resulting consequence is the introduction of a multiplicative factor of $1/(1-r)$ in the equation for the log likelihood ratio. The effect of this multiplicative constant, in turn, is simply to require a shift in the numerical value of the likelihood ratio cut-off, or criterion value, to compensate for the effect of the correlation; the basic optimal decision strategy otherwise remains the same.

As a second example, it is straightforward to show that when this same method of introducing a correlation of magnitude r between the two presentation intervals is applied to the same-different discrimination paradigm, it leads to a fairly simple form for the optimal decision rule in this case as well. Thus assumption (A1) does not appear to be absolutely critical for obtaining tractable results.

Of course, this particular method of modeling the correlation may not be appropriate in all instances; indeed, as Green and Swets (1974) point out, one can allow for correlations between the various stimulus occurrences and presentation intervals in multiple-interval tasks in a number of alternative ways. Unfortunately, the most general such correlated models tend to have a rather large number of free parameters, and this makes it difficult for one to say anything specific about their predictions without making at least some simplifying assumptions.

In addition, in order to apply any type of correlated model, it is necessary to assume that the underlying sensory representations of the stimuli take a specific distributional form, such as the Gaussian; this contrasts with the independence case, where one need not be committed to the specific form of the underlying distributions in order to obtain testable predictions

---

[3]One example of a case where a nonzero correlation between X_1 and X_2 is introduced by the way the experimenter presents the stimuli is the so-called "roving standard paradigm". For instance, in a roving standard 2AFC study of tone frequency discrimination, the two stimuli might on one trial have frequencies of 200 Hz and 201 Hz and on the next trial have frequencies of 201 Hz and 202 Hz, with the observer always required to judge whether the first presentation interval contained the higher or the lower of the two tones. Such "mixed designs" are not treated in this paper; rather, the experimental conditions are assumed to be blocked.

that can be applied to experimental data. Thus the notion of independence is one that is very convenient to retain, as long as experimental evidence in the given task does not specifically rule it out.

Many of these same considerations are also relevant with respect to assumption (A2), which states that there is no effect of the presentation intervals per se. This assumption in effect specifies that if two presentation intervals yield exactly the same observation value — so that, for example, $x_1 = x_2$ — then the odds that the first presentation interval contained an A rather than a B presentation must be exactly the same as the corresponding odds for the second presentation interval. This, in turn, is the same as requiring that the likelihood ratio function that maps observation values x into odds values, or likelihood ratios $L(x)$, must be invariant over temporal variables, so that there is but one such function for all presentation intervals.

Of course, if the observer were aware of a systematic effect across the presentation intervals — say, perhaps, a primacy effect due to temporal fatigue or a recency effect due to memory loss — he could conceivably take this into account by, in effect, computing separate likelihood ratio functions for each presentation interval, thus compensating for any change in the quality of the sensory information from interval to interval. Mathematically this does not generally lead to complications in the form taken by the optimal decision rule. That is, the overall approach taken in this paper will involve using the independence assumption (A1) to express the likelihood ratio $L(x)$ for a complex paradigm, consisting of n presentation intervals, as a function of likelihood ratios $L(x_1), L(x_2), \ldots, L(x_n)$ associated with each of the presentation intervals; thus one could just as well take these various likelihood ratios as resulting from different likelihood ratio functions rather than from instances of a single likelihood ratio function (e.g. one could subscript them in the form $L_1(x_1), L_2(x_2), \ldots, L_n(x_n)$).

However, as soon as one allows effects of the presentation intervals per se, it then becomes of interest to consider sequential effects that involve interactions between the presentation intervals and the particular A or B stimuli that appear in the intervals, and at some point the level of generality inherent in the analysis leaves little in the way of specific results that can be stated — or at least little in the way of explicit constraints that enable an observer's decision-making performance to be interrelated across a number of tasks, which is a major goal of this paper. Thus the approach taken here will involve utilizing (A1) and (A2) to simplify analysis of the various complex paradigms that will be considered; experimental tests of optimality will be interpreted as, in effect, tests of the optimal decision strategy conjoined with assumptions (A1) and (A2). From the standpoint of attempting to find the most restricted (and therefore specific) model that will apply to a given paradigm, one would presumably wish to reject such a conjoined model before proceeding to a more general model that makes less specific predictions by relaxing one or both of these restrictions.

The optimal decision rule. Thus, use of assumptions (A1) and (A2) allows the joint density associated with obtaining an observation vector $x = (x_1, x_2)$ to be written as the product of the densities associated with each of its components, x_1 and x_2, so that

$$f(x|S_1) = f[(x_1, x_2)|<AB>] = f(x_1|A)\, f(x_2|B)\,,$$
$$f(x|S_2) = f[(x_1, x_2)|<BA>] = f(x_1|B)\, f(x_2|A)\,. \tag{5}$$

By taking the ratio of these two equations, we can express the likelihood ratio for the 2AFC task in terms of the likelihood ratios associated with the two individual presentation intervals, so that

$$L(x) = \frac{f(x|S_1)}{f(x|S_2)} = \frac{f(x_1|A)/f(x_1|B)}{f(x_2|A)/f(x_2|B)} = \frac{L(x_1)}{L(x_2)}\,. \tag{6}$$

Finally, this expression can be substituted in Decision Rule 1 to obtain the optimal decision rule for the 2AFC paradigm in the form of

DECISION RULE 4. *Respond "R_1" iff*

$$L(x) = \frac{L(x_1)}{L(x_2)} \geq \frac{\pi_2}{\pi_1}\,.$$

Otherwise, respond "R_2".

According to this decision rule, the observer should calculate a likelihood ratio for each of the two intervals, and then compare the two by, in turn, forming their ratio. For example, if the odds were 3 to 1 that the first interval contained an A rather than a B presentation, and 2 to 1 that the second interval contained an A rather than a B presentation, then whenever π_2/π_1 is less than or equal to $(3/1)/(2/1) = 3/2$ the observer should make response "R_1". Assuming symmetry in pay-offs, this translates into the requirement that the observer should report that the A stimulus occurred in the first presentation interval whenever the initial, a priori probability of it being there, π_1, is greater than or equal to 2/5.

The Gaussian case. The most common assumption made in applying the foregoing analysis to experimental data obtained from a 2AFC task is that the two conditional density functions in Equation (5) are Gaussian and share a common variance — in other words, the sensory representations of the A and B stimuli satisfy the assumptions of the Gaussian case considered earlier. In this special case, the exponential function in Equation (4) relating the Gaussian likelihood ratio $L(x_i)$ to the perceived sensory strength x_i $(i = 1,2)$ can be substituted in Equation (6) to obtain the 2AFC likelihood ratio in the form

$$L(x) = \frac{e^{dx_1}}{e^{dx_2}} = e^{d(x_1 - x_2)}\,. \tag{7}$$

The interesting feature in this result is that the likelihood ratio for the Gaussian forced-choice case is again an exponential function, exactly as for the yes-no case, but with the substitution of $x_1 - x_2$ for x. This in turn implies that the optimal decision rule for the Gaussian 2AFC task is identical to the optimal decision rule for the Gaussian yes-no task, provided the value of $X_1 - X_2$ in 2AFC is substituted for the value of the decision variable X in yes-no; in other words, the optimal decision strategy is for the observer to take the Gaussian random variable $Y = X_1 - X_2$ as his decision variable and otherwise perform exactly as he would in a yes-no task (that is, use Decision Rule 3). By forming the difference in sensations between the two presentation intervals, the observer in effect converts the required 2AFC decision into a simple, yes-no judgment on the difference in sensory strengths between the two presentation intervals.

Given this reduction of the forced-choice task to a yes-no judgment on the difference $X_1 - X_2$, it is of interest to determine the relation between this derived yes-no task and a simple yes-no task involving the same A and B stimuli. Because a single interval of the 2AFC task corresponds to the single interval of a simple yes-no task in all respects, and because in the 2AFC task the observer receives a so-called "second chance" at identifying the presented stimulus condition as a consequence of the complementary stimulus presentations in I_1 and I_2 (i.e. on any given trial definitive information from either interval would suffice to correctly identify the appropriate response), one would expect to find an improvement in performance from a yes-no to a 2AFC task, and indeed it is well known (e.g. Green and Swets, 1974, pp. 67-68) that the sensitivity measure d' is expected to increase by a factor of $\sqrt{2}$.

It is relatively straightforward to see why this $\sqrt{2}$ improvement is predicted. When $S_1 = <AB>$ is the presented stimulus sequence in the 2AFC task, the A presentation in the first interval causes the sensation random variable X_1 to have a mean value of $d/2$ (cf. Figure 2), the B presentation in the second interval causes the sensation random variable X_2 to have a mean value of $-d/2$, and thus the random variable representing the sensation difference $X_1 - X_2$ has a mean value of $(d/2) - (-d/2) = d$. Since an analogous computation for the case when the presented stimulus condition is $S_2 = <BA>$ gives a value of $-d$ for the mean of $X_1 - X_2$, the total distance between the S_1 and S_2 distributions on the difference continuum is $d - (-d) = 2d$. To convert this distance between the two distributions into a measure of d', it is necessary to scale it in terms of the common standard deviation underlying the two difference distributions; since X_1 and X_2 by assumption (A1) are independent random variables, their unit variances will add to produce a variance of 2 — or, equivalently, a standard deviation of $\sqrt{2}$ — for the difference random variable $X_1 - X_2$. Hence, the d' measure for the 2AFC task takes the form

$$d'_{2AFC} = \frac{2d}{\sqrt{2}} = \sqrt{2}\, d'_{YN} \; . \tag{8}$$

We will later show in Section 6 that this same $\sqrt{2}$ improvement factor plays an important theoretical role in several other psychophysical contexts.

4. The same-different paradigm. Figure 4 shows the trial structure of the so-called "same-different" or "variable standard AX" discrimination paradigm, which has received widespread use both in studies of speech perception (for a review, see Macmillan, Kaplan and Creelman, 1977) and in studies employing a variety of other psychophysical stimuli (for reviews, see Krueger, 1978, and Vickers, 1979, Chapter 4). As shown in the figure, a single trial of the same-different task is comprised of two presentation intervals, in which A or B stimuli can appear, as in the 2AFC task; however, unlike the 2AFC task, the two intervals need not involve complementary stimulus presentations, since all four possible sequences involving the two stimulus types can occur. The observer's task in a same-different experiment is to use the response "R_1" to report that the stimuli in the two presentation intervals are the "same" if either an $<AA>$ or a $<BB>$ stimulus sequence is presented, and to use the response "R_2" to report that the stimuli in the two presentation intervals are "different" if either an $<AB>$ or a $<BA>$ stimulus sequence appears. As usual, we assume that the experimenter is free to vary the frequency with which response "R_1" versus "R_2" is appropriate, by controlling the probabilities π_1 and π_2.

THE "SAME - DIFFERENT" DISCRIMINATION PARADIGM

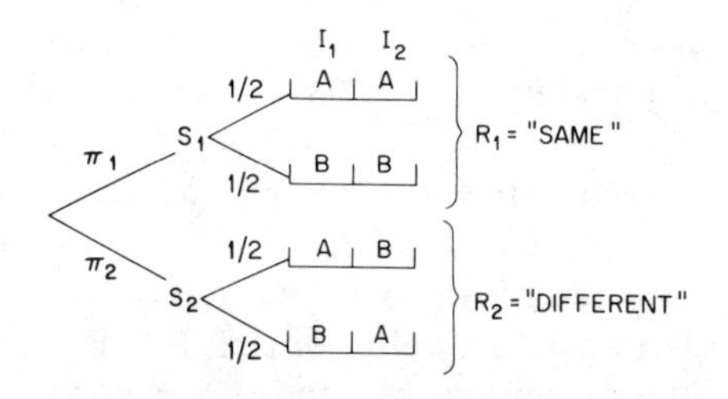

FIGURE 4. Structure of the "same-different" or "variable standard AX" paradigm. The observer's task is to say "same" if the stimuli that appear in intervals I_1 and I_2 are both A presentations or both B presentations, and to say "different" if one is an A presentation while the other is a B presentation. The experimenter controls the probabilities π_1 and $\pi_2 = 1 - \pi_1$, which determine the relative frequencies with which the two stimulus conditions appear.

The optimal decision rule. The optimal decision rule for the same-different paradigm may be derived by means of essentially the same procedure that was used earlier with the 2AFC task, provided that allowance is made for the fact that each of the stimulus conditions S_i $(i = 1,2)$ now consists of a mixture of two presentation sequences. Taking this factor into account, we can express the probability of obtaining the observation vector $x = (x_1, x_2)$, given a particular stimulus condition S_i, as a mixture defined over the sequences comprising S_i in the form

$$\begin{aligned} Pr(x|S_1) &= Pr(S_1 = <AA>)Pr(x|S_1 = <AA>) \\ &\quad + Pr(S_1 = <BB>)Pr(x|S_1 = <BB>)\,, \\ Pr(x|S_2) &= Pr(S_2 = <AB>)Pr(x|S_2 = <AB>) \\ &\quad + Pr(S_2 = <BA>)Pr(x|S_2 = <BA>)\,. \end{aligned} \tag{9}$$

In these equations, the mixture coefficients — i.e. the unconditional probabilities associated with the various stimulus sequences — are all equal to 1/2, due to the design of the paradigm (cf. Figure 4). The terms involving conditional probabilities, on the other hand, can be directly translated into expressions involving conditional probability density functions. When we make these substitutions, the equations in (9) take the form

$$f(x|S_1) = 1/2\, f(x| < AA >) + 1/2\, f(x| < BB >)\,,$$
$$f(x|S_2) = 1/2\, f(x| < AB >) + 1/2\, f(x| < BA >)\,. \tag{10}$$

The likelihood ratio for the same-different task consists of the ratio of these two equations, or

$$L(x) = \frac{f(x|S_1)}{f(x|S_2)} = \frac{f(x| < AA >) + f(x| < BB >)}{f(x| < AB >) + f(x| < BA >)}\,. \tag{11}$$

The four probability density functions that appear in this equation correspond to the four different stimulus sequences that are presented in the task. The utility of assumptions (A1) and (A2) that were introduced earlier in the discussion of the 2AFC task is that they allow these four probability density functions to be rewritten in terms of their corresponding marginal probability density functions in the form

$$L(x) = \frac{f(x_1|A)f(x_2|A) + f(x_1|B)f(x_2|B)}{f(x_1|A)f(x_2|B) + f(x_1|B)f(x_2|A)}\,. \tag{12}$$

By dividing both the numerator and the denominator of this expression by $f(x_1|B)f(x_2|B)$, we can express the likelihood ratio $L(x)$ for the two-interval task in terms of likelihood ratios $L(x_1)$ and $L(x_2)$ associated with the individual intervals of the task, so that

$$L(x) = \frac{\dfrac{f(x_1|A)}{f(x_1|B)}\dfrac{f(x_2|A)}{f(x_2|B)} + 1}{\dfrac{f(x_1|A)}{f(x_1|B)} + \dfrac{f(x_2|A)}{f(x_2|B)}} = \frac{L(x_1)L(x_2) + 1}{L(x_1) + L(x_2)}\,. \tag{13}$$

Thus the likelihood ratio for same-different discrimination is a fairly simple function of the likelihood ratios associated with the two presentation intervals. When this simple function is substituted in Decision Rule 1′, we obtain the optimal decision rule for the same-different paradigm in the form of

DECISION RULE 5. *Respond "R_1" iff*

$$L(x) = \frac{L(x_1)L(x_2) + 1}{L(x_1) + L(x_2)} \geq \frac{\pi_2}{\pi_1}\,.$$

Otherwise, respond "R_2".

To apply this decision rule, the decision-maker first computes a likelihood ratio for each of the two presentation intervals, then combines the two likelihood ratios in the form indicated by Equation (13), and finally compares the result to the inverse ratio of the presentation probabilities in order to determine a response. Given that he uses this strategy, his response is guaranteed to be optimal in the sense defined earlier, namely it will be consistent with a decision objective of minimizing his long-run number of incorrect responses or maximizing his expected reward value according to some payoff matrix defined over the various stimulus-response outcomes.

An important alternative form of this optimal decision strategy can be obtained by noting that the combination of terms comprising the likelihood ratio in Decision Rule 5 suggests a straightforward factorization, and hence separation, of terms involving $L(x_1)$ and $L(x_2)$. That is, the inequality in Decision Rule 5 can be rewritten as the constraint that

$$L(x_1)L(x_2) - \frac{\pi_2}{\pi_1} L(x_1) - \frac{\pi_2}{\pi_1} L(x_2) + 1 \geq 0 ,$$

which, in turn, is equivalent to the requirement that

$$\left(L(x_1) - \frac{\pi_2}{\pi_1}\right)\left(L(x_2) - \frac{\pi_2}{\pi_1}\right) \geq \left(\frac{\pi_2}{\pi_1}\right)^2 - 1 .$$

When this inequality is substituted for the inequality in Decision Rule 5, the result is a slightly different form of the decision rule, which may be written as

DECISION RULE 5′. *Respond* "R_1" *iff*

$$\left(L(x_1) - \frac{\pi_2}{\pi_1}\right)\left(L(x_2) - \frac{\pi_2}{\pi_1}\right) \geq \left(\frac{\pi_2}{\pi_1}\right)^2 - 1 .$$

Otherwise, respond "R_2".

This form of the optimal decision rule is of considerable interest, since it makes apparent an important simplification that results when $\pi_1 = \pi_2$, which corresponds to "same" and "different" trials being balanced in the experimental design, as they quite frequently are in practice. For this special case, the right-hand side of the inequality in Decision Rule 5′ vanishes, and the requirement for responding "R_1" can be written as the requirement that

$$\left[L(x_1) - 1\right]\left[L(x_2) - 1\right] \geq 0 .$$

Since this requirement will be satisfied when the likelihood ratios associated with the two presentation intervals are both greater than or equal to 1, or are both less than or equal to 1, we may express the important $\pi_1 = \pi_2$ special case of Decision Rule 5′ as

DECISION RULE 5″. *When* $\pi_1 = \pi_2$, *respond* "R_1" *iff*

$$[L(x_1) \geq 1 \textit{ and } L(x_2) \geq 1] \textit{ or } [L(x_1) \leq 1 \textit{ and } L(x_2) \leq 1].$$

Otherwise, respond "R_2".

A verbal translation of this decision rule gives what would seem to be a rather intuitive result. When "same" and "different" trials are equally likely to occur, the observer should respond "same" ("R_1") if both intervals are more likely to contain instances of A stimuli than B stimuli, or if both intervals are more likely to contain instances of B stimuli than A stimuli; on the other hand, he should respond "different" ("R_2") if one interval is more likely to contain an instance of an A stimulus while the other interval is more likely to contain an instance of a B stimulus.

In terms of the probability density functions $f(x_i|A)$ and $f(x_i|B)$ associated with the A and B stimuli, this translates into the following procedure: For those regions where the probability density $f(x_i|A)$ exceeds the probability density $f(x_i|B)$, the observer should assign a categorization label "A", representing the fact that the observation value corresponding to any given point in such a region is more likely to result from a presentation of stimulus A than from a presentation of stimulus B. Likewise, where $f(x_i|B)$ exceeds $f(x_i|A)$ he should assign a categorization label "B"; cases where the two densities coincide may be arbitrarily assigned to either categorization class, since the observer derives no information from a sensory state that is equally likely to have resulted from either an A or a B stimulus presentation, and thus he can do no better than chance performance on a trial where such a sensory state occurs. Then, after using the labels associated with the various regions to independently categorize the observation values from the two presentation intervals, he should respond "same" if the resulting categorizations consist of "AA" or "BB", and "different" if the resulting categorizations consist of "AB" or "BA".

The Gaussian case. In order to compare and contrast this decision rule with those employed by other models that have been proposed for same-different discrimination, it is useful to examine the form that the optimal decision strategy takes when the assumptions of the Gaussian case are appropriate. For this special case, the exponential relation between likelihood ratio and sensory strength given in Equation (4) can be used to rewrite Decision Rule 5 as

DECISION RULE 6. *Respond* "R_1" *iff*

$$L(x) = \frac{e^{d(x_1+x_2)} + 1}{e^{dx_1} + e^{dx_2}} \geq \frac{\pi_2}{\pi_1}.$$

Otherwise, respond "R_2".

The equal-likelihood contours determined by this decision rule are shown in Figure 5. In general, these contours are curvilinear in form, although for a likelihood ratio value of 1 they are linear and coincide with the X_1 and X_2 axes. This latter special case corresponds to the result contained in Decision Rule 5″, which states that when $\pi_1 = \pi_2$ the observer should make response "R_1" if and only if the likelihood ratios associated with the two presentation intervals are both greater than or equal to 1, or are both less than or equal to 1. For the Gaussian case, the likelihood ratio $L(x)$ is a strictly increasing function of the sensory strength x, and assumes the value 1 at the point labeled $x = 0$ in Figure 2, where the density functions corresponding to A and B stimulus presentations intersect; thus the requirement in Decision Rule 5″ can be phrased in terms of sensory strengths as the requirement that x_1 and x_2 both be greater than or equal to 0 or both be less than or equal to 0. This gives

DECISION RULE 6′. *When* $\pi_1 = \pi_2$, *respond* "R_1" *iff* x_1 *and* x_2 *have the same sign. Otherwise, respond* "R_2".

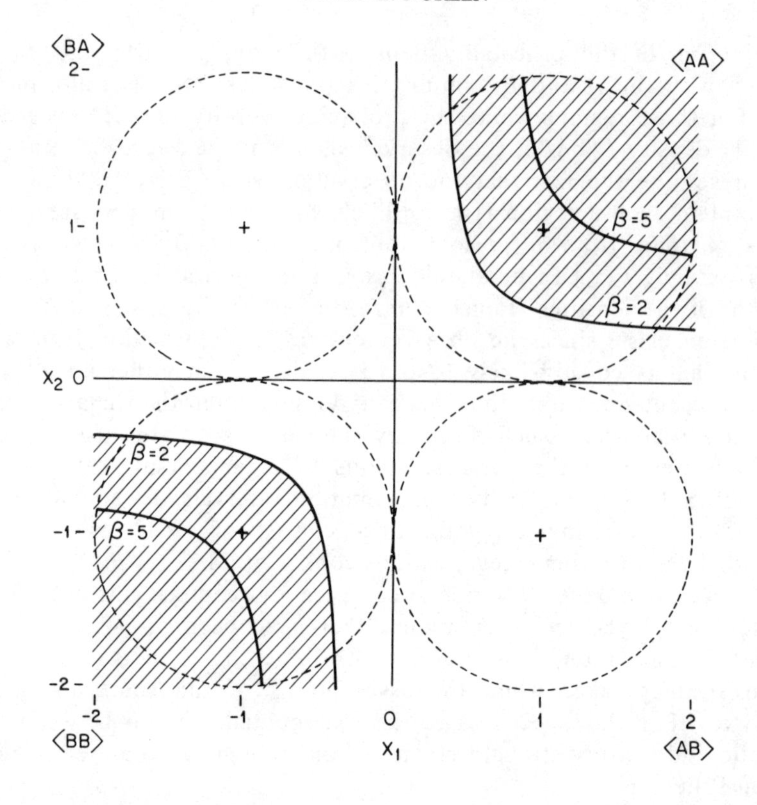

FIGURE 5. Optimal decision regions in the $X_1 \times X_2$ sensory space for the Gaussian special case of the same-different paradigm (yes-no $d' = 2.0$). The shaded area is the locus of all observations that would lead to a "same" response if a likelihood ratio criterion of 2.0 were adopted by the observer.

In summary, in the Gaussian $\pi_1 = \pi_2$ case the zero point of the sensory strength scale acts as a natural boundary for classifying an observation as an A versus a B. Accordingly, pairs of observation values falling on the same side of the zero point are classified as resulting from "same" stimuli, while pairs of observation values falling on opposite sides of the zero point are classified as resulting from "different" stimuli.

It is straightforward to use this decision rule to interrelate yes-no and same-different performance, so that data from a yes-no task can be used to predict an observer's performance in a same-different task that employs the same A and B stimuli. To construct such a relation, let p be the integral of the stimulus B density function from $x = -\infty$ up to $x = 0$; thus p is the probability that an instance of stimulus B is correctly classified as such in a yes-no task where the observer is unbiased. By symmetry about $x = 0$ of the Gaussian distributions corresponding to A and B stimulus presentations (cf. Figure 2), p also represents the probability that an instance of stimulus A is correctly classified as such, and hence p can be taken as simply the percentage of correct responses in the unbiased yes-no paradigm.

On any given trial of a same-different task, there are two distinct ways the observer can be correct. One such way is by accurately identifying both presentations, which has probability p^2 and which occurs when both observation values fall in regions where they are correctly classified. Alternatively, he can be correct by misidentifying both presentations (i.e. classifying $<AA>$ as $<BB>$, or $<AB>$ as $<BA>$), which has probability $(1-p)^2$; this latter case represents a situation where the two misclassification errors, in effect, cancel each other out and spuriously produce a correct response due to the nature of the experimental paradigm. Therefore, letting p_{SD} denote the probability of a correct response in the same-different task and rewriting p as p_{YN} to emphasize its role as the percentage of correct responses in the unbiased yes-no task, we have for the predicted relation between same-different and yes-no performance

$$p_{SD} = p_{YN}^2 + (1 - p_{YN})^2 \,. \tag{14}$$

For some purposes, it may be useful to eliminate p_{YN} from this equation and express p_{SD} directly in terms of the underlying d' measure that represents the scaled distance between the Gaussian distributions corresponding to the sensory representations of the A and B stimuli. In such cases, p_{YN} can be written in the form

$$p_{YN} = \Phi(d/2), \tag{15}$$

where the right-hand side represents the area accumulated up to the point $x = d/2$ under a standardized normal density function having a mean of 0 and a standard deviation of 1. When this expression for p_{YN} is substituted in (14), the resulting equation gives p_{SD} directly as a function of d.

In summary, Equations (14) and (15) jointly specify the predicted percentage of correct responses in a same-different task where the underlying A and B distributions satisfy the assumptions of the Gaussian case. Equation (14) actually holds somewhat more generally, for an arbitrary choice of stimulus A and B distributions, as long as the integral of the stimulus A probability density function over regions where the likelihood ratio exceeds 1 is equal in magnitude to the integral of the stimulus B probability density function over regions where the likelihood ratio is less than 1 — in other words, as long as the overall probability of an A presentation being correctly classified is the same as the overall probability of a B presentation being correctly classified.

Relation to categorical perception. In fact, the use of Equation (14) goes even somewhat beyond that prescribed by this more general context, since it is exactly the expression that one would write down on the basis of a two-state, categorical-perception-based notion of performance. That is, suppose that although a set of auditory stimuli is constructed by successively varying the value of some physical parameter in equal steps, individuals listening to the stimuli nevertheless seem to hear only two categories of sounds, which we may label, following our earlier convention, as perceptual categories A and B.

When only two perceptual categories exist, the probability of a correct "same" judgment can be expressed as the probability that both members of the pair are heard as instances of the *A* perceptual category, together with (i.e. summed with) the probability that both members of the pair are heard as instances of the *B* perceptual category. In other words, the two-state, categorical perception approach specifies that the probability of a correct "same" response takes exactly the form given by Equation (14), where p_{YN} represents the probability that a presented stimulus instance is categorically perceived as an *A*. In actual practice, the value of p_{YN} is usually taken from the labeling function obtained in a two-choice identification experiment, where the observer classifies each presented stimulus as an instance of perceptual category *A* or perceptual category *B*. For a given stimulus *X*, then, p_{YN} represents the proportion of times that *X* is labeled as an instance of perceptual category *A*, and Equation (14) gives the probability of a correct "same" response when the stimulus sequence <*XX*> is presented in a same-different discrimination task.

It is important to note, however, that this correspondence between the optimal decision strategy and the categorical decision rule holds only for the special $\pi_1 = \pi_2$ case that we have just been considering. Figure 5 shows that when $\pi_1 \neq \pi_2$ the contours of equal likelihood ratio values determined by the optimal strategy are no longer linear and orthogonal, and thus the information from the two presentation intervals can no longer be neatly decomposed into the two independent decisions required by the categorical decision rule. Thus the optimal decision strategy and the categorical decision rule coincide only in a special type of same-different task, although the very common tendency for experimenters to balance "same" and "different" trials in the experimental design means that the $\pi_1 = \pi_2$ special case plays a particularly important role in the empirical literature.

Sensory difference theory. The theoretical approach that we will consider as an alternative to the optimal decision strategy is one in which the observer is assumed to use the sensory difference $X_1 - X_2$ as his decision variable. This sensory difference decision rule is illustrated in Figure 6, where underlying Gaussian distributions are assumed for convenience in illustration.

For both the <*AB*> and <*BA*> stimulus sequences, the situation is exactly the same as in the 2AFC paradigm: The <*AB*> stimulus sequence gives rise to a sensory difference distribution centered at $+d$, while the <*BA*> stimulus sequence gives rise to a sensory difference distribution centered at $-d$. However, unlike the 2AFC case, there is now a third distribution centered at 0, representing the sensory difference distribution that results from a presentation of either stimulus sequence <*AA*> or <*BB*>.

To reach a decision, the observer sets two response criteria, one at $+k$ and the other at $-k$, and responds "same" if and only if the sensory difference $X_1 - X_2$ falls between these two values. In other words, as long as the *absolute* sensory difference is sufficiently small, the observer responds "same"; otherwise, he responds "different". Through adjustment of the criterion parameter k, he is free to vary the overall frequency with which these two responses occur.

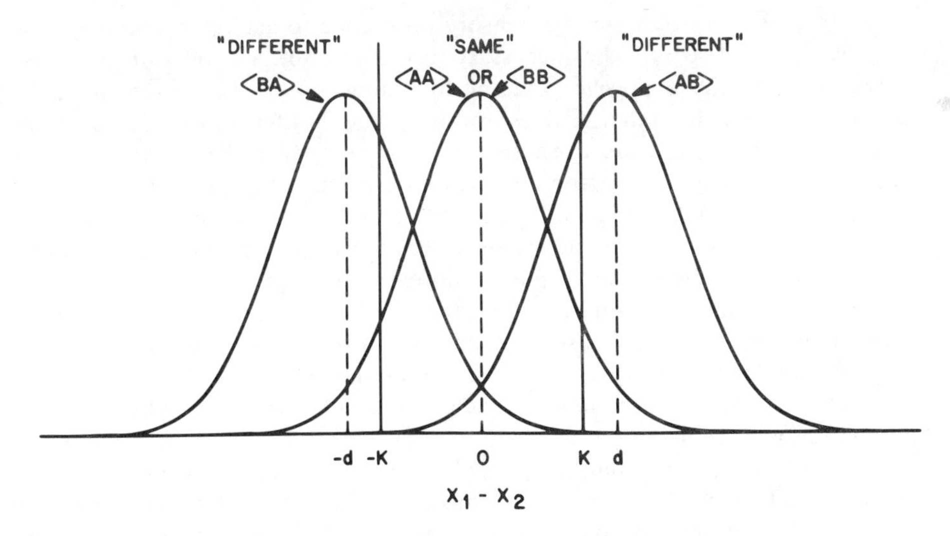

FIGURE 6. Illustration of the sensory difference decision rule in the same-different paradigm. Each distribution has a variance of 2.

This sensory difference model has been proposed by a number of different investigators and in a variety of experimental contexts. Its history is reviewed by Macmillan, Kaplan and Creelman (1977) and Krueger (1978) for the case of same-different judgments, and by Vickers (1979) for the closely-related case of three-category judgments (where the observer must distinguish between $<AB>$ and $<BA>$ stimulus presentations, rather than combine them together into a single "different" response category).

Perhaps one factor playing a prominent role in the widespread use of the model consists of the common tendency for the experimenter to instruct the observer that he must compare the two presented stimuli and decide if they are sufficiently similar to merit a judgment of "same", or, conversely, if they appear sufficiently dissimilar to merit a judgment of "different". Given the use of such instructions, it may seem natural to translate "comparing the two presented stimuli" into "forming the difference between the sensory impressions associated with the two presentation intervals", and this leads naturally to the sensory difference decision rule.

A second reason for the sensory difference model's widespread popularity may derive from the role of $X_1 - X_2$ as the optimal decision variable in the Gaussian 2AFC task; given this role, it may seem straightforward to preserve the same decision axis for same-different judgments as that used to model 2AFC judgments, and to simply add a third distribution centered at 0 for $<AA>$ and $<BB>$ stimulus presentations. On the basis of this straightforward extension, one might even expect a sensory difference decision rule to retain the same optimal status in the same-different paradigm that it enjoys for forced-choice judgments; indeed, Macmillan et al. (1977, pp. 470-471) pointed out that a monotonic relation exists between $y = x_1 - x_2$ and $L(y)$, an expression for the likelihood ratio computed under the assumption that the decision variable is distributed as depicted in Figure 6.

However, in order for the sensory difference to act as an optimal decision variable, it would be necessary for a monotonic relation to exist between y and the expression for the likelihood ratio $L(x)$ given in Equation (13), and such a relation does not in fact hold. One way to demonstrate this is to consider the decision regions that the sensory difference model defines in the $X_1 \times X_2$ decision space: A given value of the decision criterion k defines a region bounded by two parallel lines of unit slope and with y-intercepts at $+k$ and $-k$ as the region where the response "same" is appropriate; outside this region the response "different" is appropriate. This picture contrasts considerably with that depicted by Figure 5, where the decision regions in the $X_1 \times X_2$ decision space are shown for the optimal decision rule; in this latter case, linear decision contours result only when $\pi_1 = \pi_2$, and even in this instance they do not correspond to those specified by sensory difference theory.

To illustrate the nonoptimality of the sensory difference decision rule in somewhat more detail, we may consider a hypothetical decision-making task involving judgments of heights, where the decision variable no longer represents an unobservable sensory dimension, but rather corresponds to an observable dimension of numerical values, consisting of individuals' heights. That is, suppose that we have a population of females whose distribution of heights has a mean value of 5'3", and that we likewise have an equally numerous population of males whose height distribution has a mean value of 5'7". On any given trial of this hypothetical task, two heights are chosen randomly from the combined population of males and females and presented to the decision maker, who must decide if the two heights represent individuals of the same sex or of different sexes. The fact that the two populations are equally numerous, together with random sampling of individuals, ensures that $\pi_1 = \pi_2$; for illustrative purposes, we assume that the two distributions of height are Gaussian and share a common variance.

In the optimal decision strategy for this task, the decision maker first sets a cut-off midway between the means of the two distributions, at 5'5"; he then declares that the two individuals are of the same sex if they are both taller than or both shorter than this 5'5" criterion height, and declares that they are of different sex if one individual is taller than 5'5" while the other individual is shorter than this criterion height. In effect, then, the 5'5" criterion acts as a category boundary for classifying an individual on the basis of height, with individuals taller than it being classified as males and individuals shorter than it being classified as females. To perform optimally, the decision maker first independently categorizes the two individuals as males or females, then responds "same" if the categorizations agree and "different" if they disagree.

It is straightforward to construct specific examples that both demonstrate the use of this optimal decision strategy and illustrate how it differs from one based on the decision variable X_1-X_2. For example, if the two heights presented are 5'7" and 5'3", then a decision-maker using the optimal decision rule categorizes the two individuals as a male and female, respectively, and accordingly responds that they differ in sex; on the other hand, if the two heights are 5'11" and 5'7", then he categorizes both individuals as males, and thus determines that a "same sex" response is appropriate. However, note that in each of these two hypothetical decision-making situations there is a 4" height difference between the two individuals involved. Thus a decision maker using the height difference X_1-X_2 would be forced to assign an identical response to the two situations — e.g. if a 4" difference in height is sufficient to evoke a "different" response in the case of the individuals who are 5'7" and 5'3" tall, then the same 4" height difference must trigger a "different" response in the case of individuals who are 5'11" and 5'7" tall. In short, by considering only the *relative* difference in the two individuals' heights, a decision maker using X_1-X_2 as his decision variable ignores the exact location of the two individuals along the height continuum, and thus he fails to make use of the *absolute* information that plays an important role in the categorization process in the optimal specification of a response.

The multidimensional case. Similar examples that show the importance of absolute location information in optimal decision-making can be constructed for situations other than the unidimensional Gaussian case assumed here for the distribution of heights. For instance, Hefner (1958) proposed an extension of Thurstone's law of comparative judgment in which each of two stimulus presentations is assumed to give rise to a point drawn from a multivariate normal discriminal process distribution, with the *distance* between the points being used by the observer to compute a judgment of "same" versus "different" (a brief description of this model also appears in Sorkin, 1962). If r is the dimensionality of the space in which these two points are embedded, then the case $r = 1$ corresponds to the sensory difference model illustrated in Figure 6, since the unidimensional distance between the two points is just the absolute sensory difference $|x_1-x_2|$. The case where r is greater than 1, then, can be thought of as a straightforward extension of sensory difference theory to situations where there may be multidimensional sensory representations of the stimuli.

However, it is clear that this "sensory *distance* model" will face exactly the same problems with nonoptimal decisions for the case when r is greater than 1 as it does when r is equal to 1. To see this, consider once again the case where $\pi_1 = \pi_2$; if one of the sampled stimulus points is more likely to represent an A presentation while the other is more likely to represent a B presentation, then the optimal decision rule specifies that the observer should respond "different" regardless of how small a distance there is between the two sampled points. Conversely, if both presentations are more likely to be A's than B's (or vice versa), then the optimal response is for the observer to report "same", regardless of how large a distance there is between the two sampled points. Thus one may construct examples where a given interpoint distance should sometimes be mapped to the "same" response and other times be mapped to the "different" response, depending upon exactly where in the space the two sampled points happen to fall. For instance, if the two sampled points coincide with the means, or centroids, of the two multivariate density functions, then there is very good evidence that the two presented stimuli differ, and the optimal decision procedure specifies a "different" response. However, if the same interpoint distance is associated with two points sampled from a tail region of the density functions where, say, the stimulus A probability density uniformly exceeds the stimulus B probability density, then the optimal decision procedure specifies a "same" response, since an $<AA>$ stimulus presentation is the most likely source of the sensory information. Thus, just as was the case with the unidimensional, sensory difference model considered earlier, the multidimensional, relative distance model produces nonoptimal decisions through its failure to incorporate absolute location information in the decision-making process.

Of course, there may be cases where individuals for one reason or another do not use absolute location information in their judgments, and in such cases sensory difference theory, or its extensions, may provide a useful descriptive model of the performance that observers actually exhibit, despite its use of a nonoptimal decision rule. Hence it is of interest for the optimal decision strategy and sensory difference theory to be compared in experimental settings, so that one may determine which theoretical approach is more in accord with the obtained results for a given type of task. Toward this end, a study of tone frequency discrimination by Creelman and Macmillan (1979) may provide a useful example for illustrating one way that the two models may be applied in an experimental context.

An empirical illustration. In Creelman and Macmillan's study, nine different psychophysical procedures were used to investigate the discrimination of tone frequencies (data were also collected for phase discrimination, but for present purposes we will consider only the frequency discrimination data). Their stimuli consisted of sinusoidal signals of either 200 Hz or $200+f$ Hz, where for a given experimental run f assumed a fixed value chosen from a range of .25 Hz to 1.5 Hz. Included among the nine psychophysical procedures were a standard yes-no task, where the observer was required to identify which of two prespecified frequencies was presented on each trial, and a same-different or variable standard AX task of the form shown in Figure 4.

Creelman and Macmillan reported their data in terms of d' measures, but it is straightforward to convert these measures into unbiased percentage correct scores (which represent the percentage correct that an observer achieves when he has no bias for favoring response "R_1" over response "R_2"), and the latter may be somewhat more convenient for illustrative purposes here. When these conversions are made using the data from Table 2 of Creelman and Macmillan (1979), the obtained levels of performance are 73% correct in the yes-no task and 60% correct in the same-different task.

The challenge facing a mathematical model of the decision process is to account for this 13 percentage point decline in obtained performance scores from the yes-no to the same-different task. For both the optimal decision strategy and the sensory difference model, it is a straightforward procedure to calculate the expected decline and compare it to the decline that was actually obtained. While there are a number of different ways this can be done, perhaps one of the simplest is to use the obtained yes-no percentage correct score to predict the corresponding same-different score.

Accordingly, Equation (14) gives the predicted percentage of correct same-different judgments for the optimal model — it is simply the probability that the stimuli in both presentation intervals are identified correctly, or $(.73)^2$, summed with the probability that both stimuli are misidentified, or $(1-.73)^2$. This gives a predicted value of 61% correct, which is reasonably close to the 60% level of correct performance that was obtained.

For the sensory difference model, on the other hand, the d' value of 1.24 that Creelman and Macmillan obtained for the yes-no task can be translated into a same-different unbiased percentage correct score by using the tables provided by Kaplan, Macmillan and Creelman (1978, p. 806). The result is a predicted value of 58% correct responses, which again is reasonably close to the obtained performance score of 60% correct.

It may seem somewhat surprising that the optimal model and the sensory difference model give such similar predictions, given the various differences between the two theoretical approaches that were pointed out earlier. One might have expected, for example, that the nonoptimality inherent in the sensory difference decision rule would have led to more than a 3% reduction in the predicted percentage of correct responses from the value of 61% that was predicted by the optimal decision rule.

However, Table 1 shows that the predictions of the two models are very similar over their entire range of performance values, reaching a maximum discrepancy of only about 8 percentage points (cf. last column of Table 1) when the unbiased yes-no percentage correct value is approximately 95%. This close similarity means that it may be quite difficult to obtain experimental evidence that strongly favors one model over the other. To

TABLE 1

Comparison of the Sensory Difference Model and the Optimal Decision Rule on the Basis of Predicted Percentage Correct Scores in the Same-Different Task

		Predicted Value of p_{SD}		
p_{YN}	d'_{YN}	*Sensory Difference Model*	*Optimal Decision Rule*	Δ
50%	0	50%	50%	0
55	0.25	50	51	1
60	0.51	51	52	1
65	0.77	53	55	2
70	1.05	56	58	2
75	1.35	59	63	4
80	1.68	63	68	5
85	2.08	69	75	6
90	2.57	75	82	7
95	3.29	83	91	8
100	∞	100	100	0

contrast them empirically, one would presumably select the stimuli so that observers gave approximately 95% correct responses in a yes-no task, then look for a consistent pattern across blocks of trials or across observers that favored one model over the other.[4]

5. The ABX paradigm. The ABX discrimination paradigm has received widespread use both in studies designed to assess categorical perception of speech and speech-like sounds (see Macmillan, Kaplan and Creelman, 1977, for a review) and in experiments utilizing more traditional types of psychophysical stimuli, such as pure tones (e.g. Creelman and Macmillan, 1979). Perhaps one factor playing an important role in the widespread use of the paradigm is the fact that it can be taken as a variant of the two-stimulus complete identification paradigm where the two stimulus alternatives are externally presented as reference stimuli on each trial.

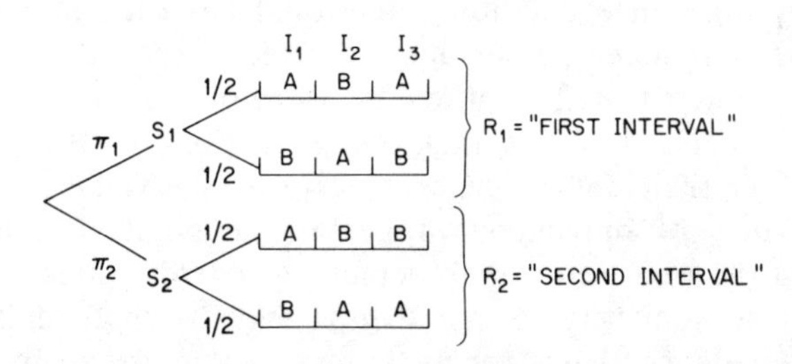

FIGURE 7. Structure of the "ABX" discrimination paradigm. The observer's task is to indicate whether the test stimulus in I_3 represents a replication of the standard presented in I_1 or the standard presented in I_2. The experimenter controls the probabilities π_1 and $\pi_2 = 1-\pi_1$, which determine the relative frequencies with which the two stimulus conditions appear.

Thus, as shown in Figure 7, a single trial consists of the presentation of the two reference stimuli, or standards, followed by a third, test stimulus. The observer's task is to judge whether the third, test stimulus represents a replication of the first-presented standard or a replication of the second-presented standard. Accordingly, for stimulus presentations $<ABA>$ and $<BAB>$ he should indicate that the test stimulus is identical to the first-presented standard by making response "R_1", while for stimulus presentations $<ABB>$ and $<BAA>$ he should indicate that the test stimulus is identical to the second-presented standard by making response "R_2". As

---

[4]A somewhat more general method for testing between the two models might involve varying stimulus presentation probabilities or costs and payoffs in order to generate a receiver-operating-characteristic or isosensitivity curve. For A and B distributions that satisfy the assumptions of the Gaussian case, the two models make contrasting predictions: The sensory difference model predicts that the isosensitivity curves should be asymmetric about the minor diagonal of the unit square (Sorkin, 1962, p. 1747; Macmillan, Kaplan and Creelman, 1977, pp. 456-457), while the optimal decision rule analysis predicts that the isosensitivity curves should be symmetric.

usual, we assume that the experimenter may vary the relative frequencies with which the two responses are appropriate, by controlling the stimulus presentation probabilities π_1 and π_2.

The optimal decision rule. The optimal decision rule for the ABX paradigm may be derived by means of essentially the same procedure as was used earlier. If in the present case we interpret assumption (A1) as specifying that the decision variables X_1, X_2, and X_3 are mutually independent, then we can use it together with assumption (A2) to express the likelihood ratio for ABX discrimination in the form

$$L(x) = \frac{f(x_1|A)f(x_2|B)f(x_3|A) + f(x_1|B)f(x_2|A)f(x_3|B)}{f(x_1|A)f(x_2|B)f(x_3|B) + f(x_1|B)f(x_2|A)f(x_3|A)} . \quad (16)$$

When both the numerator and the denominator of this expression are divided by $f(x_1|B)f(x_2|B)f(x_3|B)$, we obtain

$$L(x) = \frac{L(x_1)L(x_3) + L(x_2)}{L(x_1) + L(x_2)L(x_3)} . \quad (17)$$

Finally, this expression for the likelihood ratio can be substituted in Decision Rule 1′ to obtain the optimal decision rule for the ABX discrimination task in the form of

DECISION RULE 7. *Respond "R_1" iff*

$$L(x) = \frac{L(x_1)L(x_3) + L(x_2)}{L(x_1) + L(x_2)L(x_3)} \geq \frac{\pi_2}{\pi_1} .$$

Otherwise, respond "R_2".

As was the case with the same-different paradigm considered earlier, there are a number of alternative forms in which this decision rule can be written. One particularly useful such form is obtained by dividing both the numerator and the denominator of the likelihood ratio in Decision Rule 7 by $L(x_2)$. This gives

DECISION RULE 7′. *Respond "R_1" iff*

$$L(x) = \frac{[L(x_1)/L(x_2)]L(x_3) + 1}{[L(x_1)/L(x_2)] + L(x_3)} \geq \frac{\pi_2}{\pi_1} .$$

Otherwise, respond "R_2".

The interesting feature in this result is that the ABX likelihood ratio now takes exactly the same form as the expression for the same-different likelihood ratio in Equation (13), but with $L(x_1)$ and $L(x_2)$ in (13) replaced by, respectively, $[L(x_1)/L(x_2)]$ and $L(x_3)$ in the present case. This, in turn, is of interest, since the first of these latter two terms, $[L(x_1)/L(x_2)]$, acts as the optimal decision variable for a two-alternative forced-choice task (cf. Equation (6) and Decision Rule 4). Therefore, taken together these two results imply that the observer can perform optimally in the ABX task by initially treating the first two presentation intervals as a 2AFC task, where he obtains information relevant for deciding whether the standards occurred in an $<AB>$ or $<BA>$ presentation order, and by then combining this 2AFC information with the information from the third, test stimulus exactly as if he were making a same-different judgment based on the two sources of information.

The optimality of this decision strategy is apparent when one considers the structure of the ABX paradigm. That is, note that if the second presentation interval (I_2) in the decision-tree representation of the ABX paradigm in Figure 7 is ignored, then the paradigm corresponds exactly to the same-different paradigm of Figure 4. Thus, in principle, an observer could simply disregard I_2 and treat the ABX task as requiring a same-different judgment between the first and third presentation intervals, I_1 and I_3. However, because I_2 always contains the stimulus that is complementary to the stimulus appearing in I_1, it conveys a redundant source of information that may be used to help identify the stimulus that appeared in I_1 in exactly the same fashion as was discussed earlier in the two-alternative forced-choice task. Thus, the optimal decision strategy for the ABX task specifies that the observer should use the information from I_2 in exactly the same manner as is specified by the optimal 2AFC decision strategy, to augment his information of the first-presented stimulus; he should then combine this augmented information regarding the identity of the first stimulus with the information derived from the third, test stimulus in the same manner as is specified by the optimal decision procedure for the same-different task. The result is a decision that satisfies the optimality criteria discussed earlier, namely it minimizes the probability of an incorrect response or maximizes the expected value of the observer's payoff, given some payoff matrix defined over the stimulus-response outcomes.

Several examples that illustrate how the forced-choice information from the two standards may be combined with the information from the third, test stimulus will be considered presently, when we discuss the Gaussian special case of the ABX paradigm and also when we discuss a height judgment task analogous to the one introduced earlier. For the moment, it is of interest to note that as a consequence of the similarity in form between the expression for the ABX likelihood ratio in Decision Rule 7′ and the expression for the same-different likelihood ratio in Equation (13) we can apply a number of other properties of the same-different optimal decision strategy to the ABX discrimination task by simply replacing the likelihood ratios $L(x_1)$ and $L(x_2)$ in the same-different results with the terms $[L(x_1)/L(x_2)]$ and $L(x_3)$, respectively. For instance, the special $\pi_1 = \pi_2$ case of the same-different paradigm, described in Decision Rule 5″, can be rewritten for the ABX discrimination task as

DECISION RULE 7″. *When* $\pi_1 = \pi_2$, *respond* "R_1" *iff*

$$[L(x_1) \geq L(x_2) \text{ and } L(x_3) \geq 1] \text{ or } [L(x_1) \leq L(x_2) \text{ and } L(x_3) \leq 1] .$$

Otherwise, respond "R_2".

In the ABX task, this special $\pi_1 = \pi_2$ case corresponds to the commonly-encountered situation in which experimental conditions are arranged so that the a priori probability of the third, test stimulus matching the first-presented standard is the same as the a priori probability of it matching the second-presented standard. Under these circumstances, Decision Rule 7″ specifies that the observer should respond "R_1" (e.g. "test stimulus matches the first-presented standard") if and only if: (i) I_1 is more likely than I_2 to have contained the A standard and the third, test stimulus is more likely to represent an A than a B presentation; or (ii) I_1 is more likely than I_2 to have contained the B standard and the third, test stimulus is more likely to represent a B than an A presentation. Thus we once again have a decision strategy that can be interpreted as involving several categorization processes on the part of the observer; in effect, the two standards are categorized as representative of an $<AB>$ or a $<BA>$ presentation order, the third, test stimulus is categorized as an $<A>$ or a $<B>$ stimulus presentation, and then finally the outcomes of these two categorization processes are combined to determine an "R_1" or "R_2" response.

Given this decision strategy, it is straightforward to express the probability of a correct ABX judgment in terms of the probabilities associated with each of these categorization subprocesses. To do so, let p_{2AFC} denote the probability of a correct response by an unbiased observer in a two-alternative forced-choice task and let p_{YN} denote the corresponding probability correct for a yes-no task that utilizes the same A and B stimuli. Then the probability that the unbiased observer responds correctly when the two tasks are concatenated to form an ABX discrimination task takes the form

$$p_{ABX} = p_{2AFC}\, p_{YN} + (1 - p_{2AFC})(1 - p_{YN}) \ . \qquad (18)$$

It is of interest to note that this same expression can be derived using a discrete-state, categorical-perception-based model of the type discussed by Liberman, Harris, Hoffman and Griffith (1957), Pollack and Pisoni (1971), Creelman and Macmillan (1979), and others. However, as Creelman and Macmillan (1979, pp. 153-154) note, such a model predicts that the yes-no and 2AFC paradigms should yield the same percentage correct and also that the ABX and same-different paradigms should yield the same percentage correct (to derive this latter prediction, compare our Equations (18) and (14) with $p_{YN} = p_{2AFC}$), and neither of these predictions is supported by their frequency discrimination data. Rather, as will be discussed later in the paper, their data were in approximate agreement with the predictions of the optimal decision rule analysis, although the value of d' in the 2AFC task appeared to be a bit larger than one would expect on the basis of the obtained yes-no performance.

The Gaussian case. As was the case with the paradigms we considered earlier, the optimal decision rule for the ABX paradigm takes a particularly simple form when the assumptions of the Gaussian case are appropriate. For this special case, the exponential relation between likelihood ratio and sensory strength given in Equation (4) can be used to rewrite Decision Rule 7′ as

DECISION RULE 8. *Respond "R_1" iff*

$$L(x) = \frac{e^{d(x_1 - x_2 + x_3)} + 1}{e^{d(x_1 - x_2)} + e^{dx_3}} \geq \frac{\pi_2}{\pi_1} \ .$$

Otherwise, respond "R_2".

Macmillan, Kaplan and Creelman (1977) have previously commented upon various aspects of the decision strategy associated with this special case of the optimal decision rule; their Figure 5, for example, illustrates the equal likelihood contours that Decision Rule 8 determines and their Equation (4) expresses the likelihood ratio that appears in the decision rule in a somewhat different form, based on an alternative choice of origin for the decision axes. Accordingly, we will forgo a detailed discussion of the Gaussian special case of the optimal decision strategy in this subsection, and instead merely highlight some of its more important aspects.

One important feature associated with Decision Rule 8 is that it illustrates how the 2AFC information from the sensory strengths corresponding to the two standards should be combined with the information from the third, test stimulus in order to determine a response. That is, the $x_1 - x_2$ term that appears in the likelihood ratio represents the forced-choice information associated with the presentation order of the two standards, in accord with the fact demonstrated earlier that the sensory difference acts as the optimal decision variable in the Gaussian 2AFC task. A comparison of Decision Rule 8 with Decision Rule 6 shows that the forced-choice information in $x_1 - x_2$ should be combined with the test-stimulus information in x_3 exactly as if the observer were making an optimal same-different judgment on the basis of the two sources of information. Thus, the sensory difference information from the first two presentation intervals in effect acts as a single stimulus, or "equivalent standard", that is matched against the third, test stimulus in order to determine a response.

When the stimulus conditions are balanced so that $\pi_1 = \pi_2$, the categorical strategy that was discussed earlier is appropriate. The decision rule for this special case of the Gaussian ABX task can be obtained by referring to the corresponding special case of the Gaussian same-different task in Decision Rule 6′ and substituting $x_1 - x_2$ for x_1 and x_3 for x_2, respectively, to obtain

DECISION RULE 8′. *When $\pi_1 = \pi_2$, respond "R_1" iff $x_1 - x_2$ and x_3 have the same sign. Otherwise, respond "R_2".*

When this decision rule applies, the probability of a correct ABX judgment can be written as the sum of two terms, one corresponding to the joint probability that both $x_1 - x_2$ and x_3 are positive, and the other corresponding to the joint probability that they both are negative. This, of course, is essentially a restatement of Equation (18), which expresses ABX performance in terms of yes-no and 2AFC performance, although the parametric form of the special Gaussian case allows the predicted ABX performance to be stated entirely in terms of the distance parameter d. To obtain such an equation, we may use the expression that Equation (15) defines for p_{YN} as a function of d, together with a similar expression for p_{2AFC} that incorporates the $\sqrt{2}$ improvement factor distinguishing the 2AFC paradigm from the yes-no task; when these two expressions are substituted in (18), the result is a predicted level of ABX performance given by

$$p_{\text{ABX}} = \Phi(d/\sqrt{2})\Phi(d/2) + [1 - \Phi(d/\sqrt{2})][1 - \Phi(d/2)] . \tag{19}$$

This result agrees with Equation (3) of Macmillan, Kaplan and Creelman (1977), where it was obtained by means of a somewhat different argument.

Sensory difference theory. The theoretical approach that we will consider as an alternative to the optimal decision strategy is one that was originally proposed by Pierce and Gilbert (1958) and that, like the sensory difference model of the same-different paradigm, is based on the notion that the observer reaches a decision on the basis of differences in sensations. That is, the Pierce and Gilbert model assumes that the observer compares the first-presented standard to the test stimulus by forming the sensory difference $|X_1 - X_3|$, compares the second-presented standard to the test stimulus by forming the sensory difference $|X_2 - X_3|$, and then chooses response "R_1" or "R_2" depending upon whether the first or the second of these respective differences is the smaller. Thus the model can be interpreted as a mathematical representation of the instructions that are often given to an observer in an ABX task, where he is asked to state whether the first- or the second-presented standard is more similar to the third, test stimulus; the model simply incorporates the seemingly quite reasonable assumption that the similarity judgment is based on the difference in the sensory strength values that correspond to the two stimuli being compared, just as is assumed in the sensory difference model of the same-different task.

This sensory difference interpretation of the ABX paradigm can be contrasted with the optimal decision rule analysis by means of the height example introduced earlier. That is, suppose once again that we have both a population of females with mean height 5'3" and an equally numerous population of males with mean height 5'7"; also, assume as before that the two distributions of height are Gaussian and share a common variance. On any given trial of this hypothetical decision-making task, a male and a female are randomly selected to serve as standards, and the order in which their heights will be presented to the decision-maker is determined by, say, the flip of a coin; in addition, an individual is selected at random from the combined population of males and females to provide a height that will serve as the third, test stimulus. The decision-maker is then given the heights of these three individuals and asked to judge whether the sex of the third, test individual is the same as the sex of the first individual (i.e. the first-presented standard) or the second individual (the second-presented standard).

While there are some triads of heights for which the sensory difference model and the optimal decision rule analysis agree in their response classification, one height triad that leads to a divergence of the two approaches is the triad $<5'7", 5'9", 5'6">$. For this particular stimulus presentation, a decision-maker using the sensory difference model compares the heights of the first and third individuals and finds that the 1" height difference between 5'7" and 5'6" is smaller than the 3" height difference between the second and the third individuals; thus he responds "R_1" to indicate that the first-presented standard more closely resembles the third, test presentation than does the second-presented standard. On the other hand, a decision-maker using the optimal decision rule takes advantage of the fact that one of the first two heights must represent a male while the other must represent a female, since the two standards are chosen to be complementary in sex; accordingly, he categorizes the shorter, 5'7" individual as the female and the taller, 5'9" individual as the male, thereby obtaining a <female, male> classification for the presentation order of the two standards. Next, he categorizes the third, test individual as a male, since the 5'6" height that appears in the third presentation interval exceeds the 5'5" criterion that serves as a boundary for categorizing an individual as a male versus a female. Thus the final result is a judgment that the presented stimulus sequence corresponded to a <female, male, male> presentation, and this leads to a response of "R_2", in contrast to the "R_1" response determined by the sensory difference model.

The reason the two models determine opposite responses is apparent when one considers the structure of the example. That is, the three heights presented to the decision-maker all represent individuals that are relatively tall with respect to the combined population of males and females; indeed, without any knowledge of the particular structure of the ABX paradigm, one would be inclined to classify all three individuals as males, since they all exceed the 5'5" criterion height that acts as a response boundary in the yes-no task. However, the decision strategy prescribed by the optimal decision rule takes into account the complementary nature of the two standards — that regardless of how tall the first two individuals are, one or the other must be a female — and specifies that the second and third individuals are the two individuals who are most likely to be of the same sex, even though the first and third individuals are more similar on the basis of height per se.

In short, the optimal model uses both *relative* information regarding the presentation order of the two standards, with the taller of the first two individuals being classified as the male regardless of absolute height, and *absolute* information concerning the value of the third, test stimulus. The sensory difference model, on the other hand, considers only the *relative* information derived from a comparison of each of the standards to the test stimulus, and thereby fails to provide a means for taking into account the *absolute* information that is necessary for the optimal specification of a response.

The multidimensional case. A similar problem with absolute information is encountered when one attempts to generalize the sensory difference approach by modifying the unidimensional sensory representation considered by Pierce and Gilbert (1958). That is, one could easily suppose, in analogy with the approach of Hefner (1958) that was considered earlier in the same-different paradigm, that the sensory effect of the stimulus appearing in any given presentation interval is represented by a point in some abstract, multidimensional sensory space. A sensory *distance* extension of the sensory difference decision rule would then specify that the observer should respond "R_1" if and only if the distance between the points corresponding to the first and third stimulus presentations is less than the distance between the points corresponding to the second and third stimulus presentations.

However, to see that this decision rule produces nonoptimal decisions, suppose that on a particular trial the three points that represent the sensory effects associated with the three presentation intervals all happen to fall in a region of the space where, say, the stimulus A probability density uniformly exceeds the stimulus B probability density; in addition, assume that the point that is more likely to represent the B standard lies closer to the point corresponding to the third, test stimulus than does the point that is more likely to represent the A standard. In this case, the sensory distance model specifies that the observer should indicate as his response the presentation interval corresponding to the point that is more likely to represent the B standard, while an observer using the optimal decision rule chooses instead the presentation interval corresponding to the point that is more likely to represent the A standard. The sensory distance model's specification of a nonoptimal response in this case can again be traced to the model's failure to take into account the absolute information associated with the location of the points in the multidimensional sensory space.

Of course, while a theoretical approach based upon a nonoptimal decision rule can be immediately ruled out as a normative model of how decisions ideally ought to be made, such an approach may still serve a useful role as a descriptive model of how individuals actually make decisions, provided empirical evidence does not suggest otherwise. Fortunately, the Creelman and Macmillan (1979) study of tone frequency discrimination that was mentioned earlier included an ABX discrimination task that can be used as an empirical illustration of the theoretical approaches considered here. In the next subsection, we consider the Creelman and Macmillan data first from the theoretical standpoint provided by the optimal decision rule analysis, and then from the theoretical standpoint provided by sensory difference theory.

An empirical illustration. There are several different ways the optimal decision rule approach can be used to generate a predicted percentage correct score in the ABX task. One straightforward method employs Equation (18), which expresses the probability of a correct ABX judgment in terms of the corresponding correct response probabilities in the yes-no and 2AFC tasks. In this method, the d' values of 1.24 and 2.41 that Creelman and Macmillan report for obtained yes-no and 2AFC performance, respectively, are translated into unbiased percentage correct scores of 73% and 88%; when these two percentage correct scores are substituted in (18), the result is a predicted ABX performance level of 67% or 68%, depending upon how many significant digits are retained before rounding, and this compares favorably with the 68% level of unbiased correct responding that is associated with the reported d' value of .95 for the ABX task.

A second method for using the optimal decision rule analysis to generate a predicted performance score employs only the yes-no data. In this method, the obtained d' value of 1.24 in the yes-no task is substituted in Equation (19) to obtain a predicted value of 64% correct responses in the ABX paradigm. The fact that this predicted value is 4 percentage points lower than the obtained value of 68% may seem somewhat puzzling in view of the essentially exact correspondence between predicted and observed values that was obtained when both the yes-no and 2AFC data were used in the first method; however, the two methods agree in their predictions only when the observed 2AFC performance is exactly predicted by the yes-no data, and, as was noted earlier, the observed performance level for the 2AFC task was found to be a bit better than would be expected on the basis of the yes-no performance. Thus, it is this discrepancy between the obtained yes-no and 2AFC performance levels that is responsible for the 4 percentage point difference in the two methods.

As an alternative to considering tests of models that depend directly on the numerical values of predicted percentage correct scores, one can consider tests that specify only a particular ordering of conditions. One such ordinal prediction of the optimal decision rule analysis is an immediate consequence of results that were pointed out earlier — ABX performance should exceed same-different performance for exactly the same reason that 2AFC performance is predicted to exceed yes-no performance, namely the observer has a second chance at identifying the stimulus in the first presentation interval as a consequence of the complementary stimulus appearing in the second presentation interval. This prediction is indeed satisfied by the Creelman and Macmillan data, since the 68% level of correct performance for the ABX task exceeds the 60% level of correct performance for the same-different task. Thus the optimal model correctly orders the four paradigms we have discussed so far; in terms of increasing performance levels, this ordering of paradigms takes the form: same-different (60% correct), ABX (68%), yes-no (73%) and 2AFC (88%).

Given this reasonable success of the optimal decision rule analysis in ordering the various experimental paradigms and in providing at least a first-order approximation to the numerical percentage correct values, it is of interest to examine the corresponding predictions associated with the sensory difference approach. Like the optimal decision rule analysis, the sensory difference approach provides a straightforward method for using yes-no data to generate a predicted ABX performance score; the equation that Pierce and Gilbert (1958, p. 595) derived that interrelates the two tasks can be expressed in the form

$$p_{\mathrm{ABX}} = \Phi(d/\sqrt{2})\Phi(d/\sqrt{6}) + [1 - \Phi(d/\sqrt{2})][1 - \Phi(d/\sqrt{6})]\ , \qquad (20)$$

where d is again the d' measure, representing the distance in standard deviations between the means of the two Gaussian distributions that are assumed to underlie performance in the simple yes-no task. Thus to generate a predicted ABX performance score, one can simply use the yes-no data to estimate d' in the usual way, then substitute this estimate of d' in Equation (20) to obtain the predicted percentage correct.

It is interesting to note, however, that a comparison of this equation with Equation (19) shows that the only difference between the two is that an argument of $d/2$ in the equation based on the optimal decision rule analysis has been replaced by an argument of $d/\sqrt{6}$ in the equation derived from the sensory difference model; since $\sqrt{6}$ is approximately 2.45, it appears that the two theoretical approaches will often be quite close in their predicted values of ABX performance, given of course that the same yes-no data are used to generate predictions in both cases. In the case of the Creelman and Macmillan data, the obtained yes-no d' value of 1.24 can be substituted in Equation

(20) to obtain a predicted ABX performance level of 62%, and this is only two percentage points less than the 64% estimate provided by the optimal decision rule analysis.

6. Discussion. Thus despite the considerable conceptual differences between the sensory difference and optimal decision rule strategies, the two theoretical approaches often lead to very similar predictions. In particular, for the Creelman and Macmillan (1979) study of tone frequency discrimination both theoretical approaches correctly predict that the percentage of correct responses will increase as one successively examines same-different, ABX, yes-no and 2AFC tasks, and both approaches provide at least a first-order approximation to the numerical values of the obtained percentage correct scores in these tasks. Accordingly, it would be difficult to make a strong argument for favoring one theoretical approach over the other on the basis of the analyses we have considered thus far.

In the long run, however, evaluations of the overall usefulness of a particular theoretical approach will most likely involve not only the several paradigms that we have examined here, but also a variety of additional experimental paradigms that are used in the study of psychophysical discrimination performance. An important feature of the optimal decision rule analysis in this regard is that its application to new paradigms is usually quite straightforward, even in cases where it is not immediately obvious exactly how a corresponding sensory difference model would be formulated. In the remainder of this section, we will outline some predictions that the optimal decision rule analysis makes for several additional experimental paradigms, and then conclude by briefly commenting upon some possible limitations of the basic method of analysis.

The AXA paradigm. The AXA discrimination paradigm is a variant of the same-different task that is often used to study auditory phase discrimination, although Creelman and Macmillan (1979) have used it to study frequency discrimination as well. On any given trial of an AXA experiment, two identical stimuli are presented as standards in the first and third presentation intervals of a three-interval presentation sequence. The observer's task is to report whether the second, test stimulus is the same as, or differs from, these outer two standards — that is, he should respond "same" if either the stimulus sequence $<AAA>$ or $<BBB>$ is presented, and "different" if either $<ABA>$ or $<BAB>$ appears.

The optimal decision rule analysis of this task gives an interesting prediction for the situation in which the A and B stimuli satisfy the assumptions of the Gaussian case. Under these circumstances, the observer's unbiased percentage correct score in the AXA task should be identical to the unbiased percentage correct score that he achieves when the same A and B stimuli are used in an ABX task. In other words, the observer is predicted to derive the same benefit from two identical standards that he derives from two complementary standards.

The values of d' that Creelman and Macmillan (1979) report for the AXA and ABX paradigms are .87 and .95, respectively. In terms of unbiased percentage correct scores, these d' measures translate into values of 67% and 68% correct responses, so the prediction of equivalent performance levels would seem to be supported in this particular study.

The 4I2AFC paradigm. To introduce our next paradigm, suppose that observers treat a pair of stimulus presentations as a single perceptual event; then the 2AFC task is in effect a yes-no task in which the stimulus pairs <*AB*> and <*BA*> act as the two alternatives. Under these circumstances, the equivalent of a two-alternative forced-choice task would be a four-interval task in which both <*AB*> and <*BA*> occur on each trial, in a randomly determined order. Creelman and Macmillan (1979) briefly mentioned this task, but noted that it did not seem to have been used in previous psychophysical work; we will refer to it as the "four-interval, two-alternative forced-choice task", or the "4I2AFC" paradigm, since there are four intervals in which stimulus presentations occur, although the observer is only required to make a forced-choice judgment between two alternatives, by responding "R_1" to the presentation sequence <*ABBA*> and "R_2" to the presentation sequence <*BAAB*>.

In the optimal decision rule analysis, this 4I2AFC task reduces in a formal sense to a 2AFC task in which the usual <*A*> and <*B*> stimuli are replaced by the stimulus pairs <*AB*> and <*BA*>. This, of course, is not surprising given the structure of the task — indeed, this formal reduction is in some sense a restatement of the procedure that we used to design the task. However, it is of some interest from an empirical standpoint, since it means that familiar relations that are predicted to hold between the yes-no and 2AFC tasks are also predicted to hold between the 2AFC and the 4I2AFC paradigms.

For example, in one such familiar relation the optimal decision strategy specifies that the area under an observer's yes-no isosensitivity curve (also known as a "receiver-operating-characteristic" or "ROC" curve) should equal his unbiased percentage correct score in a 2AFC task that is constructed from the same A and B stimuli, regardless of the form taken by the underlying stimulus A and B distributions. The formal statement of this prediction is often referred to as the so-called "Area Theorem", and a proof of it is given by Green and Swets (1974, pp. 45-48). If their proof is modified, however, so that the two stimulus alternatives in the yes-no task correspond to the <*AB*> and <*BA*> stimuli of the 2AFC task, then an analogous "Area Theorem" is obtained for performance in a 4I2AFC paradigm. In other words, the optimal decision rule analysis predicts that the area under an observer's 2AFC isosensitivity curve will equal his unbiased percentage correct score in a 4I2AFC task.

Another set of testable predictions is expected to hold when the stimulus A and B distributions satisfy the assumptions of the Gaussian case. Under these circumstances, just as the optimal decision strategy for the 2AFC task involves a yes-no judgment on the sensory difference $X_1 - X_2$, the optimal decision strategy for the 4I2AFC paradigm involves a yes-no judgment on the decision variable $(X_1 - X_2) - (X_3 - X_4)$. This in turn implies that the value of the response measure d' for the 4I2AFC task should exhibit the same $\sqrt{2}$ improvement factor relative to the value of d' in the 2AFC case that the 2AFC case is predicted to show relative to the yes-no paradigm; in other words, one would expect d' to improve by a total factor of 2 when the 4I2AFC paradigm is contrasted directly with the yes-no task. In summary, the optimal decision rule analysis provides a theoretical framework of empirically-testable predictions that may play a useful role in subsequent experimental investigation of the 4I2AFC task.

The 4IAX paradigm. The last paradigm that will be noted here is the four-interval same-different or "4IAX" paradigm, which has been used by Pisoni and Lazarus (1974) to study synthetic speech stimuli and by Creelman and Macmillan (1979) to study tone frequency discrimination. On any given trial of this paradigm, a "same" pair and a "different" pair of stimuli are presented, and the observer is required to specify whether the first two or the last two stimuli represent the pair of stimuli whose members are the "same". For example, if the presented stimulus sequence is $<AABA>$, the observer should indicate that the first pair of stimuli are the "same", while if the presented stimulus sequence is $<BAAA>$, he should indicate that the second pair of stimuli are the "same".

The optimal decision rule analysis shows that the following three-step decision procedure represents an optimal decision strategy for the 4IAX task:

(1) First, Equation (13) is used to compute a same-different likelihood ratio $L(x_1, x_2)$ that expresses the degree of similarity between the stimuli that were presented in the first two presentation intervals.

(2) Next, Step 1 is repeated with the third and fourth presentation intervals, in order to obtain a likelihood ratio $L(x_3, x_4)$ for the second pair of stimuli.

(3) Finally, Decision Rule 4, the optimal decision rule for the 2AFC paradigm, is used to determine which likelihood ratio is more likely to represent the pair of "same" stimuli. That is, the observer should respond "R_1" to indicate that it is the first pair of stimuli that are the "same" if and only if $L(x_1, x_2) \geq (\pi_2/\pi_1)L(x_3, x_4)$; otherwise, he should respond "R_2". (Here π_1 denotes the a priori probability that the "same" stimulus pair will appear in the first pair of presentation intervals.)

Thus in the optimal decision rule analysis the 4IAX task reduces to a 2AFC task in which the usual $<A>$ and $<B>$ stimuli are replaced by "same" and "different" stimulus pairs. Once again, this may seem a rather unsurprising result, given the structure of the task; however, it is important from an empirical standpoint, since it suggests that familiar relations that are predicted to hold between the yes-no and 2AFC tasks may also be predicted to hold between the same-different and 4IAX paradigms.

For example, one such familiar relation is predicted to hold when the underlying A and B distributions satisfy the assumptions of the Gaussian case. In this case, the four possible stimulus sequences of the same-different task determine but two distributions of likelihood ratio values, one for "same" trials and another for "different" trials, and the same-different paradigm reduces to the equivalent of a yes-no task. Under these circumstances, when a modified version of the proof of the "Area Theorem" is applied to this equivalent yes-no task, a testable prediction that interrelates same-different and 4IAX performance is obtained, namely the area under the same-different isosensitivity curve is predicted to give the observer's unbiased percentage correct score in the 4IAX task.[5]

It would be convenient for model-testing purposes if the $\sqrt{2}$ relation that is predicted to hold between the d' measures in the Gaussian yes-no and 2AFC tasks were also predicted to hold between corresponding measures in the same-different and 4IAX tasks. However, when the stimulus A and B distributions satisfy the assumptions of the Gaussian case, the same-different task reduces to a yes-no task in which the underlying distributions are non-Gaussian. Thus, if the Gaussian assumptions introduced earlier in the paper are appropriate, one would not expect an exact $\sqrt{2}$ relation to hold between d' measures for the same-different and 4IAX tasks.

Of course, it may be the case that under certain conditions the $\sqrt{2}$ relation, though not exact, will still serve as a useful approximation to the predictions derived from the non-Gaussian representation. In this regard, it is interesting that from a purely empirical standpoint one can use the value of $d' = .51$ that Creelman and Macmillan (1979) obtained in the same-different task, which corresponds to an unbiased percentage correct score of 60%, to predict a value of $d' = (\sqrt{2})(.51) = .72$, or an unbiased percentage correct score of 64%, in the 4IAX paradigm, and this is not too different from the value of $d' = .67$, or an unbiased percentage correct score of 63%, that was actually obtained. It remains for a more detailed investigation of the nature of the approximation to determine whether this empirical correspondence is one that is predicted by the optimal decision rule analysis, or whether it instead represents an entirely fortuitous result.

Some possible limitations. One could continue to create new discrimination paradigms indefinitely by interpreting existing paradigms as yes-no tasks, and constructing tasks that act as the corresponding two-alternative forced-choice equivalents; however, it is clear that as the number of resulting

---

[5]It is straightforward to generalize this result beyond the Gaussian case we have considered here to an important class of models characterized by reflection symmetry about the origin of the A and B probability density functions. That is, for a class of models defined by the requirement that $f(x_i|A) = f(-x_i|B)$ $(i=1,2)$, the four stimulus sequences of the same-different task yield but two distributions of likelihood ratio values, and it therefore follows through a modified proof of the "Area Theorem" that the area under the same-different isosensitivity curve gives the predicted unbiased percentage correct score in the 4IAX task. This reflection-symmetry condition, which holds for the important equal-variance logistic case among others, also imposes a variety of other testable predictions that can be examined in other paradigms. For example, for all the paradigms discussed in this paper, the reflection-symmetry condition implies that the isosensitivity curves determined by the optimal decision rule analysis should be symmetric about the minor diagonal of the ROC space (see Green and Swets, 1974, pp. 48-49, for a discussion of this property with respect to the 2AFC paradigm). As a second example, the condition implies that the isosensitivity curves generated by the optimal decision rule analysis in the ABX and AXA paradigms will exactly coincide, provided that the same A and B stimuli are used in both tasks. Finally, it may be noted that if the decision variable X is unidimensional, then all the above results hold somewhat more generally, for any pair of distributions for which a monotonic transformation of the decision axis will render the reflection-symmetry condition satisfied.

presentation intervals becomes large, it becomes less compelling to assume that there is no memory loss associated with the observer's retention of the stimulus information that is derived from each presentation. Thus it would seem that at some point an optimal decision rule analysis based on the assumptions used throughout this paper must encounter serious difficulties as a descriptive model of an observer's actual performance.

Indeed, such potential difficulties are by no means limited to paradigms having a large number of presentation intervals, since it seems likely that the assumptions used here will be a bit too strong at times for even some of the simple paradigms that were discussed earlier. For instance, one rather strong prediction associated with the assumptions is that in an ABX discrimination task it should not matter which of the three presentation intervals contains the X test stimulus, as long as the observer is made aware of the structure of the paradigm beforehand. Hence, performance is predicted to be identical in ABX, AXB and XAB tasks, regardless of the form of the underlying stimulus A and B distributions. One factor that might argue against the likelihood of this prediction holding true is the presence of superficial differences in the structure of the three tasks that could lead observers to perform differentially in the three circumstances — for example, in the AXB task there is always one pair of successive stimuli that are the "same" and a second pair of successive stimuli that are "different", while in the other two tasks either the "same" pair or the "different" pair is separated by an intervening standard, as a consequence of the test stimulus appearing first or last. In some cases the presence or absence of this intervening standard may turn out not to make much of a difference in the observer's obtained performance levels in the three tasks; however, it would be rather surprising if sufficiently powerful experimental designs could not turn up at least a few instances where reliable differences among the three types of tasks were found, and such cases would argue against the specific form of the optimal decision rule analysis considered here.

In the final analysis, however, the interesting question is not whether there exist tasks for which the predictions of the optimal decision rule analysis are inaccurate, since certainly such tasks exist. Rather, the interesting question is whether some sort of characterization can be made of the types of experimental conditions and the types of response requirements for which the analysis is versus is not likely to be appropriate. That is, the optimal decision rule analysis is basically a theoretical framework that, under some circumstances, may provide a useful summary of an observer's performance in a variety of different tasks. The relevant problem now is to define the precise nature of these circumstances.

7. Summary. The major points of this paper can be summarized as follows:

(1) The optimal decision rule for a given psychophysical paradigm can often be written in a rather simple form by expressing the likelihood ratio $L(x)$ for an observation vector x in terms of individual likelihood ratios $L(x_i)$ $(i = 1, \ldots, n)$ associated with each of the n presentation intervals of the paradigm.

(2) In an important special case of the same-different task when "same" and "different" trials are balanced in the experimental design, the optimal decision rule specifies that the observer should separately categorize the stimuli appearing in the two presentation intervals, and then give his response based on the resulting categorical information.

(3) In the optimal decision strategy for the ABX discrimination task, the observer matches forced-choice information from the two standards against information from the third, test stimulus exactly as if he were making an optimal same-different judgment on the basis of the two sources of information.

(4) For both the same-different and ABX discrimination tasks, it is straightforward to construct examples that show how the optimal decision strategy differs from a decision strategy in which the observer uses differences in sensations. Nevertheless, in many cases the two theoretical approaches give very similar values for the predicted percentage of correct responses. For the tone frequency discrimination data of Creelman and Macmillan (1979), both of these approaches correctly predict that the percentage of correct responses should increase as one successively examines same-different, ABX, yes-no and two-alternative forced-choice tasks, and the numerical values of the models' predictions provide at least a first-order approximation of the obtained values.

(5) The optimal decision rule analysis can be used to generate predictions of an observer's performance in a variety of other paradigms. For instance, it predicts identical percentage correct scores for important special cases of ABX and AXA discrimination tasks, and this prediction was found to be approximately true for the data of Creelman and Macmillan (1979). A number of other new predictions await detailed experimental tests.

References

Creelman, C. D. and Macmillan, N. A. Auditory phase and frequency discrimination: A comparison of nine procedures. *Journal of Experimental Psychology: Human Perception and Performance*, 1979, **5**, 146-156.

Green, D. M. and Swets, J. A. *Signal detection theory and psychophysics*. Huntington, New York: Robert E. Krieger Publishing Company, 1974.

Hefner, R. A. *Extensions of the law of comparative judgment to discriminable and multidimensional stimuli.* Unpublished doctoral dissertation, University of Michigan, 1958.

Kac, M. A note on learning signal detection. *IRE Transactions on Information Theory*, 1962, **IT-8**, 126-128.

Kac, M. Some mathematical models in science. *Science*, 1969, **166**, 695-699.

Kaplan, H. L., Macmillan, N. A., and Creelman, C. D. Tables of d' for variable-standard discrimination paradigms. *Behavior Research Methods and Instrumentation*, 1978, **10**, 796-813.

Krueger, L. E. A theory of perceptual matching. *Psychological Review*, 1978, **85**, 278-304.

Laming, D. *Mathematical psychology*. New York: Academic Press, 1973.

Lappin, J. S. The relativity of choice behavior and the effect of prior knowledge on the speed and accuracy of recognition. In N. J. Castellan, Jr. and F. Restle (Eds.), *Cognitive Theory, Volume 3.* Hillsdale, N.J.: Lawrence Erlbaum Associates, 1978, pp. 139-168.

Liberman, A. M., Harris, K. S., Hoffman, H. S., and Griffith, B. C. The discrimination of speech sounds within and across phoneme boundaries. *Journal of Experimental Psychology*, 1957, **54**, 358-368.

Luce, R. D. Detection and recognition. In R. D. Luce, R. R. Bush and E. Galanter (Eds.), *Handbook of Mathematical Psychology, Volume I.* New York: Wiley, 1963, pp. 103-189.

Macmillan, N. A., Kaplan, H. L., and Creelman, C. D. The psychophysics of categorical perception. *Psychological Review*, 1977, **84**, 452-471.

Pierce, J. R. and Gilbert, E. N. On AX and ABX limens. *Journal of the Acoustical Society of America*, 1958, **30**, 593-595.

Pisoni, D. B. and Lazarus, J. H. Categorical and noncategorical modes of speech perception along the voicing continuum. *Journal of the Acoustical Society of America*, 1974, **55**, 328-333.

Pollack, I. and Pisoni, D. B. On the comparison between identification and discrimination tests in speech perception. *Psychonomic Science*, 1971, **24**, 299-300.

Sorkin, R. D. Extension of the theory of signal detectability to matching procedures in psychoacoustics. *The Journal of the Acoustical Society of America*, 1962, **34**, 1745-1751.

Vickers, D. *Decision processes in visual perception.* New York: Academic Press, 1979.

BELL LABORATORIES, MURRAY HILL, NEW JERSEY 07974

Copy for this paper was prepared from computer-composed galleys provided by the author.

SIAM-AMS Proceedings
Volume **13**
1981

Mathematical Models of Binocular Vision[1]

GEORGE SPERLING

Overview. When I accepted the invitation to participate in this symposium, I promised to discuss some representative models of perception. It soon became apparent that the mathematical models currently being proposed are so numerous and so complex that it would require volumes to do justice to even one subspeciality, such as color vision or binocular vision. For example, simple linear matrix operations sufficed for the theory of color mixture for 100 years (Wyszecki and Stiles [**1967**]); today's more comprehensive theories of color perception invoke additional nonlinear operations (e.g. Pugh and Mollon [**1979**]).[2] There are new developments in theories such as factor analysis, multidimensional scaling, and cluster analysis which have been used to describe the mapping of physical stimuli (such as simple color patches but also much more complex stimuli) into psychologically significant space (Shepard [**1980**]). Measurement theory has evolved as a branch of mathematics to describe the mapping of physical stimulus dimensions (most often intensity) into psychological dimensions (Krantz, Luce, Suppes, and Tversky [**1971**], Roberts [**1979**]).

In sensory psychology, particularly in the study of vision and of hearing, linear equations have been used with considerable success to describe the receptors, i.e., the optical properties of the lens of the eye (Fry [**1955**], Krauskopf [**1962**]) and the transducer properties of the outer, middle, and inner ear (basilar membrane, Allen [**1977a**], [**1977b**]). Sine wave stimulus patterns and linear theory are widely used to describe the first order psychological properties of sensory systems (Licklider [**1961**], Graham [**1981**]). But in these domains, familiar engineering methods have been all but exhausted; current theories deal with complex, nonlinear sensory mechanisms (Schroeder [**1975**], Victor, Shapley, and Knight [**1977**]). In fact, process models (such as models of the presumed

1980 *Mathematics Subject Classification.* Primary 92A25; Secondary 58C28.

[1]For reprints write to Professor G. Sperling, Department of Psychology, New York University, 6 Washington Place, New York, N. Y. 10003.

[2]It is impractical to give here complete reference lists for the various subjects mentioned in the text; only a few representative sources are cited.

neural processes that underly perception) can quickly become mathematically intractable; only a few workers have reduced the corresponding sets of nonlinear differential equations to something understandable (Sperling and Sondhi [**1968**], Grossberg [**1972**], [**1973**], Luce and Green [**1972**], [**1974**]).

Visual illusions that result from viewing "impossible" figures are better understood via topological analysis of the stimuli (Cowan [**1974**]); errors in the perception of a sequence in a string of characters are better understood by means of combinatorial algebraics (Sperling and Melchner [**1976**]); Riemannian geometry is used to describe the binocular perception of space (Luneberg [**1947**]); and Markov, random walk, and diffusion models occur in the study of reaction time, decision making, and other processes (Norman [**1972**], Link and Heath [**1975**]).

In the study of selective attention (e.g., to one part of the visual field or to one kind of character symbol in a visual search task) new concepts such as the attention operating characteristics have developed (Sperling and Melchner [**1978**], Navon and Gopher [**1979**]) that are intimately related to the receiver operating characteristics of signal detection theory (Green and Swets [**1966**], Egan [**1975**]), as well as to important concepts of economics. Relevant mathematical specialties are utility theory and linear programming.

In a study of visual motion perception in ambiguous displays, functional equations was the mathematical tool used to develop a process model (Burt and Sperling [**1981**]). In fact, as many kinds of mathematics seem to be applied to perception as there are problems in perception. I believe this multiplicity of theories without a reduction to a common core is inherent in the nature of psychology (Sperling [**1978**]), and we should not expect the situation to change. The moral, alas, is that we need many different models to deal with the many different aspects of perception.

The perceptual problems I will be concerned with in this paper involve the dynamics of resolving perceptual ambiguities in general; examples are given from binocular vision and depth perception. The relevant areas of mathematics are nonlinear differential equations and catastrophe (or potential) theory. I have chosen these examples because the possibility of further mathematical development is most appealing.

Path-dependence in binocular vision: The phenomena. We begin with the general problem of path-dependent perceptual states: how do we understand the cases where our perception of the present stimulus depends on our perception of the immediately preceding stimuli?[3]

Multistability of vergence. A classical example of this kind of phenomenon is described by Helmholtz [**1924**, p. 58]. Helmholtz viewed a binocular stereogram, that is, two identical or nearly identical images, so that each eye saw only its

[3]Much of the subsequent discussion is based on Sperling [**1970**], which the reader should consult for details, additional references, and interaction models.

corresponding image (Figure 1). First, he achieved vergence-fusion of the sterogram, that is, he arranged for the left and right half-images to fall on corresponding retinal points, and they were perceived as a single, unitary image (Figure 1(b)). At this point, the stereo images were arrayed so that the lines of sight of his two eyes were parallel. Then, Helmholtz began to increase slowly the separation in physical space between the two half-images so that, in order to maintain fusion, his eyes were forced to diverge. He found that by pulling the images apart in this way he was able to cause the lines of sight of his eyes to diverge by as much as eight degrees to follow the images. When the divergence angle required to maintain fusion increased beyond eight degrees, Helmholtz lost fusion; that is, his eyes returned to a parallel position and he saw two partially overlapping stereo half-fields (Figure 1(c)). To restore fusion, Helmholtz had to bring the stereo half-images much closer together than the point at which fusion broke.

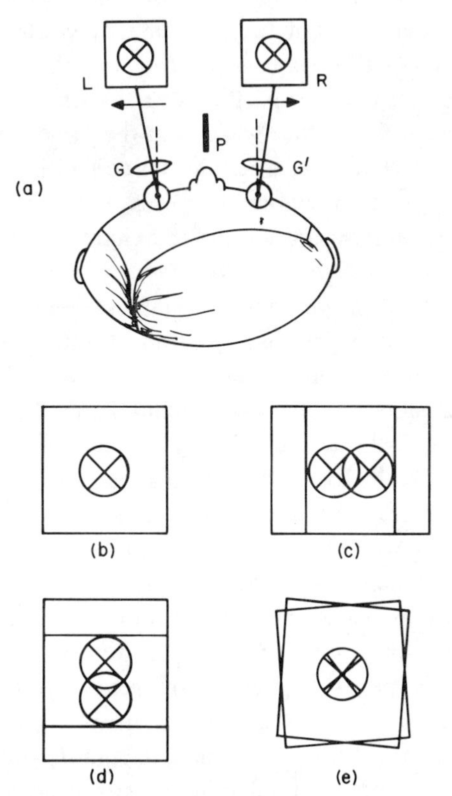

FIGURE 1.(a) A sterogram showing the left L and right R half-images viewed by the left and right eyes, respectively. The lenses G, G' enable the observer to focus at the short viewing distance. The partition P isolates the views of the two eyes. The arrows indicate the direction in which the stimuli are moved in Helmholtz's horizontal vergence demonstration. The broken lines indicate parallel lines of sight; the solid lines indicate the actual lines of sight when the observer is able to verge accurately at the angle of divergence shown. (b) Representation of the cyclopean perception when the eyes are correctly verged. The perception after vergence fails for (c) horizontal vergence, (d) vertical vergence, and (e) torsional vergence.

This example illustrates path-dependence for a particular stereogram separation, say, eight degrees. If the eyes were diverged prior to the eight degree stereogram–perhaps because the previous stereogram had a seven degree separation and they were able to fuse it–the eyes remained fused. If the immediately preceding stereogram had a zero or a 12 degree separation, the eyes would not fuse. Whether the eyes were fused or not depended on their history. Thus, we have two stable states, fused or not fused, in response to the same present stimulus, and the particular state that is achieved depends on the sequence of the preceding stimuli.

This example exploits the most familiar of the eyes' vergence systems–the system that controls the convergence and divergence of the lines of sight of the two eyes. Many people have voluntary control over lateral vergence, particularly convergence, and that control can considerably complicate actual demonstrations and theory. However, the demonstration works equally well with vertical vergence which, so far, has not been voluntarily controlled. Helmholtz found that when one stereo half-image was raised and the other lowered, the eyes would diverge vertically–one up, the other down–to maintain fusion. The angular range of vertical vergence was about the same vertically as the range of horizontal divergence. Similarly, when the stereo half-images are rotated around their centers in opposite directions, the eyes will follow these rotations with corresponding torsional rotations around their axes.

Multistability of accommodation (*focusing*). The accommodation (focusing) system of the eyes provides perhaps the simplest demonstration of path-dependence.[4] Take a piece of paper that has some printing on it and make a hole about $\frac{1}{2}$ to 1 cm diameter in the center. Close one eye and bring the hole as close to the other as possible, all the while maintaining the paper in sharp focus. Direct the line of sight through the middle of the hole at a distant object while maintaining the paper in focus. The hole should appear to be filled with an extremely blurred image while the surrounding paper is sharp (Figure 2). Now move the paper away while maintaining the line of sight. The eye will focus on the distant object, which now appears clearly. Without changing the line of sight, reintroduce the paper to the position it occupied just previously. It should be possible to now look through the hole and to see the distant object in focus and the surrounding paper blurred. (Younger subjects have a much better demonstration because of their greater range of accommodation. Once the accommodative range becomes too small, voluntary–rather than a stimulus–control of accommodation can predominate.) However, by adjusting the parameters of the demonstration–aperture size, distance, contrast, etc.–bistability can usually be restored. The demonstrations show path-dependence in accommodation. Precisely the same visual stimulus (distant view through a near aperture) can lead to two states of accommodation: near focus or far focus, depending on

[4]The demonstration of path-dependence in accommodation was reported by Sperling [**1970**].

the previous stimuli, i.e., on where the eye happened to be focused before looking through the hole.

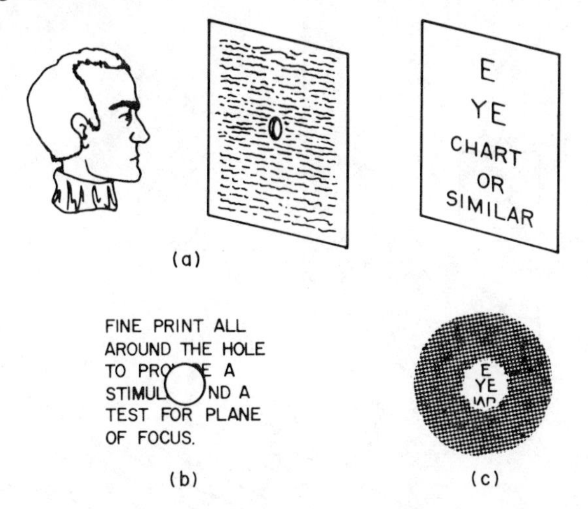

FIGURE 2. (a) Arrangement for observing two states of accommodation (focusing of the eye). The observer looks through the aperture with one eye at a distant object. (b) The view, accommodated on the aperature. (c) The view, accommodated on the distant object.

Multistability of fusion. Perhaps the most interesting instance of path-dependence in this class of visual phenomena occurs in perceptual fusion. To demonstrate it, the eyes view a stereogram through a complex optical system that cancels the effect of eye movements; that is, the images are optically stabilized on their respective retinas irrespective of where the eyes happen to point. Now, let the images be moved apart on the retinas just as, in the first example, they were moved apart in physical space. Eye movements cannot compensate for the image slip; nevertheless, perceptual fusion can be maintained for an appreciable image separation before fusion suddenly "breaks" and the two half-images appear to be partially overlapping instead of a fused unitary image (Diner [**1978**], Fender and Julesz [**1967**]). Again, we have two perceptual states–fused and unfused–in response to the same physical stimulus on the retina.

Path-dependence, multiple stable states, hysteresis. Insofar as we consider only deterministic theories, path-dependence and multiple stable states are equivalent. Different states can be reached only via different paths, and path-dependence implies that the same stimulus will lead to different responses (multiple stable states) depending on the path taken. Hysteresis refers to a particular kind of path-dependence, namely, the tendency of a state to perseverate even after inducing conditions have changed to be more appropriate for some other state. This is, in fact, the kind of path-dependence that is most frequently observed, and the kind that has been the subject of all the examples.

Path-dependence: *Theory*. To begin to describe these phenomena mathematically, we start with one particular example–vergence. Let the vergence angle that would be required to place the two half-images of a stereogram onto exactly

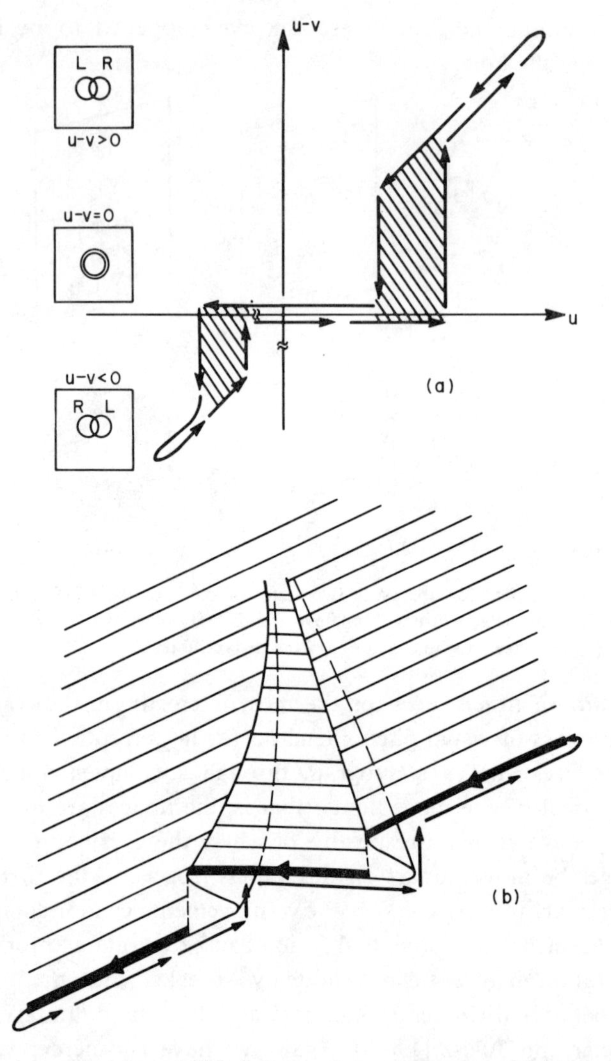

FIGURE 3. Graphical representation of Helmholtz's vergence experiment. (a) Abscissa u represents the stimulus disparity between images to the left and right eyes; v is the eyes' vergence angle, and the ordinate is $u - v$ the perceived disparity, which is shown, as perceived in the inserts at the extreme left. In the inserts, L and R indicate left-eye and right-eye stimuli, respectively. The arrows indicate the sequence of successive events (stimuli presented and the resulting perceived disparity) in an experiment. Right part of graph represents divergence, the left part convergence; the break in the axes indicates that the breakaway vergence angles are different in the two directions. The shaded area represents hysteresis–the difference between perceived disparity depending upon the path taken. (b) A catastrophy theory representation of the experiment in (a). The arrows represent the path of an experiment; heavy arrows, the stimulus converging, and light arrows, the stimulus diverging. See text for details.

corresponding retinal points be u. Let the actual vergence angle between the eyes be v. Then the (perceived) disparity is $u - v$. Helmholtz's experiment can be represented in a graph of $u - v$ versus v, as in Figure 3. We start with the eyes parallel ($v = 0$) and the stimulus positioned on corresponding retinal points

($u - v = 0$), represented at the point (0, 0) in Figure 3(a). As u is increased by pulling the stereo half-images apart (vertically or horizontally), the eyes follow at first: thus the perceived disparity $u - v$ is 0. When the critical angle is reached, the eyes fall back to their parallel position and the perceived disparity equals the actual disparity $u - v$ between the two stereo half-images. For all larger values of u, the perceived disparity $u - v$ remains equal to u since the eyes remain parallel ($v = 0$). When the angle between the stereo half-images is again reduced, another critical angle, smaller than the breakaway angle, is reached where vergence-fusion is again restored. These events are represented by following the arrows around one half of Figure 3(a). As the separation between the stereo half-images is reduced further and becomes negative, a similar sequence of events is produced in the other half of the graph. Thus, if the initial separation (right side of Figure 3(a)) represents divergence, the left side represents convergence. If the right side represents displacement of the left image above the right image, the other side represents the opposite displacement.

Catastrophe theory description. Figure 3(b) represents a catastrophe theory (Thom [**1974**], Susman and Zahler [**1978**]) description of the events of Figure 3(a). The path of Figure 3(a) is traced on the surface of Figure 3(b). When a fold in the surface is reached, a catastrophe, that is, a jump to another level occurs. In this representation, the third dimension (depth into the page) is not essential. It could have an interpretation as the level of illumination upon or as the contrast of the stimulus, with the unfolded rear section of sheet representing subthreshold amounts of illumination or contrast.

Potential theory description. Elegant though catastrophe theory may be, this particular representation of a path-dependent phenomenon within that framework is merely descriptive; it does not yield any insight into the nature of the underlying processes. A theory for the class of phenomena in question is needed. The theory proposed here is a kind of potential theory in which an electrical potential field governs the movement of a charged particle. I will use a gravimetric analog of potential theory, in which a marble rolls on a bumpy surface–the energy surface–under the influence of gravity, because this analogy is particularly concrete and intuitively accessible.[5]

Two kinds of factors designated as internal and external factors control the shape of the energy surface. For example, the basic surface governing horizontal vergence is bowl-shaped, the shape being determined by internal factors. The marble tends to roll to the center of the bowl, representing the tendency of the eyes to verge to a neutral position in the absence of an external stimulus. Specifically, let the displacement vergence energy surface $g(v)$ be a concave-up function of v. Let the movement of the marble along this surface be governed by

[5]In a potential field $e'(x)$, the force acting on a particle is proportional to e'. An energy field $e(x)$ is the integral of a potential field so that the force is proportional to $e' = \partial e(x)/\partial x$. In the gravimetric analog, force $f = -ke'/(1 + e'^2)$, an irrelevant complication, which becomes negligible as $e' \to 0$. For appropriately scaled units, the gravimetric and potential models are equivalent.

a simple differential equation:

$$\frac{d^2v}{dt^2} + k\frac{dv}{dt} = \frac{-dg(v)}{dv}, \qquad k > 0. \tag{1}$$

For concreteness, think of the marble as rolling in a smooth bowl filled with salad oil. The dynamic properties of motion are given by equation (1), but it is evident that so long as there is nonzero friction in the medium, the marble must eventually come to rest at the equilibrium point, the bottom of the bowl. The horizontal projection of the marble's position represents the instantaneous vergence position of the eyes. For convenience, let the value of v at the minimum of $g(v)$ be zero; this point represents the neutral position of eyes in the absence of any visual stimulus, i.e., in the dark. The motion of the marble when it is placed up on the side of the bowl and released represents the vergence movements of the eyes when they are verged on some particular stimulus, say, a book, and the light is suddenly turned off. In the dark, the eyes return to their neutral position (Figure 4).

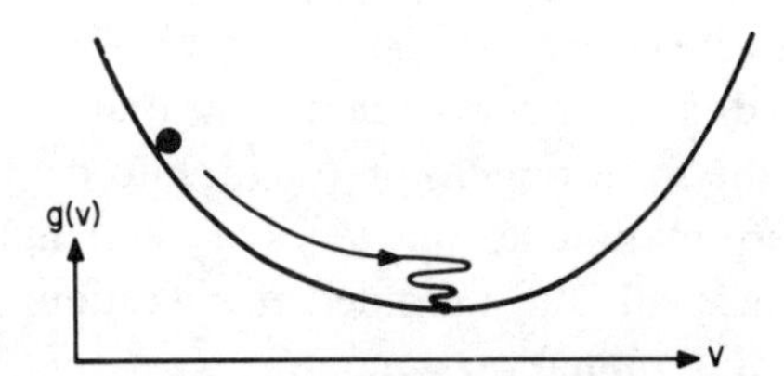

FIGURE 4. Vergence displacement energy $g(v)$ as a function of vergence angle v. The eye's vergence position is represented by the projection onto the abscissa of a marble rolling on the surface. This surface governs the eyes' return to their neutral vergence positon when they are somehow displaced from it; a typical path is indicated.

To show how external factors–the external stimuli impinging on the eyes–can introduce new equilibrium states, we need some definitions. Let the stimulus to the left retina be described as a luminance distribution $l_L(x, y)$ which is a function of x and y, the horizontal and vertical coordinates expressed in units of visual angle. Let the stimulus to the right eye be $l_R(x, y)$. Define $h(v)$, the vergence image-disparity energy as a function of vergence angle v,

$$h(v) = -\int\int_{x,y}\left[l_L\left(x - \frac{v}{2}, y\right) - l_R\left(x + \frac{v}{2}, y\right)\right]^2 dx\, dy. \tag{2}$$

Except for the minus sign, the vergence image-disparity is essentially the unnormalized cross-correlation between the images to the left and right eyes; h has a minus sign so that it has a minimum when the correlation has a maximum (Figure 5). Vergence image-disparity energy $h(v)$ expresses how well the images on the two retinas would match if the eyes were to assume a vergence angle of exactly v.

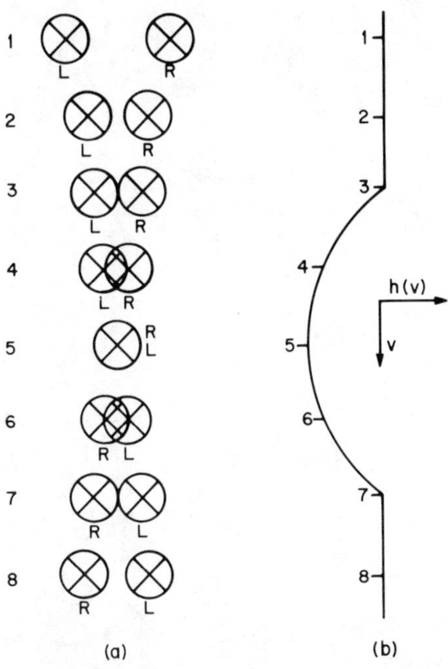

FIGURE 5. Illustration of the computation of vergence image-disparity energy. (a) Luminance distributions l_L and l_R on the left and right retinas as a function of vergence shift v for the stimuli illustrated in row 4. (b) Vergence image-disparity energy $h(v)$ for the stimulus in row 4. The numbers on the graph of $h(v)$ indicate the corresponding row in (a). The axes are placed at 0, 0; that is, v is zero in row 4. The eyes would have to diverge slightly from their initial position at 5 to achieve vergence-fusion of stimulus 4, as indicated by the corresponding minimum in $h(v)$. The figure can be rotated 90 degrees to place the left side on the bottom; in this case, it represents vertical vergence.

The internal and external factors controlling vergence are added to obtain the net vergence energy:

$$e(v; l_L, l_R) = g(v) + h(v; l_L, l_R). \tag{3}$$

Figure 6(a) shows an example of a unistable vergence stimulus. That is, a stereogram is viewed in which the separation of the two half-images is so slight that the eyes inevitably fuse on it. The vergence energy surface corresponding to it has a single minimum.

Now let the stereo half-images be pulled apart. This has the effect of moving the minimum of g and the minimum of h further apart. Sooner or later a point is reached where the two minima are so far apart that when they are added to form $e(v)$, it has two minima (Figure 6(b)). These minima correspond to the two stable equilibrium points for such stereograms. The eyes achieve the off-center equilibrium point when the stereo half-images are drawn apart slowly, all the while maintaining vergence-fusion. This corresponds to drawing the minima of $e(v)$ apart slowly, all the while keeping the marble in the bottom of the moving minimum. If the marble were initially at $v = 0$ when the energy surface was suddenly formed to have two minima, it would remain in the minimum at zero. Thus, there are two equilibrium points for the marble, depending on the sequence of events leading up to the two-minimum surface.

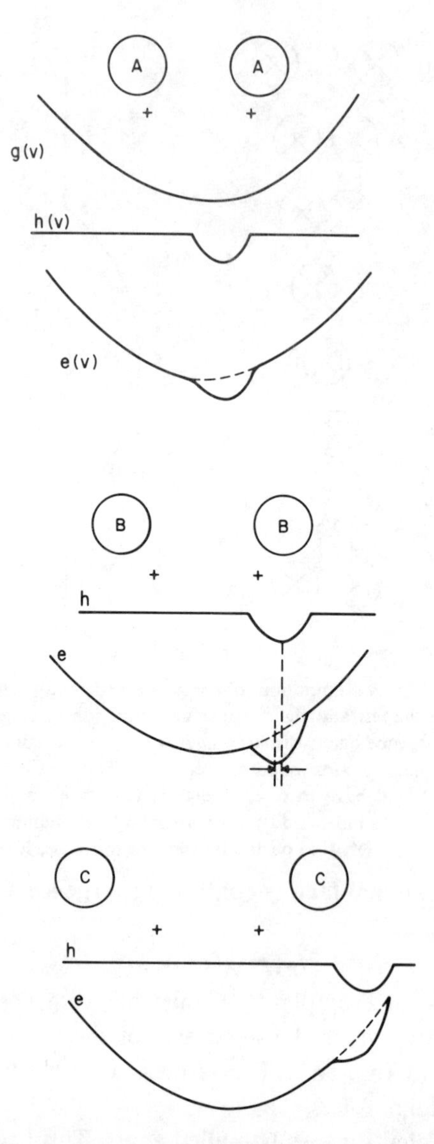

FIGURE 6. Potential theory model. (a) Two halves of a stereogram (A, A), fixation points are indicated by $+$. Vergence displacement energy $g(v)$ adds to vergence image-disparity energy $h(v)$ to produce (net) vergence energy $e(v)$, which has a single minimum slightly displaced in the direction of divergence. (b) Two halves of a stereogram requiring a greater divergence than (a) to achieve fusion. The corresponding vergence energy surface has two minima. The arrows indicate "vergence disparity": the displacement of the minimum of $e(v)$ [which determines the actual vergence position] away from the minimum of $h(v)$ [which minimizes L/R image disparity]. (c) A stereogram which is at the limit of vergence fusion. The $e(v)$ surface has only one minimum corresponding to the neutral vergence position.

Figure 6(c) illustrates what happens when the separation between the stereo half-images is increased further. The minimum produced by $h(v)$ eventually becomes too shallow to hold the marble; beyond this critical separation there is again only one stable equilibrium point.

This qualitative model of multistable path-dependent states is in agreement with the major experimental phenomena (multiple stable states, hysteresis, etc.) described in the examples and summarized in Figure 3. The minima of the energy function $\partial e/\partial v = 0$ correspond to the points on the catastrophe surface in Figure 3(b). The advantage of the potential theory representation is that it provides a mechanism for the system dynamics, not merely a representation of the equilibrium states, which is all that is provided by a catastrophe theory representation.

Vergence disparity. The potential theory representation also provides an explanation for some second order phenomena, such as "vergence disparity". When the eyes are induced to verge at a position other than the neutral point, they do not move away far enough from the neutral point to produce perfect left/right image registration. There is a disparity between the optimum and the actual vergence position (Ogle **[1950]**, Fincham and Walton **[1957]**). In the theory, this "vergence disparity" is represented by the fact that the minimum of $e(v)$ does not lie exactly over the minimum of $h(v)$; because of the concaveness of $g(v)$, the minimum of $e(v)$ is displaced towards the intrinsic neutral point (minimum of $g(v)$). A closely related observation is that detailed, high contrast, large area stimuli are most effective in inducing the eyes to depart from their neutral position and in minimizing "vergence disparity". According to the computation of $h(v)$ in the theory, such stimuli have larger, steeper minima and therefore these minima are less deviated by the addition of $g(v)$ to produce $e(v)$ (reducing "vergence disparity") and their steeper sides preserve minima further from the neutral point.

Optimization theory. The vergence theory can also be thought of as an error theory. There is an error function $h(v)$ that represents the error in image registration as a function of vergence angle v, and an "error" function $g(v)$ that represents the displacement v of the eyes from their preferred position as an error. The equilibrium vergence position is the one that minimizes the summed errors.

A theory for accommodation can be generated according to principles that are quite analogous to those of the theory for vergence. There is an error function [an accommodation displacement function $g(v)$] that represents the deviation v of accommodation from its preferred position as an error. And, there is an error function [image defocus function, $h(v)$] that represents image blur. The equilibrium value of accommodation minimizes the sum of these two "errors". The main phenomena and conclusions about vergence can be transposed directly to accommodation.

When we consider accommodation and vergence together, we observe that where the eyes are induced to accommodate determines where they prefer to verge and vice versa. The interaction between these two systems produces very interesting phenomena and theories to account for them (Sperling **[1970]**). Complex though such theories may be, they are elementary in another sense.

The eyes can be verged at only one angle at one time. They can be focused only at one distance. To reproduce the path-dependent phenomena of vergence and accommodation, a neural model need compute only one value, $\partial h(v_i)/\partial v$ [at $v = v_i$, the eyes' current (instantaneous) vergence or accommodation position] and add this value to a reference signal $\partial g(v)/\partial v$ [at $v = v_i$] to produce the motor control signal. But in a neural system for perceptual fusion, fusion image disparity must be calculated for a range of possible z-depth values $(z_{\min}, z_{\max})$ because for any fixed vergence position, perceptual fusion is possible over a range of depths. Furthermore, at each visual direction x, y, fusion at a different depth z is possible. This leads to an interaction between all points in the x, y plane, instead of an interaction between just two points as in the interaction of vergence and accommodation.

In fact, in one visual direction, only one fusion plane is observable.[6] This is a manifestation of a very general property of perceptual systems; they cannot simultaneously perform two antagonistic responses. Therefore, there is no reason to preserve information that would ultimately lead to antagonistic responses. Some of these kinds of decisions occur early in perceptual processing.

Two illustrative examples are figure/ground interaction (perceiving one part of a scene as figure precludes simultaneously perceiving it as ground) and binocular rivalry (seeing the image from one eye precludes simultaneously seeing a contradictory image from the other eye). Here, we will develop a particular theory of the neural processes that subserve binocular fusion, but the principles are widely applicable.

Neural theory.

Neural binocular field. The neural binocular field (NBF) is a representation inside the organism of the depth and space relations outside, literally, a reflection of the world in the brain of the beholder. The basic theory (Figure 7) usually is attributed to Kepler (Boring [**1933**], Kaufman [**1965**]). The x, y dimensions of the NBF represent, approximately, the dimensions perpendicular to the observer's visual axes; the z dimension represents depth in space. Signals from the left and right retinas enter the NBF at an "angle" so that as z increases, these signals are represented at successively greater x translations relative to each other. The middle z-plane of the field represents the horopter, the surface in space that projects to precisely corresponding retinal points.[7]

---

[6]Sperling [**1970**]. The possible counterexample to the rule of one depth to one point is a partially transparent screen that partially obscures an object behind it. It appears, however, that each surface (screen and object) is represented by a patchwork of many small areas in the visual field; the two kinds of patches do not intersect.

[7]In fact, in the vertical direction, the horopter is not even approximately perpendicular to the visual axes (Helmholtz [**1924**]). Points along the midline in space (horizon) project to the horizontal retinal midline, but points along the vertical midline in space actually project to a retinal line that intersects the vertical retinal midline at an angle of several degrees. However, these practical complications in the precise shape of the horopter are irrelevant for the general principles being considered here, particularly insofar as we restrict our attention to a horizontal meridian.

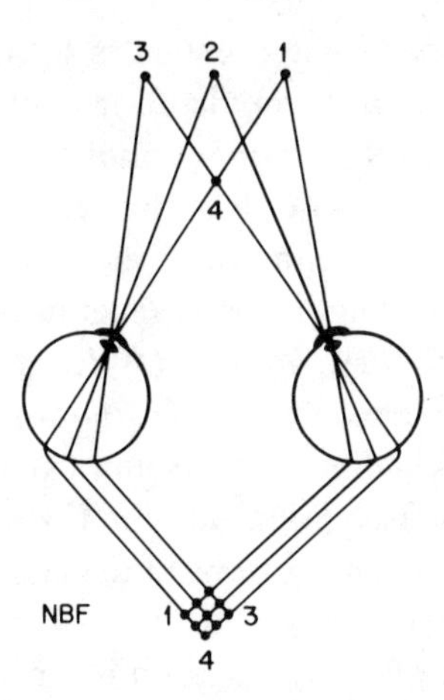

FIGURE 7. Keplerian projection theory. Objects in space, 1, 2, 3, 4, project through the retinas onto a neural binocular field, NBF, that mirrors the external spatial relationships. Intersections of signals from the two eyes (indicated by dots) represent possible matches. Three objects produce nine possible matches in the NBF; only three of these are "correct", and the remainder are "ghosts". For example, it is locally ambiguous whether the correspondence at point 4 in the NBF represents a real object 4 or a ghost match of 3 with 1.

Given a Keplerian NBF, the neural problem of stereoscopic depth perception reduces to three subproblems: (1) detecting correspondences (intersections) in the NBF; (2) determining the "distance" of such correspondences from the midline of the NBF; and (3) utilizing this depth information, i.e., using depth information by concatenating it with pattern information about an object to reach a determination of object shape or identity, or using depth information directly to control vergence movements of the eyes and other motor responses. The present summary focuses on the first of these problems– which is challenging enough. The other problems are treated in Sperling **[1970]** and elsewhere.

The most obvious complication in detecting correspondences in the NBF is that depth information is by its very nature *local* (that is, it is concerned with computing the depth plane of very small areas, x, y), and the information from any small area is inherently ambiguous. For example, if the neural inputs to the NBF carry information simply about whether a small area is lighter or darker than its immediate surround, then the "true" matches will represent a small fraction of all the matches, most of which will be accidental false matches–or ghost matches as they are called. If the inputs to the NBF carry highly specific information, then we are not likely to find the specific information coded in a small area, (so that area's depth cannot be computed). Furthermore, to convey depth in small areas with specialized codes would require an enormous proliferation of inputs so that there would be a reasonable chance of finding not only a highly localized input but also an appropriately-specialized one. The solution to this problem is to use information from the larger context to assist in the local decision–a solution with very interesting mathematical properties.

There is an even more basic principle the visual system seems to have adopted *en route* to its solution of the ghost matching problem. At any given x, y

coordinate (representing one visual direction in the cyclopean field of view) only one surface is perceived and only one depth magnitude is computed. What kind of neural organization in the NBF can accomplish this?

The mathematical neuron. In order that this section be self-sufficient, we begin with the description of a neuron, the basic element of the nervous system, with apologies to the readers for whom this is repetitious. A neuron receives excitatory inputs $[x_i(t)]$ and inhibitory inputs $[z_j(t)]$ from many other neurons, $[i]$, $[j]$. It combines all its inputs (all of which are non-negative) to produce a single non-negative output $y(t)$, which in turn becomes an excitatory and/or inhibitory input to other neurons. The exceptions are the first-order neurons, which receive direct sensory input, and the last neurons in the chain, which produce outputs to muscles, glands, etc.

From a functional point of view, a neuron is a mass of electrically conducting jelly surrounded by a thin membrane which is characterized by high resistance R and a capacitance C. An excitatory input to the neuron consists, essentially, of injections of electrical charge (in the form of charged ions) into the neuron, i.e., an excitatory input is a current $x(t)$.

In modeling a neuron, the excitatory input currents are assumed to add linearly. The accumulated charge decays (exponentially) through the RC membrane, and the output is proportional to the resulting voltage. Thus, to a first approximation, a real neuron is modeled by an RC integrator; its response to an excitatory impulse $\delta(t)$ at time $t = 0$ is simply $e^{-t/(RC)}$.

The most common response of neurons to inhibitory inputs is a proportional increase in membrane conductance $(1/R)$. Thus, inhibition is not a subtractive process–the opposite of excitation; inhibition is modeled as a *shunting* process (Furman [**1965**]) that divides excitation. A nonlinear differential equation to describe these relations is

$$\frac{d}{dt}y(t) + \frac{y(t)}{CR(t)} = X(t) \tag{4a}$$

where

$$R^{-1}(t) = R_0^{-1} + Z(t). \tag{4b}$$

Capital letters X, Z are used to designate the sums over all excitatory and all inhibitory inputs, respectively. This system was proposed and investigated by Sperling and Sondhi [**1968**].

Closer examination of typical real neurons implies complications to the seemingly simple equations (4). Some of these can be incorporated gracefully; others produce intractable complications. Saturation of excitatory inputs is incorporated into (4a) by replacing the right-hand term $x(t)$ with $x(t)(B - Y(t))$. [The resulting systems have been elegantly dealt with by Grossberg [**1973**]. The feedforward shunting inhibition proposed by Sperling and Sondhi [**1968**] yielded a formulation that is virtually equivalent to Grossberg's.] Other complications are: the equilibrium voltage to which inhibition drives the neuron interior generally is somewhat lower than the resting voltage for zero excitatory inputs.

Many neurons have a threshold $\varepsilon > 0$ below which there is no output. This is modeled by replacing $y(t)$ by $y'(t) = \max(0, y(t) - \varepsilon)$. The conducting interior of the neuron has significant resistance, so that–depending on the shape of the neuron–the exact placement of excitatory and inhibitory inputs can have important effects. This effect has been dealt with by computer simulations (Rall [**1964**]).

Functional model. The major psychophysical findings for which we seek a neural theory are: (1) multiple stable states/path-dependence in perceptual fusion (i.e., in fusion with stabilized retinal images); (2) only one perceivable depth at any one time in any one visual direction; (3) context dependency of fusion–a solution for the ghost-ambiguity problem. We propose operations that are carried out on the information in the NBF, and then a neural model that could accomplish these operations. Uniqueness is not claimed either for the operations or the model.

The functional model begins with the assumption that there is a structure like the NBF (Figure 7) receiving spatially labelled inputs, and that there are elements capable of detecting binocular correspondences. The outflow from the NBF is generated by the correspondence-detecting elements, and this outflow determines object perception, stereoscopic depth perception, vergence, etc. The binocular correspondence-detecting elements interact among themselves by competition and by cooperation to determine the outflow.

Competition, cooperation/competition. First consider a thin column in the NBF, occupying a small area $dxdy$ and extending in the z direction. The column is assumed to function so as to never allow more than one z-level to be active at one time ("monoactivity", Sperling [**1970**]). This monoactivity function simply associates with each small visual area one and only one perceived depth. In the neural model, monoactivity is achieved by a purely competitive interaction. The second function is cooperative/competitive. Activity at one level z in one column cooperates with (does not inhibit) activity in adjacent columns at the same level z and inhibits (competes with) everything else. The restriction of activity to one level in a z-depth column is obviously a trivial but perfectly satisfactory way to account for the restriction of perceived fusion to a single depth plane at any x, y point in space.

The cooperative/competitive principle offers a solution to the ghost-ambiguity problem since, on the one hand, it restricts activity to one level of the NBF at any one point, and on the other, it encourages promising areas of activity to expand and to dominate weaker areas. The advantage of a cooperative/competitive principle for the solution to the binocular matching problem has been widely appreciated since the principle was first proposed by Sperling [**1970**]. The principle has subsequently been adopted as the basis of their binocular models by Julesz [**1971**], Dev [**1975**], Nelson [**1975**], and Marr and Poggio [**1976**].

Monoactive neural model: A case of pure competition. A simple neural network to achieve competitive selection of a single neuron in a column is illustrated in

Figure 8. Each neuron i receives an external input $x_i(t)$, produces an output $y_i(t)$, and receives its net inhibitory input Z_i from every other neuron j in the column $Z_i = k \sum_{j \neq i} y_j(t)$.

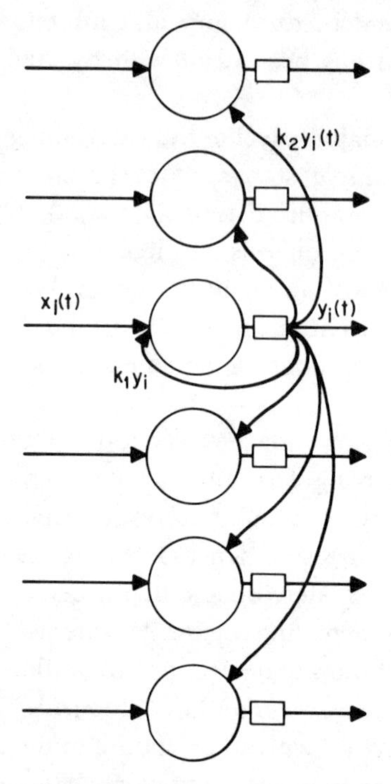

FIGURE 8. A monoactive column of model neurons. Output connections are shown for one neuron, i. It receives an external input $x_i(t)$ that sums with an output-produced feedback excitatory input $k_1 y_i(t)$. It sends shunting inhibitory signals $k_2 y_i(t)$ to all other cells. The small rectangular boxes in the input path represent thresholding operations: when the input is y, $y > 0$, the output is $\max(y - \varepsilon, 0)$.

A neural network with merely recurrent inhibition would be a very tame "filter" until three additions were made: (1) Positive feedback: each neuron feeds back a portion of its output as an excitatory input; (2) Threshold: each neural output y is subjected to a thresholding operation such that $y_t = \max(y - \varepsilon, 0)$; (3) Saturation: each neuron has a maximum output, B. The effect of positive self-feedback is to destabilize the system greatly and to increase the intensity of interactions. If the strength of inhibitory connections k_y is sufficient so that an active neuron with output B can reduce all other outputs to below their threshold, then it alone will have an output and all other neurons will be silent. This situation obtains until another neuron captures the system. The system thus appears to have the desired property of *monoactivity*–only one active neuron at any one time (Sperling [**1970**, Appendix B]).

Although the above argument may sound believable, it was (Grossberg [**1973**], [**1978**]) who first provided an adequate mathematical analysis of competitive systems of this kind (excitatory self-feedback, widespread inhibitory feedback).

Let us re-examine the assumptions. As has been pointed out earlier (in the section entitled *The mathematical neuron*) output saturation is an inherent part of Grossberg's theory so it does not require a special assumption. In order for the system to have the multistable monoactive property, Grossberg found that the form of the inhibitory feedback return function was critical; that is, the amount of signal fed back had to be an S-shaped function of input intensity. [Sperling's output function is zero for inputs below threshold and rises linearly to B when inputs are above the threshold; thus, threshold and saturation create an angular member of the S-function class.]

The point of this discussion is that monoactivity of a column of model neurons can be achieved under a wide variety of conditions and assumptions. Sufficient conditions are: inhibitory feedback onto the other neurons within the monoactive column, excitatory self-feedback, and an S-shaped feedback return function (or output threshold and saturation).

Monoactive systems perform an exceedingly universal and useful function: selecting one from many inputs. For example, in motivational systems, when an animal is hungry, thirsty, and sleepy, and can satisfy these drives at three different locations, it does not walk to the mean location. It selects these activities one at a time. When there are several objects of possible interest in the visual field, the eyes do not point at the blank space between them. They fixate the objects successively. The same is true of vergence movements when objects at different depth planes are present in the field of view.

In the case of binocular fusion, the theory proposes that there is a monoactive system at each x, y column of the NBF. This is the correlate in the NBF of the restriction that depth can be perceived at only one depth at any one visual direction. Another possible example is the ambiguous figure: perceiving one perceptual organization precludes simultaneously perceiving the other possible organizations. In general, perceptual and motor functions that involve categorization are likely to be accomplished by means of a monoactive network.

Double-coupled neural model: *cooperation/competition*. The monoactive neural model essentially made single decisions: which neuron will be dominant at any one time. It had only one kind of interaction and may be considered to be single-coupled. The second neural model uses context to resolve ambiguity. Thus, many neurons are active at every time and the model deals with determining the *combination* of active neurons. It is double-coupled in the sense that it exhibits one kind of interaction with members of its column and another kind between columns.

The basic building block is a monoactive column of many strongly competitive neurons. Such a column represents the z dimension, depth, in the NBF. Many of these columns are densely packed together in the x, y plane, perpendicular to the z axis. The position in the *column* of the one active neuron *column* represents a decision about depth locally. How does this decision affect the decision arrived at in the adjacent columns? The proposal put forward by

Sperling [**1970**] is that a neuron at level z in one column facilitates neurons at and near to level z in adjacent columns, by inhibiting neurons elsewhere in the adjacent columns. In effect, the neuron attempts to influence adjacent columns to respond more-or-less in the same way as its own column. It makes inhibitory feedback connections to neurons at different levels in adjacent columns as in its own column. Lateral excitatory connections between same-level neighbors strengthen the cooperativity beyond that achieved by inhibiting-the-inhibition. But it is necessary that lateral interactions between neighbors be different from vertical interactions within a column. Otherwise, the whole NBF would function like a single column. The lateral connections to adjacent columns are not so strong that they override strong input signals. When inputs are weak or ambiguous, the effect of cooperation/competition interaction is to build "planes of influence" at the z level of the model neurons that happen to have the strongest input signals. When there is a depth-ambiguous area between two unambiguous areas, the boundary between the two planes of influence can be quite unstable. [In stereograms constructed to satisfy these conditions, it is sometimes possible actually to see the rapid sweep of one depth plane over another (Sperling [**1970**]).]

The interaction between adjacent model neurons that represent different depth planes can result in far-reaching effects that travel like a wave. In other words, local cooperation/competition can result in global interactions. This is one of the features that makes cooperation/competition models so interesting and attractive to theorists. While systems like the one proposed here have been simulated, I do not know of a comprehensive mathematical treatment; it seems to be a worthwhile area of investigation.

The cooperation/competition mechanism that serves binocular vision quite probably serves other perceptual functions that involve local/global interactions, such as figure/ground relations and depth perception. Interpreting a portion of a scene as figure causes the area interpreted as figure to spread and to extend its boundaries to the limits of the area interpreted as ground. Similarly, seeing an ambiguous motion stimulus (composed of many points) moving in one direction in one part of a stimulus causes the motion interpretation to be applied to all other parts.

Summary and conclusion. Simple demonstrations of path-dependent perceptual states led to simple, useful descriptive models (potential theory) and to formidable process models (cooperation/competition networks) that still present substantial mathematical challenges.

REFERENCES

Allen, J. B., 1977a. *Two-dimensional cochlear fluid model: New results*, J. Acoust. Soc. Amer. **61**, 110–119.

———, 1977b. *Cochlear micromechanics–a mechanism for transforming mechanical to neural tuning within the cochlea*, J. Acoust. Soc. Amer. **62**, 930 – 939.

Boring, E. G., 1933. *The physical dimensions of consciousness*, Century Co., New York.

Burt, P. and G. Sperling, 1981. *Time, distance, and feature trade-offs in visual apparent motion*, Psych. Rev. **88**, 171–195.

Cowan, T. M., 1974. *The theory of braids and the analysis of impossible figures*, J. Math. Psych. **11**, 190–212.

Dev, P., 1975. *Perception of depth surfaces in random-dot stereograms: a neural model*, Internat. J. Man-Machine Studies **7**, 511–528.

Diner, D. B., 1978. *Hysteresis in human binocular fusion: a second look*, doctoral thesis, California Institute of Technology.

Egan, J. P., 1975. *Signal detection theory and ROC analysis*, Academic Press, New York.

Fender, D. and B. Julesz, 1967. *Extension of Panum's fusional area in binocularly stabilized vision*, J. Optical Soc. Amer. **57**, 819–830.

Fincham, E. F. and J. Walton, 1957. *The reciprocal actions of accommodation and convergence*, J. Physiology (London) **137**, 488–508.

Fry, G. A., 1955. *Blur of the retinal image*, Ohio State University Press, Columbus, Ohio.

Furman, G. G., 1965. *Comparison of models for subtraction and shunting lateral-inhibition in receptor-neuron fields*, Kybernetika **2**, 257–274.

Graham, N., 1981. *The visual system does a crude Fourier analysis of patterns*, these PROCEEDINGS.

Green, D. M. and J. A. Swets, 1966. *Signal detection theory and psychophysics*, Wiley, New York.

Grossberg, S., 1972. *Pattern learning by functional-differential neural networks with arbitrary path weights*, Delay and Functional-Differential Equations and their Applications (K. Schmitt, ed.), Academic Press, New York, pp. 121–160.

______, 1973. *Contour enhancement, short term memory, and constancies in reverberating neural networks*, Stud. Appl. Math. **52**, 217–257.

______, 1978. *Competition, decision, and consensus*, J. Math. Anal. Appl. **66**, 470–493.

von Helmholtz, H. L. F., 1924. *Treatise on physiological optics*, 3rd ed. (J. P. C. Southall, trans.), Optical Soc. Amer., Rochester, N. Y., 1924; reprint, Dover, New York, 1962.

Julesz, B., 1971. *Foundations of cyclopean perception*, Univ. of Chicago Press, Chicago, Ill.

Kaufman, Lloyd, 1965. *Some new stereoscopic phenomena and their implications for the theory of stereopsis*, Amer. J. Psych. **78**, 1–20.

Krantz, D. H., R. D. Luce, P. Suppes and A. Tversky, 1971. *Additive and polynomial representation*, Foundations of Measurement, vol. 1, Academic Press, New York.

Krauskopf, J., 1962. *Light distribution in human retinal image*, J. Optical Soc. Amer. **52**, 1046–1050.

Licklider, J. C. R., 1961. *Basic correlates of the auditory stimulus*, Handbook of Experimental Psychology (S. S. Stevens, ed.), Wiley, New York, pp. 985–1039.

Link, S. W. and R. A. Heath, 1975. *A sequential theory of psychological discrimination*, Psychometrika **40**, 77–195.

Luce, R. D. and D. M. Green, 1972. *A neural timing theory for response times and the psychophysics of intensity*, Psych. Rev. **79**, 14–57.

______, 1974. *Counting and timing mechanisms in auditory discrimination and reaction time*, Contemporary Developments in Mathematical Psychology, vol. 2, Measurement, Psychophysics, and Neural Information Processing (D. H. Krantz, R. C. Atkinson, R. D. Luce and P. Suppes, eds.), Freeman, San Francisco, Calif.

Luneburg, R. K., 1947. *Mathematical analysis of binocular vision*, Princeton Univ. Press, Princeton, N. J.

Marr, D. and T. Poggio, 1976. Science **194**, 283–287.

Navon, D. and D. Gopher, 1979. *On the economy of the human-processing system*, Psych. Rev. **86**, 214–255.

Nelson, J. I., 1975. *Globality and stereoscopic fusion in binocular vision*, J. Theoret. Biol. **49**, 1–88.

Norman, F., 1972. *Markov processes and learning models*, Academic Press, New York.

Ogle, K. N., 1950. *Researches in binocular vision*, W. B. Saunders, Philadephia, Pa.

Pugh, Jr., E. N. and J. D. Mollon, 1979. *A theory of the* Π_1 *and* Π_3 *color mechanisms of Stiles*, Vision Res. **19**, 293–312.

Rall, W., 1964. *Theoretical significance of dendritic trees for neuronal input-output relations*, Neural Theory and Modeling (R. F. Reiss, ed.), Stanford Univ. Press, Stanford, Ca.

Roberts, F. S., 1979. *Encyclopedia of mathematics and its applications*, vol. 7, *Measurement theory*, Addison-Wesley, Reading, Mass.

Schroeder, M. R., 1975. *Models of hearing*, Proc. IEEE, **63**, 1332–1350.

Shepard, Roger N., 1980. *Multidimensional scaling, tree-fitting, and clustering*, Science **210**, 390–398.

Sperling, G., 1970. *Binocular vision: a physical and a neural theory*, Amer. J. Psych. **83**, 461–534.

______, 1978. *The goal of theory in experimental psychology*, unpublished technical memorandum, Bell Laboratories.

Sperling, G. and M. J. Melchner, 1976. *Estimating item and order information*, J. Math. Psych. **13**, 192–213.

______, 1978. *Visual search, visual attention, and the attention operating characteristic*, Attention and Performance. VII (J. Requin, ed.), Erlbaum, Hillsdale, N. J., pp. 675–686.

Sperling, G. and M. M. Sondhi, 1968. *Model for visual luminance discrimination and flicker detection*, J. Optical Soc. Amer. **58**, 1133–1145.

Sussman, H. J. and R. S. Zahler, 1978. *Catastrophe theory as applied to the social and biological sciences: a critique*, Synthese **37**, 117–216.

Thom, T., 1974. *La théorie des catastrophes: état présent et perspectives*, Dynamical Systems (A. Manning, ed.), Lecture Notes in Math., vol. 468, Springer-Verlag, Warwick, pp. 366–372.

Victor, J., R. Shapley and B. Knight, 1977. *Nonlinear analysis of cat retinal ganglion cells in the frequency domain*, Proc. Nat. Acad. Sci. U.S.A. **74**, 3068–3072.

Wyszecki, G. and W. S. Stiles, 1967. *Color Science*, Wiley, New York.

Department of Psychology, New York University, New York, New York 10003

SIAM-AMS Proceedings
Volume **13**
1981

Reaction Time Distributions Predicted by Serial Self-terminating Models of Memory Search[1]

DIRK VORBERG

1. Introduction. The use of reaction time (RT) measurement in the study of cognitive processes is one of the oldest ideas in experimental psychology, but still one of the most active ones in the field of human information processing of today. It is essentially due to the Dutch physiologist F. C. Donders [**1969**]. He advanced the assumption that cognitive processes proceed in stages or *serially*, and proposed to measure the duration of a cognitive process by comparing the average RT's from a task which does and another one which does not require the execution of that process for the solution of the task.

Much of the current popularity of the RT method and also of its theoretical foundations is due to the work of S. Sternberg [**1966**], [**1969**], [**1975**]. As an example, let us consider Sternberg's [**1966**] memory scanning experiment which has since become a classic in the field. In the experiment, he studied how long it takes us to scan the contents of short-term memory.

On a given trial of the experiment, the subject was to memorize a short list of digits which were displayed sequentially. After a brief delay, a test digit was shown, and the subject was to decide as fast as possible whether the digit was one of the digits in memory. For example, the memory list might be (3, 1, 5, 7, 4). On a *positive* test trial, the test digit '7' might be shown; on a *negative* test trial, '8' might be shown as test digit. The subject indicated his decision ('yes' or 'no') by operating one of two response levers. RT was defined as the time from the onset of the test digit to the occurrence of the response. The focus of interest was whether and how RT is influenced by the number of items

1980 *Mathematics Subject Classification.* Primary 92A25.

[1]This work was done during my sabbatical year at the Human Information-Processing Research Department, Bell Laboratories, Murray Hill, N. J. I am grateful to the Bell Labs and Saul Sternberg for the hospitality, and to him, Jan van Santen, and Dave Noreen for many helpful discussions. Thanks are due to Wolfgang Hell for a critical reading of the manuscript.

in short-term memory. Figure 1 shows the results found by Sternberg [**1966**], which have been confirmed since by numerous other studies (see Sternberg [**1975**] for a review of the literature). Mean RT increases linearly with list length, n, which implies that there is a fixed increment in RT as list length changes from n to $n + 1$. Surprisingly, this seems to be true for both types of test trials. To account for his findings, Sternberg proposed a simple model which assumes that the representations of the list elements in short-term memory are compared in a serial fashion, i.e., one by one, to the representation of the test digit. If the mean time per comparison is constant, it follows that RT should increase linearly with list length, n, on negative test trials, with slope equal to the mean comparison time. On positive test trials, however, the number of comparisons required to find the digit that matches the test digit may be less than n since the subject may find the critical digit before the search has exhausted the whole list. In fact, only $(n + 1)/2$ comparisons on the average will be required to find the critical digit on positive trials. If the search through short-term memory were *self-terminating* in this fashion, i.e., were terminated and a response initiated as soon as the critical digit is found, mean RT should increase linearly on positive test trials as well, but with half the slope of the RT function on negative test trials. This prediction clearly fails, as is evident from Figure 1. To account for these results, Sternberg proposed that the search through short-term memory is *exhaustive* rather than self-terminating, i.e., proceeds always to the end of the list, even if the critical element has been found before.

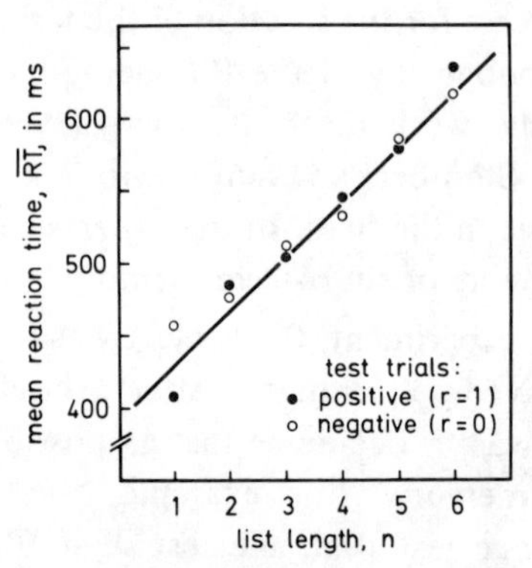

FIGURE 1. Mean memory scanning time as a function of list length, from Sternberg [**1966**].

Sternberg's distinction between exhaustive and self-terminating search processes has become extremely influential, although his proposal that short-term memory scanning is exhaustive has not been universally accepted in spite of his findings. In fact, serial self-terminating (SST) models of RT have become almost ubiquitous in cognitive psychology (see Sternberg [**1973**] for many recent examples of SST models in short-term memory scanning, choice reaction times, semantic memory, recognition of pictures, visual search, etc.). It is easy to see why SST models seem so attractive. They permit one to account for observed changes in mean RT with experimental conditions by assuming that, whereas the durations of the component processes themselves remain unaffected, only the distribution of the required number of components changes with conditions. Particularly intriguing examples of such models to account for a diversity of findings in choice-reaction tasks and memory-scanning have been proposed by Falmagne, Cohen and Dwivedi [**1975**] and Theios and his coworkers (Theios [**1973**], Theios, Smith, Haviland, Traupman and Moy [**1973**]).

With very few exceptions, the analysis and test of SST models has been in terms of mean RTs. Moreover, the predictions used to test the models are often derived under the restrictive assumptions of constant means and stochastic independence of the component processing times. However, Sternberg [**1973**] has shown that strong testable predictions can be derived from SST models under much weaker assumptions, if attention is shifted from the mean RTs to properties of the whole RT distribution. The purpose of this paper is to present theoretical results which generalize and extend Sternberg's method to an experimental paradigm that I shall call (n, r)-experiment. These results provide powerful tools for testing whole classes of SST models without invoking specific distributional assumptions. The basic idea behind Sternberg's method and its generalizations is to exploit the mixture properties that the RT distribution must have if it is to be accounted for by an SST model: Since RT is assumed to be determined by the nonobservable number of operations required before the search is terminated, the RT distribution can be represented by a mixture of conditional distributions which depend on the actual number of search steps, but ought to be independent of the maximum number of steps which would be needed for an exhaustive search. Before I proceed, let me illustrate what I mean by an (n, r)-experiment. Consider the following experiment reported by Bamber [**1969**]:

On a trial, subjects were shown two letter strings and asked to decide whether these were the same or different. A trial began with the presentation of a stimulus card which contained from one to four consonants in a row. After inspection of the letter string the subject initiated the tachistoscopic presentation of a test string containing the same number of consonants. On 'same' test trials, the letters were identical to those shown before and in the same order; on 'different' test trials, the n-letter test string contained r letters which were different from the n letters of the first string. 'TBLN' – 'TBLN' is an example of the strings displayed on a 'same' trial, and 'RVD' – 'RTD' illustrates a 'different' trial with one letter different. Subjects responded 'same' or 'different' by pressing one of two buttons; RT was measured from the onset of the test string. Of particular interest is how RT changes with the length of the letter string, n, and the number of differences, $r = 0, 1, \ldots, n$. Figure 2 shows mean RTs on 'different' trials as a function of n and r.

As will be described below, the results of Bamber's experiment agree nicely with the predictions of an SST model; the theoretical predictions are given by the unfilled dots labeled 'predicted $E_{rn}(\text{RT})$'.

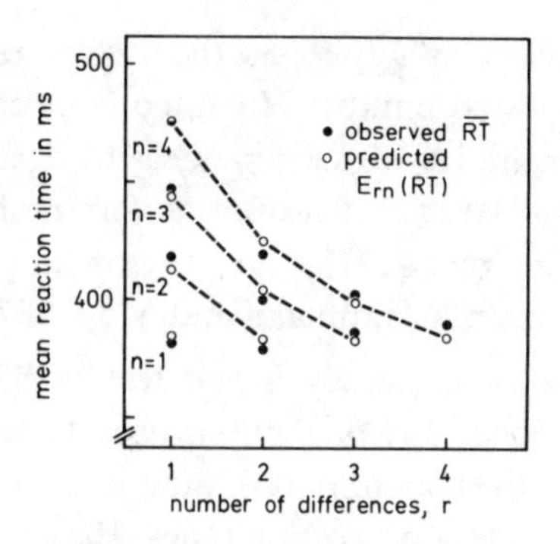

FIGURE 2. Mean reaction time in same-different comparisons as a function of list length, n, and number of differences, r from Bamber [**1969**].

The outline of the paper is as follows. In the next section, I shall specify the assumptions which define an SST model for an (n, r)-experiment. Then I shall present a representation theorem for the average RT distribution of any SST model, from which testable predictions will be derived for three families of (n, r)-experiments. The final section sketches a few experimental results bearing on the predictions. As a further extension of the method, a generalized representation theorem will be given which holds under less restrictive termination assumptions.

It should be noted that the representation theorem (equation (2)) and the distributional inequality (8) were first derived by Sternberg [**1973**] for the special case of $(n, 1)$-experiments. His paper also contains valuable suggestions concerning statistical tests of the predictions as well as ingenious applications of the distributional inequality to various SST models proposed in the literature.

2. Definitions and assumptions. An (n, r)-experiment is an experimental task in which the subject is presented with a list of n items and is to give a response as quickly as possible which depends on some r critical items or *targets* embedded in the list.

EXAMPLES. In Sternberg's [**1966**] memory scanning experiments, the list to be scanned contains either one target ($r = 1$) on positive trials, namely, that the item which matches the test item, or none ($r = 0$) on negative trials. In Bamber's [**1969**] same-different experiment, the targets are the differences between paired letters in the two strings to be compared; 'same' trials correspond to $(n, 0)$-experiments, and 'different' trials to (n, r)-experiments, where $1 \leqslant r \leqslant n$. Note that, logically at least, both tasks are solved if only one target has been identified, even if $r > 1$. This is the assumption I shall concentrate upon in the following; the final section discusses an extension to situations where the subject is to base his response on $m > 1$ out of r targets.

In the following, I shall investigate some general predictions about the reaction time distributions from (n, r)-experiments with $r > 0$ which must hold if the underlying search for a target is serial and self-terminating (SST). To recapitulate, SST models are characterized by two general assumptions. First, information processing is assumed to proceed serially in the sense that at most one cognitive operation is performed at any point in time. Moreover, each operation is performed at most once, which means that the information processing system keeps track of what has already been done. Second, processing is assumed to be self-terminating in the sense that no further items are processed after the first target item is encountered.

No assumptions will be made about the order in which the serial processes are performed; in particular, it is not assumed that the same, fixed processing order occurs on every trial, nor that all processing orders are equally likely. One of the remarkable properties of SST models is that they permit quite general predictions which must hold regardless of processing order, and can thus be tested even if the experimenter does not know whether subjects use processing strategies which lead to pronounced preferences for some orders.

Let $\mathbf{x}_n \equiv (x_1, \ldots, x_n)$, and $\mathrm{Perm}(\mathbf{x}_n) = \mathrm{Perm}(x_1, \ldots, x_n)$ denote the set of all permutations of the vector $\mathbf{x}_n$; in particular, I use $\mathrm{Perm}(1, \ldots, n)$ for the permutations of the integers 1 through n. Let $J \equiv \{\mathbf{j}_r | 1 \leqslant j_1 < \cdots < j_r \leqslant n\}$ be the set of all ordered r-tuples of integers between 1 and n; usually, $\mathbf{j} \in J$ will denote the positions of the r targets in the input list on a given trial. In order to account for a particular search strategy used by a subject leading to preferences for some processing orders, I introduce the integer-valued random variables $\{S_i\}$, where $S_i = s$ if the item from input position i, $i = 1, \ldots, n$, is processed sth. It is helpful (although by no means the only possible interpretation of information processing in SST models; see Sternberg **[1973]**) to imagine the n list elements as placed in a stack memory before processing proper begins, where the elements in the stack are examined one by one and the search for a target starts at the top of the stack. Under this interpretation, S_i gives the stack position of the item that occurred in position i of the input list. Any particular search strategy then gives rise to a joint probability distribution of the $\{S_i\}$, denoted by $P_{\mathbf{j}rn}(S_1 = s_1, \ldots, S_n = s_n) = P_{\mathbf{j}rn}(\mathbf{S}_n = \mathbf{s}_n)$, where the indices are used to indicate that the subject's strategy may change as the parameters of the task are varied.

To illustrate these definitions, consider the following example. On a trial of a (5, 3)-experiment, the experimenter presents the word list {job, plate, table, truck, rabbit}. After a short delay, the subject is asked to name as quickly as possible a word from the list which contains 'b' as the third letter. Here, {job, table, rabbit} are the three target items; since they were presented as the first, third and fifth words in the list, $\mathbf{j} = (1, 3, 5)$. Figure 3 shows a hypothetical arrangement of the five items[2] in the memory stack, and illustrates the definitions of the random variables $\{S_1, \ldots, S_5\}$. In the following, interest will focus

[2]Of course, it is not the items themselves that are stored in short-term memory but some unspecified representations of them. For simplicity, however, I use the more direct and less precise language.

on the stack positions of the targets $\{S_{j_1}, \ldots, S_{j_r}\}$, and, in particular, on the highest (i.e., lowest numbered) target in the stack, $S^* = \min(S_{j_1}, \ldots, S_{j_r})$.

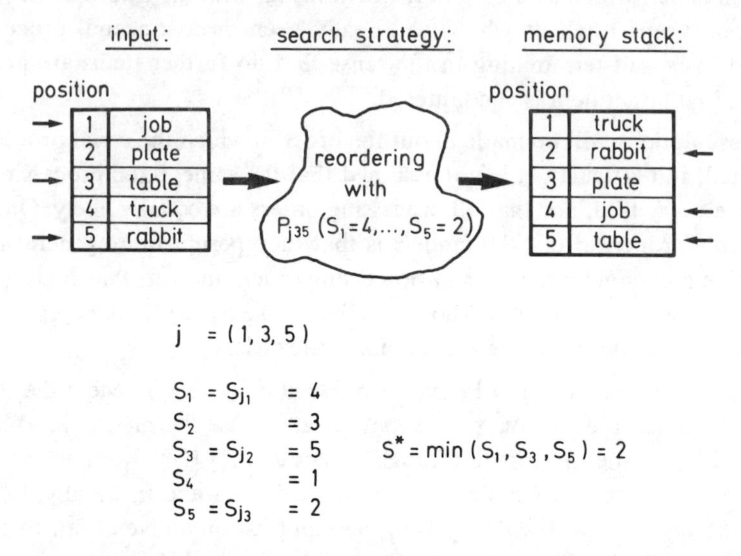

FIGURE 3. Hypothetical trial of a memory search experiment. (See text for explanation.)

Let me now try to make precise what defines an SST model for (n, r)-experiments. Essentially, each trial described in terms of list length, n, number of targets, r, and target input positions, $\mathbf{j} \in J$, is considered as a separate random experiment for which probability distributions on $\mathbf{S}_n$ and reaction time, RT, are defined. The following assumptions specify how these distributions depend on $\mathbf{j}$, r, and n, and thus permit us to relate the resulting RT distributions across experimental conditions.

(A1) *Representation of items in the stack.* For every $\mathbf{j} \in J$,

$$\sum_{\mathbf{s} \in \mathrm{Perm}(1, \ldots, n)} P_{\mathbf{j}rn}(\mathbf{S}_n = \mathbf{s}) = 1.$$

This assumption implies that, on any trial of an (n, r)-experiment, the memory stack contains n items, and each item is checked exactly once if all n stack positions are examined, whatever the order $\mathbf{S}_n$ of the representations in the stack.

(A2) *Independence of search order and target positions.* For every $\mathbf{j}, \mathbf{j}' \in J$ and $\mathbf{s} \in \mathrm{Perm}(1, \ldots, n)$,

$$P_{\mathbf{j}rn}(\mathbf{S}_n = \mathbf{s}) = P_{\mathbf{j}'rn}(\mathbf{S}_n = \mathbf{s}) \equiv P_{rn}(\mathbf{S}_n = \mathbf{s}).$$

This assumption is motivated by the fact that the identity of an item (i.e., whether it is a target or not) cannot have an effect before it is processed. Basically, the assumption implies that the order in which the elements are processed is decided upon before processing starts. (One might consider exceptions to this assumption where the detection of a target influences the processing order of the remaining items. However, by assumption (A3), this will not lead to distinguishable predictions.)

(A3) *Self-termination.* For every $\mathbf{j} \in J$, let $S^* = \min(S_{j_1}, \ldots, S_{j_r})$; then, for all $t \geqslant 0$,

(i) $P_{\mathbf{j}rn}(\mathrm{RT} \leqslant t | S_1, \ldots, S_n) = P_{\mathbf{j}rn}(\mathrm{RT} \leqslant t | S^*)$, and

(ii) $P_{\mathbf{j}rn}(\mathrm{RT} \leqslant t | S^*) = P_{rn}(\mathrm{RT} \leqslant t | S^*)$.

Part (i) makes precise the notion of a self-terminated serial search: The reaction time distribution depends only on the position of the highest target in the stack, and not on the total search order, $\mathbf{S}_n$. Of course, this is the case if the search is terminated as soon as the first target has been found. Part (ii) is analogous to (A2); it says that RT depends on the input positions of the r targets, $\mathbf{j}$, only through the highest position of any target in the stack.

To simplify the notation, I introduce

$$F_{k,rn}(t) \equiv P_{rn}(\mathrm{RT} \leqslant t | S^* = k).$$

$F_{k,rn}(t)$ gives the conditional probability of observing a RT not exceeding t on a trial of an (n, r)-experiment when the first target is found at the kth step, i.e., in stack position k.

(A4) *Serial processing.* For all $1 \leqslant k \leqslant n - r$ and $t \geqslant 0$,

$$F_{k,rn}(t) \geqslant F_{k+1,rn}(t).$$

The assumption states that completing $k + 1$ operations before time t cannot be more likely than finishing only k of them. This is certainly the case if processing is serial. In fact, the usual notion of serial processing which decomposes RT into a sum on nonnegative components, for example, $\mathrm{RT} = U_1 + \cdots + U_k + B$ if $S^* = k$, can easily be shown to imply (A4). For the purpose of this paper, however, the much weaker and more general notion of serial processing stated by (A4) is sufficient; note that (A4) does not entail any assumptions about how the conditional RT distributions depend on the number of search steps, S^*.

(A5) *Dependence on experimental conditions.* For $1 \leqslant k \leqslant n - r + 1$ and $t \geqslant 0$,

$$F_{k,rn}(t) = \begin{cases} F_k(t) & \text{(assumption (A5))}, \\ F_{k,r}(t) & \text{(assumption (A5'))}, \\ F_{k,n}(t) & \text{(assumption (A5''))}. \end{cases}$$

These three alternative assumptions specify how the RT distributions conditional on the number of steps, k, needed to detect the first target depend on the total number of items, n, and the number of targets in the list, r. Under the strongest assumption, (A5), which implies both (A5′) and (A5″), RT is conditionally independent of n and r given the number of search steps, $S^* = k$. The weaker version (A5′) assumes that RT is independent of list length but may be affected by the number of targets in the list. This may be the case under conditions where memory load, i.e., the total number of items to be stored in the stack, can be neglected; if the number of targets determines the test format, the time to encode the test and to prepare the search is likely to depend on r. Assumption (A5″) takes memory load effects into account but assumes no influence of the number of targets on the conditional RT.

3. Mixture representation of the average RT distribution. In this section, I shall show that for any SST model defined by (A1), (A2), and (A3), the average RT distribution can be written as a probability mixture of $(n - r + 1)$ conditional RT distributions, $\{F_{k,rn}(t)\}$. Intuitively, this is not surprising, since it can take any number of steps between 1 and $n - r + 1$ to locate the first of the r targets among the total n elements, and RT is assumed to depend at most on the number of steps, S^*, and on (n, r). However, it is remarkable that the mixing weights of the $F_{k,rn}(t)$ are completely determined for given (n, r). This leads to testable constraints on the RT distributions for different (n, r)-conditions. By examining the relations between the average RT distributions over a set of (n, r)-experiments, it can then be tested whether any model that is a member of the SST class can account for the data.

For the proof of the mixture representation of the average RT distribution, the following simple lemma is useful. Consider the stack positions of the r targets on a trial with target input positions $\mathbf{j} \in J$. For all target stack positions $\mathbf{s}_r = (s_1, s_2, \ldots, s_r) \in J$, the probability $P_{\mathbf{j}rn}(S_{j_1} = s_1, \ldots, S_{j_r} = s_r)$ is obtained from $P_{\mathbf{j}rn}(\mathbf{S}_n = \mathbf{s}_n)$ as the marginal distribution. Now keep $\mathbf{s}_r$ fixed. Then the following lemma holds:

LEMMA 1. *For any* $\mathbf{j}$ *and* $\mathbf{s}_r \in J$,

$$\sum_{\mathbf{k} \in J} \sum_{\mathbf{i} \in \mathrm{Perm}(\mathbf{k})} P_{\mathbf{j}rn}(S_{i_1} = s_1, \ldots, S_{i_r} = s_r) = 1. \tag{1}$$

This is easily verified by noting that the double sum represents the probability of the event that any r of the n items will be in the stack in positions $\mathbf{s}_r = (s_1, \ldots, s_r)$; however, by the assumptions, this is the certain event.

Consider the RT distribution for (n, r) that is obtained when an equally weighted average is taken over all possible input positions of the targets, $\mathbf{j} \in J$. Define

$$G_{rn}(t) \equiv \binom{n}{r}^{-1} \sum_{\mathbf{j} \in J} P_{\mathbf{j}rn}(\mathrm{RT} < t).$$

This average gives equal weight to all $\binom{n}{r}$ possible target input positions; if the targets are randomly placed in the list, $G_{rn}(t)$ corresponds just to the usual RT distribution which disregards serial position effects. If the experimenter selects target positions $\mathbf{j}$ with unequal probabilities, then $G_{rn}(t)$ corresponds to the average of the RT distributions conditional upon $\mathbf{j}$.

THEOREM 1. *For any SST model obeying* (A1), (A2), *and* (A3), *the average RT distribution is*

$$G_{rn}(t) = \sum_{k=1}^{n-r+1} \frac{\binom{n-k}{r-1}}{\binom{n}{r}} F_{k,rn}(t). \tag{2}$$

PROOF. By definition,

$$\binom{n}{r} G_{rn}(t) = \sum_{\mathbf{j} \in J} P_{\mathbf{j}rn}(\mathrm{RT} \leqslant t),$$

which, conditionalizing on the value of $S^* = \min(S_{j_1}, \ldots, S_{j_r})$, can be written as

$$\binom{n}{r} G_{rn}(t) = \sum_{\mathbf{j} \in J} \sum_{k=1}^{n-r+1} P_{\mathbf{j}rn}(\mathrm{RT} \leqslant t \mid S^* = k) P_{\mathbf{j}rn}(S^* = k)$$

$$= \sum_{k=1}^{n-r+1} F_{k,rn}(t) \sum_{\mathbf{j} \in J} P_{\mathbf{j}rn}(S^* = k)$$

by assumption (A3). For any target input positions $\mathbf{j} \in J$,

$$P_{\mathbf{j}rn}(S^* = k) = \sum_{\mathbf{i} \in \mathrm{Perm}(j)} \sum_{k < k_2 < \cdots < k_r \leqslant n} P_{\mathbf{j}rn}(S_{i_1} = k, S_{i_2} = k_2, \ldots, S_{i_r} = k_r).$$

After reordering the sums, this leads to

$$\binom{n}{r} G_{rn}(t) =$$

$$\sum_{k=1}^{n-r+1} F_{k,rn}(t) \sum_{k < k_2 < \cdots < k_r \leqslant n} \sum_{\mathbf{j} \in J} \sum_{\mathbf{i} \in \mathrm{Perm}(\mathbf{j})} P_{\mathbf{j}rn}(S_{i_1} = k, S_{i_2} = k_2, \ldots, S_{i_r} = k_r).$$

By the lemma, the rightmost two sums equal 1 for $1 \leqslant k \leqslant n - r + 1$ and all $k < k_2 < \cdots < k_r \leqslant n$. Since there are $\binom{n-k}{r-1}$ ways of selecting the ordered integers $(k_2, \ldots, k_r)$ from $k + 1$ through n, the theorem follows.

4. Properties of the mixing coefficients. As mentioned before, the usefulness of the mixture representation of the average RT distribution lies in the fact that the mixing weights or coefficients are uniquely determined by the list length, n, and the number of targets, r. In order to exploit Theorem 1 for the comparison of the average RT distributions across conditions, I derive some properties of the coefficients.

To simplify the notation, define

$$p_{k,rn} \equiv \binom{n-k}{r-1} \Big/ \binom{n}{r},$$

and $q_{k,rn} \equiv \sum_{i=1}^{k} p_{i,rn}$. Note that the weights $p_{k,rn}$ describe a probability distribution on the integers 1 through $n - r + 1$ (since $\sum_{k=1}^{n-r+1} p_{k,rn} = 1$), and that $q_{k,rn}$ gives its distribution function. The following interpretation provides some intuition for this fact:

Consider the case where the r targets and the $(n - r)$ nontarget items are placed at random into the n-slot memory stack. Distinguishing representations only with regard to whether they correspond to targets or nontargets, there are $\binom{n}{r}$ distinguishable stack arrangements. As before, let S^* be the position of the highest target in the stack. Obviously, S^* takes values between 1 and $n - r + 1$. The probability that the highest target is in stack position k, $P_{rn}(S^* = k)$, equals $p_{k,rn} = \binom{n-k}{r-1}/\binom{n}{r}$, since there are $\binom{n-k}{r-1}$ distinguishable stack arrangements which place $k - 1$ nontargets in the highest $k - 1$ positions, one target in position k, and distribute the remaining $r - 1$ targets among the lowest $n - k$ positions. Thus, Theorem 1 shows that, for any strategy of ordering the input items in the stack, the probability distribution for the highest target in the stack, when averaged across all possible target input positions, is the same as when items are placed randomly. Note in passing that the expected position of the highest target, which equals the expected number of comparisons until the first target is found, is

$$E_{rn}(S^*) = \frac{n+1}{r+1}. \tag{3}$$

The following properties, which will be needed for the applications of Theorem 1, are stated without proof. They are easily verified by inserting the definitions of $p_{k,rn}$ and $q_{k,rn}$ and by noting that $q_{k,rn} = 1 - \binom{n-k}{r}/\binom{n}{r}$.

LEMMA 2. *For all* k, $1 \leqslant k \leqslant n - r + 1$ *and* $r > 0$,

(i) $$p_{k,rn} \leqslant ((n+1)/n) p_{k,r(n+1)}, \tag{4}$$

(ii) $$q_{k,rn} \geqslant q_{k,r(n+1)}, \tag{5}$$

(iii) $$p_{k,rn} \geqslant (r/(r+1)) p_{k,(r+1)n}, \tag{6}$$

(iv) $$q_{k,rn} \leqslant q_{k,(r+1)n}. \tag{7}$$

5. Comparing RT distributions across (n, r)**-conditions.** To prove the mixture representation of the average RT distribution for a given (n, r)-condition, only assumptions (A1), (A2), and (A3) were needed. However, in order to compare the average RT distributions from experiments in which n, r, or both, vary we need assumptions (A4) and either (A5), (A5′), or (A5″) which permit us to relate the mixing distributions, i.e., the RT distributions conditional on the position of the highest target in the stack, across experimental conditions. We consider three cases separaetely.

5.1. *Experiments with fixed number of targets*. Consider experiments in which we investigate the effect of list length, n, on RT while keeping the number of targets in the list, r, constant. The positive trials from Sternberg's memory scanning **[1966]** experiment described above provide an example. Assume that the number of elements to be held in the stack does not, per se, influence RT, but only via the number of steps to locate the highest target, S^* (assumption (A5′)). From this and (A4) we derive the following testable relation which the average RT distributions have to obey in order for an SST model to hold:

THEOREM 2. *For an SST model defined by* (A1) *through* (A4) *and* (A5′) *the average RT distributions are related by*

$$G_{r(n+1)}(t) \leqslant G_{rn}(t) \leqslant ((n+1)/n) G_{r(n+1)}(t) \tag{8}$$

for all $t \geqslant 0$.

PROOF. For convenience, we drop the argument t.

(i)

$$
\begin{aligned}
G_{rn} &= \sum_{k=1}^{n-r+1} p_{k,rn} F_{k,r} \quad \text{(by Theorem 1 and (A5}'))\\
&\leqslant \sum_{k=1}^{n-r+1} \frac{n+1}{n} p_{k,r(n+1)} F_{k,r} \quad \text{(by Lemma 2(i))}\\
&\leqslant \sum_{k=1}^{n-r+1} \frac{n+1}{n} p_{k,r(n+1)} F_{k,r} + \frac{n+1}{n} p_{n-r+2,r(n+1)} F_{n-r+2,r}\\
&= \frac{n+1}{n} G_{r(n+1)}.
\end{aligned}
$$

(ii) Noting that $p_{k,rn} = q_{k,rn} - q_{k-1,rn}$ for $k > 1$ and $p_{1,rn} = q_{1,rn}$, we have by (A5$'$):

$$
\begin{aligned}
G_{rn} &= q_{1,rn} F_{1,r} + \sum_{k=2}^{n-r+1} (q_{k,rn} - q_{k-1,rn}) F_{k,r}\\
&= \sum_{k=1}^{n-r} q_{k,rn}(F_{k,r} - F_{k+1,r}) + F_{n-r+1,r} \quad (\text{since } q_{n-r+1,rn} = 1),\\
&\geqslant \sum_{k=1}^{n-r} q_{k,r(n+1)}(F_{k,r} - F_{k+1,r}) + F_{n-r+1,r} \quad \text{(by Lemma 2(ii) and (A4))}\\
&\geqslant \sum_{k=1}^{n-r} q_{k,r(n+1)}(F_{k,r} - F_{k+1,r}) + q_{n-r+1,r(n+1)}(F_{n-r+1,r} - F_{n-r+2,r})\\
&\qquad + F_{n-r+2,r} \quad (\text{since } F_{n-r+1,r} \geqslant F_{n-r+2,r} \text{ by (A4)})\\
&= G_{r(n+1)}.
\end{aligned}
$$

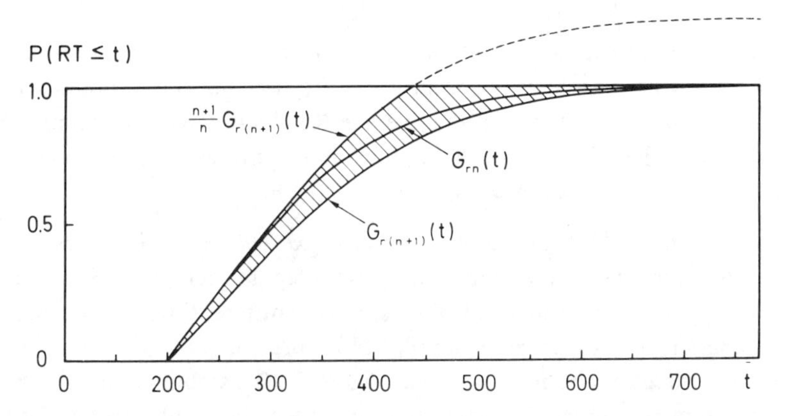

FIGURE 4. Average RT distributions from (n, r) and $(n + 1, r)$ conditions which satisfy inequality (8).

Theorem 2 generalizes the result first obtained by Sternberg [**1973**] for the special case of one target, $r = 1$. He termed the right and left part of the inequality the 'short RT' and the 'long RT' property, respectively, which must hold for neighboring values of list length in order for an SST model to be consistent with data. Figure 4 illustrates these properties and how they can be used to test the model on RT distributions obtained with list lengths n and $n + 1$ and a fixed number of targets.

By Theorem 2, the average RT distribution function observed under condition (n, r) must be bounded, for all $t \geqslant 0$, from below by that from the corresponding $(n + 1, r)$-experiment, and from above by $(n + 1)/n$ times the lower bound. Consider the proportion, p, of 'short' RT's obtained for list length n, say, $\mathrm{RT} \leqslant t_0$. Theorem 2 says that, not surprisingly, this should exceed the corresponding proportion observed with the longer list, $n + 1$; however, the 'short RT' property says that the proportion of short RT's for n items, p, must not exceed that for $n + 1$, $p' = G_{r(n+1)}(t)$, by too much, since $p = G_{rn}(t)$ can never be larger than $(n + 1)p'/n$. An equivalent way of expressing this property is that the distribution functions for n and $n + 1$ must start from the same point rather than be shifted with respect to each other. Note that the 'short RT' property does not depend on the ordering of the conditional distributions, $F_{k,r}(t)$.

The 'long RT' property says that the proportion of RT's longer than any given value t which is obtained for list length n can never be larger than that for $n + 1$. The 'long RT' property reflects the assumed stochastic ordering of the mixing distributions, $F_{k,r}(t)$. Obviously, the average distributions functions cannot cross each other since the $F_{k,r}(t)$ are uncrossed by assumption.

Note in passing that the inequality on the average distribution functions, $G_{rn}(t)$, can be expressed equivalently as an inequality in terms of the quantile functions, $G_{rn}^{-1}(t)$, which give RT as a function of cumulative probability:

$$G_{r(n+1)}^{-1}\left(\frac{n}{n+1}p\right) \leqslant G_{rn}^{-1}(p) \leqslant G_{r(n+1)}^{-1}(p), \qquad 0 \leqslant p \leqslant 1. \tag{9}$$

Sternberg [1973] has made use of this quantile inequality in empirical tests of SST models. Note also that Theorem 2 can be used to derive bounds on the relations that must hold between the average RT distributions for nonneighboring values of list length. By repeated application of the theorem, it is easily seen that $G_{r(n+m)}(t) \leqslant G_{rn}(t) \leqslant ((n + m)/n)G_{r(n+m)}(t)$, $m > 0$.

5.2. *Fixed list length (n, r)-experiments with variable number of targets*. Now consider the converse experiment where we keep list length, n, constant but systematically vary the number of targets, r. The 'different' trials from Bamber's [1969] experiment provide an example. Not surprisingly, mean RT becomes shorter as r increases from 1 to n (see Figure 2). This change in RT with the number of targets is predicted not only in the mean, but, as before, in the whole average RT distribution. There are bounds, however, which limit by how much RT can vary with r if an SST model holds.

For this situation, we replace assumption (A5′) by (A5″), which says that the RT distributions conditional on the highest target stack position do not depend on r, whereas they may depend on list length, n. Then we have the following

THEOREM 3. *For an SST model defined by* (A1)–(A4) *and* (A5″), *the average RT distribution functions are related by*

$$G_{rn}(t) \leqslant G_{(r+1)n}(t) \leqslant \frac{r+1}{r} G_{rn}(t) \quad \textit{for all } t \geqslant 0. \tag{10}$$

The proof is analogous to that of Theorem 2. The right side of the inequality follows from Lemma 2(iii) and (A5″), whereas the left side follows from the stochastic ordering of the conditional distributions (A4) and Lemma 2(iv).

As before, Theorem 3 can be used to test the model by checking whether the obtained average RT distributions for r and $r + 1$ exhibit both the 'short RT' and the 'long RT' property.

5.3. *(n, r)-experiments where both n and r are varied.* If we can justify the strong assumption (A5) that the conditional RT distributions, $\{F_{k,rn}(t)\}$, remain invariant as both list length, n, and number of targets, r, are varied, then a very powerful prediction for the relation between the average RT distributions can be generated. Compare experiments with list lengths n and $n + 1$; the theorem below says that the distribution for $(n + 1, r + 1)$ can be determined completely from those obtained under conditions (n, r) and $(n, r + 1)$. Note that the stochastic ordering assumption (A4) is not required any longer.

THEOREM 4. *For an SST model defined by* (A1)–(A3) *and* (A5) *the average RT distributions are related by*

$$G_{(r+1)(n+1)}(t) = \frac{r+1}{n+1} G_{rn}(t) + \frac{n-r}{n+1} G_{(r+1)n}(t) \tag{11}$$

for all $t \geqslant 0$.

PROOF. For convenience, drop the argument t. By Theorem 1 and (A5), we have

$$\begin{aligned}
\binom{n}{r} G_{rn} + \binom{n}{r+1} G_{(r+1)n} &= \sum_{k=1}^{n-r+1} \binom{n-k}{r-1} F_k + \sum_{k=1}^{n-r} \binom{n-k}{r} F_k \\
&= \sum_{k=1}^{n-r} \left[\binom{n-k}{r-1} + \binom{n-k}{r} \right] F_k + F_{n-r+1} \\
&= \sum_{k=1}^{n-r} \binom{n+1-k}{r} F_k + F_{n-r+1} \\
&= \binom{n+1}{r+1} G_{(r+1)(n+1)}.
\end{aligned}$$

Dividing both sides by $\binom{n+1}{r+1}$ gives the proposition. Obviously, the relation (11) between the average RT distributions reflects the well-known difference equation satisfied by binomial coefficients

$$\binom{n}{r} + \binom{n}{r+1} = \binom{n+1}{r+1}.$$

It is therefore not surprising that additional relations between the average RT distributions can be obtained which correspond to other properties of binomial coefficients. As an example, consider the following theorem which permits us to express the distribution for any (n, r) condition as a mixture of the $(n, 1)$ distributions:

THEOREM 5. *Under the conditions of Theorem 4, the average RT distribution for* $(n, r), r > 1$, *obeys*

$$G_{rn}(t) = \sum_{k=1}^{n-r+1} k \frac{\binom{n-k-1}{r-2}}{\binom{n}{r}} G_{1k}(t). \tag{12}$$

PROOF. The theorem follows as the solution of the difference equation (11) and the boundary condition $G_{nn}(t) = F_1(t) = G_{11}(t)$.

The mixture representation given by (12) again has a simple combinatorial interpretation in terms of random placement of targets and nontargets in the memory stack. With probability $k\binom{n-k-1}{r-2}/\binom{n}{r}$, the first k stack positions contain exactly one target (in any of the k possible positions) and the second target is in position $k + 1$, giving rise to the same average RT distribution as a $(k, 1)$-experiment.

Note that since Theorems 4 and 5 state identities, they can be used to generate the corresponding identities for the moments of the average RT distributions. Differentiating, multiplying by t^m and integrating with respect to t leads to

$$E_{r+1,n+1}(\mathrm{RT}^m) = \frac{r+1}{n+1} E_{rn}(\mathrm{RT}^m) + \frac{n-r}{n+1} E_{r+1,n}(\mathrm{RT}^m) \tag{13}$$

and

$$E_{rn}(\mathrm{RT}^m) = \sum_{k=1}^{n-r+1} k \frac{\binom{n-k-1}{r-2}}{\binom{n}{r}} E_{1k}(\mathrm{RT}^m), \tag{14}$$

which may be used to yield testable predictions for the way in which RT means and variances change with n and r.

6. A generalization of Theorem 1. So far, I have considered SST models only where processing always stops as soon as the first target is detected. This does not allow for imperfect processing of elements, and, obviously, cannot account for occasional failures to identify a target as such which can be observed in (n, r)-experiments. Thus in order to account for erroneous responses, the SST model will need some modification if it is to give more than a rough first approximation to the processes actually going on. Although I shall not pursue this problem in detail here, I give a generalization of Representation Theorem 1 which may serve as a possible starting point for more realistic SST models. Moreover, the generalized theorem below may also be used to generate testable predictions for (n, r)-experiments that differ from the type considered so far in that the subject is instructed to respond as soon as he has found m out of r targets.

Assumptions (A1) and (A2) are the same as before; the self-termination assumption (A3) is modified as follows. Let M be a random variable which takes on values $1, 2, \ldots, r + 1$; M indicates the number of targets that have to be detected before the search is terminated. The event $(M = r + 1)$ means that the subject exhausts the stack, i.e., examines all n items, before he initiates a response. For a given condition (n, r) the probability distribution of M is $P_{rn}(M = m) \equiv c_{m,rn}$; it is assumed not to depend on the input positions of the targets $\mathbf{j} \in J$. Let S_m^* denote the mth smallest among $S_{j_1}, S_{j_2}, \ldots, S_{j_r}$, i.e., the position of the mth highest target in the stack.

Assumption (A3) can now by replaced by

(A3′) For every $\mathbf{j} \in J$ and $t \geqslant 0$,

$$P_{\mathbf{j}rn}(\mathrm{RT} \leqslant t | S_1, \ldots, S_n, M = m) = \begin{cases} P_{rn}(\mathrm{RT} \leqslant t | S_m^*, M = m), & m \leqslant r, \\ P_{rn}(\mathrm{RT} \leqslant t | M = m), & m = r + 1. \end{cases}$$

In words, on trials where search is terminated when $M = m \leqslant r$ targets are detected, RT depends on the order of the elements in the stack only through the position of the mth highest target; if the search is exhaustive, RT does not depend on the stack order at all. Moreover, RT is conditionally independent of the input positions of the targets, $\mathbf{j} \in J$.

Define $F_{k,mrn}(t) \equiv P_{rn}(\mathrm{RT} \leqslant t | S_m^* = k, M = m)$, where $m \leqslant r$, and $F_{rn}^*(t) \equiv P_{rn}(\mathrm{RT} \leqslant t | M = r + 1)$.

THEOREM 6. *For an SST model with* (A1), (A2), *and* (A3′), *the average RT distribution equals*

$$G_{rn}(t) = \sum_{m=1}^{r} c_{m,rn} \sum_{k=m}^{n-m+1} \frac{\binom{k-1}{m-1}\binom{n-k}{r-m}}{\binom{n}{r}} F_{k,mrn}(t) + c_{r+1,rn} F_{rn}^*(t). \tag{15}$$

The proof proceeds analogously to that of Theorem 1. The mixture involves both the number of targets to be detected, m, as well as the stack position of the mth target, k. Again, the mixing weights have a simple interpretation in terms of random arrangements of the elements in the stack: Given random stack ordering, the event $(S_m^* = k)$ occurs wth probability $\binom{k-1}{m-1}\binom{n-k}{r-m}/\binom{n}{r}$ since there are $\binom{k-1}{m-1} \cdot 1 \cdot \binom{n-k}{r-m}$ ways of placing $(m - 1)$ targets among the first $(k - 1)$ positions, 1 target in position k, and the remaining $(r - m)$ among the lowest $(n - k)$ positions. Note that the expected position of the mth target, $E_{rn}(S_m^*)$, equals $m \cdot (n + 1)/(r + 1)$. By setting

$$c_{m,rn} = \begin{cases} 1 & \text{if } m = i, \\ 0 & \text{if } m \neq i, \end{cases}$$

Theorem 6 can be used to represent the appropriate mixture for (n, r)-experiments in which the subject is to respond as soon as $i \leqslant r$ targets have been found.

7. Empirical tests of SST models. In spite of a wealth of experimental studies that have been conducted to test SST models in reaction time, almost no reports exist which give the data in sufficient detail to permit an application of the test described above. This is because researchers of RT have focused their interest so far on the RT means rather than on the whole distributions. Therefore, I limit this section to some suggestions concerning possible tests of SST models and briefly mention some results obtained by Sternberg **[1973]**.

One way of testing the representation theorem (equation (2)) directly has been proposed by Sternberg **[1973]**. Keep n and r constant, and manipulate conditions which are likely to influence the subject's search strategy, e.g., by instructions or by varying the frequencies with which items from the different input positions are asked for at the test. By equation (2), the equally weighted average RT distribution must remain invariant in spite of changes in the probability distribution over the search orders. It is easy to test whether this invariance of the average RT distributions holds empirically.

Tests of SST models via equation (8), (10), (11), and (12) should be obvious. Sternberg **[1973]** has reported some results of applying equation (2) to $(n, 1)$-experiments. Two experiments involved choice reactions with a 1 : 1 mapping of stimuli and responses; the required response was either a button press or the naming of the stimulus. A possible SST model for this task assumes that the subject stores representations of all S-R pairs in memory. When a stimulus is presented, he searches his memory for the pair with the stimulus term that matches the test stimulus, and reads off the response stored with it. Assuming a serial memory search which terminates when the critical pair is located, this implies the SST model for $(n, 1)$-experiments sketched above. Sternberg tested this model by comparing the average RT distributions obtained for different numbers of stimuli, n.

The results were clearcut: Although the long RT property obtained for all conditions and all four subjects when manual responses were required, the short RT property was violated in every case. There were far too few short RT's under the $n = 4$ and $n = 8$ conditions compared to the $n = 2$ condition. Rather than fanning out from the same point, the RT distributions seemed to be shifted with respect to each other. Under naming conditions, the average distributions for $n = 2$ and $n = 8$ were compared. For each of five subjects, the short RT property was again violated. In addition, the long RT property did not hold for four subjects. Thus it seems clear that SST models have to be rejected for 1 : 1 choice reactions.

Sternberg **[1973]** also applied the test to the data from the positive trials of a memory scanning experiment in which list length was $n = 1$, $n = 2$, or $n = 4$. The data, averaged across 12 subjects, were inconsistent with the SST model. Again, the short RT property was violated, thus confirming Sternberg's **[1966]** original reaction of the SST model for memory scanning because of equal mean rates on positive and on negative test trials.

These results cast severe doubts on the adequacy of SST models for RT data in choice reaction tasks and in memory scanning. It should be noted that the distributional inequality (equation (8)) is based on weaker assumptions and is thus a test of a broader class of SST models than the usually tested predictions concerning changes in mean RT which require the assumption of a constant mean duration per processing step. It seems therefore desirable to conduct the distributional tests under conditions where SST models have been shown to be successful with respect to mean RT's, in order to see whether the model passes the more general but possibly also more stringent test.

To my knowledge no data exist by which equation (11) (which relates the distributions from (n, r) and $(n, r + 1)$ conditions to that of $(n + 1, r + 1)$) or equation (12) (which uses the distributions from $(n, 1)$-experiments to predict those from (n, r)-experiments, $r \geqslant 2$) can be tested. However, Bamber's data reported above are consistent with these predictions at least as far as the means are concerned. Bamber fitted an SST model to his data which uses the strong assumption of constant mean time per comparison, i.e.,

$$E_{rn}(\mathrm{RT}|S^* = k) = k \cdot \mu + b \quad \text{if } r \geqslant 1.$$

As Figure 2 shows, this model gives an excellent fit to the mean RT's on different-trials. It remains to be seen whether the model is also able to predict other distributional properties as well. One problem which is outside the scope of this paper is why the SST model's prediction totally breaks down on the same-trials where RT ought to be longest since the search has to be exhaustive but is actually shorter than on different-trials. This points to a shortcoming of the SST model for same-different experiments. Bamber [**1969**] proposed to fix it by assuming two separate parallel processes for the detection of sameness and of differences. This has some plausibility because of the excellent account of the mean RT's on different-trials which would be difficult to understand otherwise. At present, however, Bamber's suggestion is not widely accepted.

References

Bamber, D., 1969. *Reaction times and error rates for "same"-"different" judgments of multidimensional stimuli*, Perception and Psychophysics **6**, 169–174.

Donders, F. C., 1969. *Over de snelheid van psychische processen*, translated in Attention and Performance. II (W. G. Koster, ed.), Acta Psych. **30**, 412–431.

Falmagne, J.-C., S. Cohen, and A. Dwivedi, 1975. *Two-choice reactions and ordered memory scanning processes*, Attention and Performance. V (P. M. A. Rabbitt and S. Dornic, eds.), Academic Press, London.

Sternberg, S., 1966. *High-speed scanning in human memory*, Science **153**, 652–654.

______, 1969. *Memory-scanning: mental processes revealed by reaction-time experiments*, Amer. Sci. **57**, 421–457.

______, 1973. *Evidence against self-terminating memory search from properties of RT distributions*, paper presented at the Annual Meeting of the Psychonomic Society, St. Louis, Mo.

______, 1975. *Memory scanning: new findings and current controversies*, Quart. J. Exper. Psych. **27**, 1–32.

Theios, J., 1973. *Reaction time measurements in the study of memory processes: theory and data*, The Psychology of Learning and Motivation (G. H. Bower, ed.), vol. 7, Academic Press, New York.

Theios, J., P., G. Smith, S. E. Haviland, J. Traupmann, and M. S. Moy, 1973. *Memory scanning as a serial selfterminating process*, J. Exper. Psych. **97**, 323–336.

FACHGRUPPE PSYCHOLOGIE, UNIVERSITÄT KONSTANZ, D-7750 KONSTANZ, FEDERAL REPUBLIC OF GERMANY

BELL LABORATORIES, MURRAY HILL, NEW JERSEY 07974

ABCDEFGHIJ–CM–8987654321